DRUG METABOLISM

Drug Metabolism

Current Concepts

Edited by

CORINA IONESCU

"I. Haţieganu" University of Medicine and Pharmacy,
Cluj-Napoca, Romania

and

MINO R. CAIRA

University of Cape Town,
South Africa

Springer

A C.I.P. Catalogue record for this book is available from the Library of Congress.

ISBN-13 978-90-481-7058-6 (PB)
ISBN-10 1-4020-4142-X (e-book)
ISBN-13 978-1-4020-4142-6 (e-book)

Published by Springer,
P.O. Box 17, 3300 AA Dordrecht, The Netherlands.

www.springeronline.com

Printed on acid-free paper

Printed in the Netherlands.

Dedication

To the memory of my parents

To my beloved husband and son,
for their continuous support, understanding
and encouragement.

Corina Ionescu

CONTENTS

PREFACE

This book is intended to serve a wide audience, including students of chemistry, pharmacy, pharmacology, medicine, biochemistry and related fields, as well as health professionals and medicinal chemists. Our aim in preparing it has been threefold: to introduce essential concepts in drug metabolism (drug biotransformation), to illustrate the wide-ranging medical implications of such biological processes and to provide the reader with a perspective on current research in this area. The general intention is to demonstrate that the metabolism of a drug is a primary concern throughout its lifetime, from its inception (chemical design and optimisation) to its final clinical use, and that for any given drug, the multiple factors influencing its metabolism necessitate on-going studies of its biotransformation.

In the first chapter, the principles underlying drug absorption, distribution, metabolism and elimination are described, with drug metabolism highlighted within the context of these fundamental processes. Chapters 2 and 3 deal with the chemistry of drug biotransformation, describing both Phase I ('asynthetic') and Phase II ('synthetic') biotransformations and the enzymes that mediate them. Further details of the structural features, mechanisms of action in biotransformation, and regulation of enzymes appear in Chapter 4. Enzyme induction and inhibition, with special reference to the cytochrome P450 system, are examined in Chapter 5. This is followed, in Chapter 6, by a discussion of the influence of sex, age, hormonal status and disease state on drug biotransformation. An introduction to the relatively new discipline of pharmacogenetics, probing the effects of gene variability on drug biotransformation, is the subject of Chapter 7. This includes commentary on the implications of pharmacogenetics for the future dispensing of medicines. Chapter 8 treats two special topics that have significant clinical implications, namely drug-drug interactions and adverse reactions. Included in this chapter is an extensive tabulation of drug-drug interactions and their biological consequences. Finally, Chapter 9 attempts to demonstrate how considerations based on a sound understanding of the principles of drug metabolism (described in the earlier chapters) are incorporated into the drug design process in order to maximise the therapeutic efficacy of candidate drugs. This is of paramount interest to the medicinal chemist whose aim is to design safe and effective drugs with predictable and controllable metabolism.

The text is supported extensively by pertinent examples to illustrate the principles discussed and a special effort has been made to include frequent literature references to recent studies and reviews in order to justify the term 'current' in the title of this work.

Corina Ionescu *Mino Caira*

ACKNOWLEDGEMENTS

Prof. dr. Marius Bojiță, Rector of "I.Hațieganu" University of Medicine and Pharmacy, Cluj-Napoca, Romania, for facilitating this collaboration, his understanding and support;

Prof. dr. Felicia Loghin, Dean of the Faculty of Pharmacy, for her continuous support and encouragement;

Prof. dr. Jacques Marchand (Univ. of Rouen, France) and Prof. George C Rodgers Jr. (Univ. of Louisville, Kentucky, USA) for their encouragement and recommendations;

Prof. dr. Dan Florin Irimie, the first reader of the Romanian version of my book, for his helpful comments.

MRC expresses his gratitude to Fiona, Renata and Ariella for their infinite patience and unflagging loyalty.

Thanks are due to the University of Cape Town and the NRF (Pretoria) for supporting drug-related research projects.

We are indebted to Richard A Paselk, Abby Parrill and numerous other sources for granting us permission to reproduce numerous figures.

A special token of thanks is due to our colleagues who have assisted with technical aspects of the production of this work. They include Senior Lecturer Dr. Adrian Florea, from the Department of Cellular Biology, "I.Hațieganu" University of Medicine and Pharmacy, as well as Mr Vincent Smith and Mr Paul Dempers (both of the Department of Chemistry, University of Cape Town).

Chapter 1

DRUG METABOLISM IN CONTEXT

1.1 INTRODUCTION

There are four discrete processes in the pharmacokinetic phase during the biological disposition of a drug (or other xenobiotic), namely its absorption, distribution, metabolism and excretion – the ADME concept (Figure 1.1).

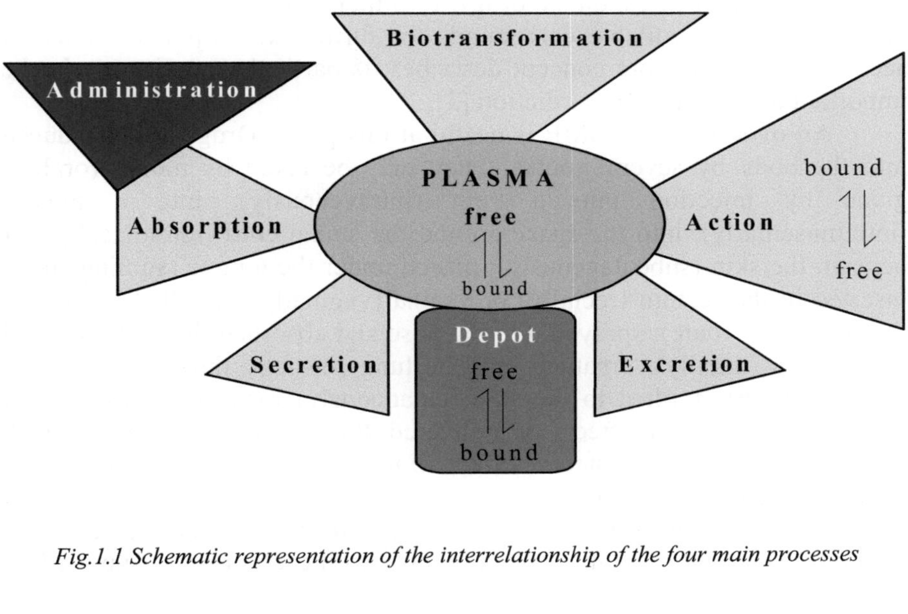

Fig.1.1 Schematic representation of the interrelationship of the four main processes

The importance of ADME in modern drug development cannot be understated. Optimisation of the performance of new drug candidates, with respect to increasing their bioavailability and controlling their duration of action, depends critically on investigation and proper exploitation of their metabolism and pharmacokinetics (PK), an activity referred to as 'early ADME studies'. Studies of metabolism and PK have accordingly evolved to be in step with innovations in modern drug-discovery, such as automated combinatorial synthetic developments, high-throughput pharmacological testing and the compilation of extensive databases. The reader is referred to a recent review highlighting a particular category of ADME investigation, namely 'metabolic stability' studies [1]. In addition to explaining the theoretical basis of metabolic stability and its relationship to metabolic clearance, the review presents some fundamental relationships between drug structure and metabolism, as well as providing examples of how metabolic stability studies have contributed to the design of drugs with improved bioavailabilities and favourable half-lives. In the final chapter of this book dealing with drug design, we return to this topic and describe further examples of the incorporation of metabolism and PK data into various strategies for increasing the therapeutic indices of new candidate drugs.

Before dealing in detail with the individual items comprising ADME, another modern aspect of drug discovery and development that merits mention here is 'ADME prediction', whose aim is to forecast the ADME behaviour of candidate drugs from their chemical structures with a view to selecting suitable compounds for further development. A recent account of the biophore concept describes its particular application in the important area of ADME prediction [2].

An overview of ADME is useful at this point. Drugs are introduced into the body by several routes. They may be taken by mouth (orally); given by injection into a vein (intravenously), into a muscle (intramuscularly), into the space around the spinal cord (intrathecally), or beneath the skin (subcutaneously); placed under the tongue (sublingually); inserted in the rectum (rectally) or vagina (vaginally); instilled in the eye (by the ocular route); sprayed into the nose and absorbed through the nasal membranes (nasally); breathed into the lungs, usually through the mouth (by inhalation); applied to the skin (cutaneously) for a local (topical) or bodywide (systemic) effect; or delivered through the skin by a patch (transdermally) for a systemic effect. Each route has specific purposes, advantages, and disadvantages.

After the drug is absorbed, it is then distributed to various organs of the body. Distribution is influenced by how well each organ is supplied by blood, organ size, binding of the drug to various components of blood and tissues, and permeability of tissue membranes. The more fat-soluble a drug is, the higher its ability to cross the cell membrane. The blood-brain-barrier restricts passage of drugs from the blood into the central nervous system and cerebrospinal fluid. Protein binding (attachment of the drug to blood proteins) is an important factor influencing drug distribution. Many drugs are bound to blood proteins such as serum albumin (the main blood protein) and are not available as active drugs.

Metabolism occurs via two types of reaction: phase I and phase II. The goal of metabolism is to change the active part of medications (also referred to as the functional group), making them more water-soluble and more readily excreted by the kidney (i.e. the body attempts to get rid of the "foreign" drug). Appropriate structural modification of drugs increases their water solubility and decreases their fat solubility, which speeds up the excretion of the drug in the urine.

Excretion occurs primarily through the urine. Fecal excretion is seen with drugs that are not absorbed from the intestines or have been secreted in the bile (which is discharged into the intestines). Drugs may also be excreted in the expired air through the lungs, in perspiration, or in breast milk. There are three processes by which drugs are eliminated through the urine: by pressure filtration of the drug through the kidney component called the Glomerulus, through active tubular secretion (like the shuttle system), and by passive diffusion from areas of high drug concentration to areas of lower concentration.

While the four processes comprising ADME were formally separated above, it should be noted that, depending on their respective pharmacokinetics, a given dose of drug may be undergoing more than one of these processes simultaneously, so that e.g. metabolism of absorbed drug may commence while part of the administered dose is still being absorbed.

1.2 ABSORPTION

Medicines may be administered to the patient in a variety of ways, but the desired therapeutic effect will be achieved only if the pharmacologically active substance reaches its site of action (the target cells in the body) in a concentration sufficient for the appropriate effect and remains there for an adequate period of time before being excreted [3-8].

Thus, to produce its characteristic effects, a drug must undergo a process of movement from the site of application into the extracellular compartment of the body and be present in appropriate concentrations at its sites of action. Absorption may therefore be defined as the sum of all processes that a drug substance may undergo after its administration before reaching the systemic circulation. Consequently, it is evident that the concentration of active drug attained depends primarily upon the extent and rate of absorption.

The extent (completeness) of absorption into the systemic circulation is sometimes defined by another parameter, designated as bioavailability. Generally, the term is used to indicate the fractional extent to which a dose of drug reaches its site of action, or a biological fluid from which the drug has access to its site of action. The amount of drug absorbed is determined by measuring the plasma concentration at intervals after dosing and integrating by estimating the area under the plasma concentration versus time curve (AUC) [3,4,7,8]. Bioavailability may vary not only between different drugs and different pharmaceutical formulations of the same drug, but also from one individual to another, depending on various factors, described in a following subsection.

The rate of absorption, expressed as the time to peak plasma concentration (T_{max}), determines the onset of pharmacological action, and also influences the intensity and sometimes the duration of drug action, and is important in addition to the extent (completeness) of absorption.

Moreover, we can define the concept of absolute bioavailability as the percentage of the drug substance contained in a defined drug formulation that enters the systemic circulation intact after initial administration of the product via the selected route. Nevertheless, it is noted that while the absolute bioavailability of two drugs may be the same (as indicated by the same AUC), the kinetics may be very different (e.g. one may have a much higher peak plasma concentration than the other, but a shorter duration) [4].

As already mentioned, drugs may be administered by many different routes, the choice of which depends upon both convenience and necessity. Under the circumstances, it is evident that knowing the advantages and disadvantages of the different routes of administration is of primary importance [3-8].

The most common, generally safe, convenient for access to the systemic circulation and most economical method of drug administration is the oral route, applicable for achieving either local or systemic effects. A number of recent reviews treat various aspects that are relevant to drug administration via this route. For example, an account has been given of the structure of oral mucosa and the factors that affect drug oral mucosal absorption and drug formulation [9]. In the case of hydrophilic drugs, the structure of the intestinal epithelium, characterised by the presence of tight junctions ('zona occludens') significantly reduces their permeability. Design of agents that are capable of increasing paracellular permeability via modulation of tight junctions has been reviewed [10]. Efflux proteins, expressed by intestinal epithelium, may limit the absorption of drugs and secrete intracellularly formed metabolites back into the intestinal lumen.

The clinical significance of the carrier-mediated efflux on intestinal absorption as well as first-pass gut wall metabolism of drugs has been highlighted [11].

Computational approaches to questions of drug absorption are also topical. We mention here a recent review that features simulation of gastrointestinal absorption and bioavailability and its application to prediction of oral drug absorption [12]. Statistical and mathematical methods were used to obtain models from which parameters for common drugs (e.g. the fraction absorbed, bioavailability, concentration-time profiles) could be predicted.

Regarding oral administration, we must mention that this route does not always give rise to sufficiently high plasma concentrations to be effective: some drugs may be absorbed unpredictably or erratically [8], or patients occasionally may have an absorption malfunction. Disadvantages of this route include also limited absorption of some drugs, determined by their physical characteristics (e.g. water solubility), destruction of some drugs by digestive enzymes or low gastric pH, irritation to the gastrointestinal mucosa, irregularities in absorption in the presence of food (or other drugs, polytherapy still being very common), as well as necessity for patient compliance. In addition, drugs in the gastrointestinal tract may be metabolised by the enzymes of the intestinal flora, mucosa, or, especially, the liver (the main location of biotransformations) before they gain access to the general circulation; so, it can be said that most orally administered drugs undergo first-pass metabolism. The extent of the latter is usually determined from a comparison of the difference in the areas under the blood concentration-time curves (AUCs) observed for oral versus i.v. drug administration, and is accurate only when drug clearance obeys first-order kinetics. Complications arising from clearance that is not first-order (e.g. that of ethanol) and errors that could result in measurements of first-pass metabolism have been reviewed recently [13].

Experimental strategies have been developed for the *in vivo* evaluation of factors affecting oral bioavailability; these can lead to estimation of the individual contributions attributable to drug absorption, losses in the gut lumen, and first-pass metabolism in the gut wall and liver [14]. The methods assume linear pharmacokinetics and constant clearance between treatments and are also appropriate for assessing metabolite bioavailability and probable sites of metabolism.

The effects of food on drug absorption also merit consideration in a discussion of oral bioavailability, since these may be quite complex. One classification of drug-food interactions includes those that cause reduced, delayed, increased and accelerated drug absorption, and those in which food does not play a role [15]. According to this account, the drug formulation is evidently also an important factor, so that 'formulation-food interactions' may be a more appropriate term than 'drug-food interactions'.

For some drugs it is sometimes assumed that parenteral administration is superior to oral administration as the latter may be impaired due to e.g. poor lipid solubility of the drug, its high molecular weight, or strongly anionic nature. This applies to the antithrombotic heparin. However, one study has shown that heparin is taken up by endothelial cells not only parenterally, but also following oral administration, despite low plasma concentrations [16]. Thus, animal experiments with unfractionated heparins (bovine and porcine) or low molecular weight heparins have yielded results supporting the thesis that heparin may well be effective when administered orally.

Drug metabolism by intestinal flora may also affect drug activity. Alteration of bowel flora (e.g. by concomitant use of antibiotics) can interrupt enterohepatic recycling and result in loss of activity of some drugs (e.g. the low oestrogen contraceptive pill).

The oral route is usually precluded only in patients with gastrointestinal (GI) intolerance or who are in preparation for anaesthesia or who have had GI surgery, as well as in situations of coma.

It is worth emphasising here that several common disorders may have an influence on the ability of the body to handle drugs. As a result, individualized therapy may become necessary for patients when e.g. gastrointestinal, cardiac, renal, liver and thyroid disorders influence drug pharmacokinetics. To avoid therapeutic failure, altering the route of administration or favourable drug co-administration (see Chapter 8) may be considered.

The rate and extent of absorption of orally administered drugs may be affected by numerous pathological factors that alter gastric emptying in instances where patients have GI diseases. Factors such as trauma, pain, labour, migraine, intestinal obstruction and gastric ulcer have been associated with decreased absorption rate, whereas conditions such as coeliac disease and duodenal ulcer may result in enhanced absorption. More specific cases include the following: reduction in the absorption of lipophilic molecules (e.g. fat-soluble vitamins) due to steatorrhoea induced by pancreatic disease; poor absorption in cardiac failure as a result of reduced GI blood flow; reduction in the absorption of ferrous sulphate and other drugs in chronic renal failure due to the buffering effect of ammonia generated by cleavage of urea; alteration in drug absorption in liver disease due to associated mucosal oedema. In the latter case, hepatic pre-systemic metabolism of drugs administered orally is impaired, resulting in significantly increased bioavailability.

Drugs that are not absorbed can have a systemic effect via an indirect action. Cholestyramine, a bile acid binding resin that lowers plasma concentrations of low-density lipoprotein cholesterol and reduces the risk of myocardial infarction in men with hypercholesterolaemia, is an example of this [6].

For systemic effects there are two main mechanisms of drug absorption by the gut: passive diffusion and active transport (a specific, carrier-mediated, energy-consuming mechanism).

Particular cases to be mentioned are controlled-release preparations [7,8,17], the Positive Higher Structures (PHS) [18], and the use of proliposomes [19], nanoparticles [20], chitosan microspheres [21], and erythrocytes [22], as potential carriers for drugs.

Controlled-release preparations are most suitable for drugs with short half-lives (< ~ 4 h) and are designed to produce slow, uniform absorption of the drug for 8 hours or longer with the obvious advantages of reduced dose frequency (improved compliance), maintenance of a therapeutic effect overnight, and lower incidence and/or intensity of undesired effects (by elimination of the peaks in drug concentration).

Opioids in a range of controlled-release preparations for oral, rectal and transdermal administration in a wide variety of pain states have been reviewed [17]. The first of these on the market (MS Contin tablets) has been in use for nearly twenty years. In contrast to short-acting immediate-release opioid preparations, which are typically administered after 4 or 6 h intervals, controlled-release preparations require considerably lower dosing frequency (e.g. once- or twice-daily doses for oral/rectal preparations, up to 7 days for transdermal preparations). With some newer preparations, analgesic therapy can begin without initial stabilisation with an immediate-release product. The primary advantages are thus sustained pain relief and patient compliance.

Nevertheless, absorption of such preparations is likely to be incomplete, so it is especially important that bioavailability be established before their general introduction. Other problems associated with slow-release preparations include the following: overdose is difficult to treat (because large amounts of drug continue to be absorbed several hours after the tablets have left the stomach); there is reduced flexibility of dosing (since sustained-release tablets should not be divided); high cost [4-8].

The PHS are bio-systems which return the enlarged molecules of a drug, resulting from attracted water molecules, to their ‘normal’ size, and thus restore their initial bioactivity [18].

Oral preparations based on liposomes as enteric-coated products can have improved in vivo stability. Reduction of toxicity and improvement in therapeutic efficacy have also resulted from the use of liposomes. Conventional liposomes may, however, present problems of stability (e.g. aggregation, susceptibility to hydrolysis, oxidation). Instead, proliposomes, which are dry, free-flowing materials that form a multi-lamellar suspension on addition of water, are devoid of such problems [19].

Pre-systemic metabolism reduces drug bioavailability. This may sometimes be overcome by using nanoparticles, in particular those that are bioadhesive (e.g. poly(methylvinylether-co-maleic anhydride) nanoparticles, that are either coated with albumin, or treated with albumin and 1,3-diaminopropane) [20].

Chitosan microspheres have been explored as carriers for drugs owing to their biocompatibility [21]. This stems from the fact that chitosan (a deacylated chitin) is a natural, non-toxic biodegradable polymer with mucoadhesive properties. Interaction with counterions such as sulphates and polysulphates, and crosslinking with glutyraldehyde lead to gel formation, a phenomenon that lends itself to pharmaceutical application. Thus, the performance of certain poorly soluble drugs has been significantly improved using this approach.

Drug pharmacokinetics can be altered significantly by encapsulation in biocompatible erythrocytes, which have been employed for delivery of drugs, enzymes and peptides [22]. Advantages include modification of release rate, enhancement of liver uptake and targeting of the reticulo-endothelial system. Targeting of particular drugs (e.g. antineoplastics such as methotrexate and carboplatin, anti-HIV peptides and nucleoside analogues) to specific organs or tissues is another important application that employs erythrocytes. (See also Chapter 9 for drug targeting using 'chemical delivery systems').

Enteric-coated formulations may also be employed in an attempt to reduce high first-pass metabolism, to achieve tissue targeting and to improve the overall safety profile of a drug, as has recently been reported for budesonide [23].

Important routes of administration that circumvent pre-systemic metabolism, providing direct and rapid access to the systemic circulation, and bypassing the intestine and liver are the buccal and sublingual routes. The sublingual route provides a very rapid onset of action (necessary, for example, in the treatment of angina attacks with nitroglycerine), while for the buccal route, the formulation ensures drug release over a prolonged period, thus giving an extended absorption and providing more sustained plasma concentrations. The drug substance must be relatively potent since the dose administered is necessarily low, and its taste must be masked (otherwise, it would result in salivation with subsequent loss of drug).

Frequently, drugs administered by these routes provide improved bioavailability compared with that from the oral route (because of the direct, rapid access to the systemic circulation). Nevertheless, we have to mention that only a few drugs may be administered successfully by these routes. Major limitations are the prerequisite for low dosage levels (generally limited to around 10 mg), the masking of taste and the risk of irritation to the mucosa, especially with prolonged treatment.

Useful especially in paediatrics (as well as in patients who are unconscious or when vomiting) is the rectal route, drugs administered in this way displaying either local or systemic effects.

Currently there is considerable interest in exploring alternative routes of administration of narcotics for the management of pain due to cancer [24]. The rectal, buccal or sublingual routes for management of acute pain syndromes have been considered as alternatives to the oral, intramuscular, intravenous and subcutaneous routes. Thus, rectal administration of morphine sulphate and chlorhydrate can lead to acceptable absorption, albeit subject to interpersonal variation. Further studies are warranted in view of the meagre pharmacokinetic data currently available for administration of narcotics via the buccal and sublingual routes in particular.

The mucosal membrane of the rectum is well supplied with blood and lymph vessels and consequently this route of drug absorption is usually high. Other advantages include avoidance of exposure to the acidity of the gastric juice and digestive enzymes, prolonged duration of action, as well as partly bypassing the portal circulation, and thus reducing pre-systemic metabolism. Usually, approximately 50% of a drug that is absorbed from the rectum will bypass the liver; the potential for hepatic first-pass metabolism is consequently less than that for an oral dose. Disadvantages are that drugs administered rectally can cause severe local irritation; in addition, rectal absorption is often irregular and incomplete. The reader is referred to a recent review on suppositories [25] describing both the pharmaceutical agents employed in rectal and/or vaginal preparations as well as novel suppositories with specific functions (e.g. suppositories that are foaming, those having localised effect, hollow suppositories).

The topical route is employed to deliver a drug at (or immediately beneath) the point of application. Therefore, this route is of limited utility. However, some success has been reported with transdermal preparations of certain drugs for systemic use (e.g. those of nitroglycerine and clonidine). The aprotic solvent dimethyl sulphoxide (DMSO) is a well-known penetrant used to enhance absorption. Topical administration using DMSO as a vehicle for systemic effects has also been investigated [26]. Other substances that enhance penetration of drugs include surface-active agents and several amides [6].

Nevertheless, it should be mentioned that systemic absorption may sometimes cause undesirable effects, as in the case of potent glucocorticoids, especially if applied to large areas (and under occlusive dressings) [6].

As with injection or buccal administration, transdermal administration bypasses pre-systemic metabolism in the gut wall or liver.

The inhalation route is one of the oldest methods of effective treatment and has been used by asthmatics for their self-medication with natural products for centuries. However, administration of drugs to the bloodstream with inhalation aerosols may be hindered by several factors. These include the limitations of dry powder inhalers, drug-excipient interactions and biological loss of the active substance in lung tissue. This subject has been reviewed recently [27].

Inhalation may be employed for delivering gaseous or volatile substances into the systemic circulation, as with most general anaesthetics or nebulised antibiotics sometimes used in children with cystic fibrosis and recurrent Pseudomonas infections. The major advantage is that the drug substance can be targeted directly to its sites of action in the lower respiratory tract with the potential for significantly reduced systemic side effects. At the same time, the large surface area of the alveoli, together with the excellent local blood supply, ensure rapid absorption, with subsequent rapid onset of action of the administered drug. More recently this route has been applied to the administration of drugs such as steroids and peptides, which are inactivated after oral administration, another major advantage of the inhalation route being avoidance of hepatic first-pass loss.

Various types of inhalers for pulmonary administration of glucocorticoids have been reviewed [28]. Details of their contents, construction and principles of operation are discussed. With inhalers, disadvantages may include incorrect use of the device itself, the existence of a high degree of coordination between breathing and activation of the device, the possibility of causing bronchoconstriction in certain cases, and toxicity of aerosol propellants.

Pulmonary absorption is also an important route of entry of certain drugs of abuse and of toxic environmental substances of varied composition and physical states. Both local and systemic reactions to allergens may occur subsequent to inhalation.

The use of liposomes in drug delivery was mentioned briefly above. A comprehensive review on the development of liposomes for local administration and its efficacy against local inflammation has appeared [29]. Topics covered include local administration to treat a number of conditions including arthritis (by intra-articular injection), pancreatitis and inflammation (of skin, airway, eye, ear, rectum, burn wounds).

Parenteral administration, currently referring to the administration of drug substances via injection, include the intravenous, intramuscular, and subcutaneous routes. These routes may be employed whenever enteral routes are contraindicated or inadequate and present main advantages such as: rapid onset of action; possibility of administration in the case of unconscious, uncooperative or uncontrollable patients; avoiding preliminary metabolism in the GI tract or liver (first-pass effect e.g. in i.v. injection) and especially in the case of intravenous infusion, facile control, enabling precise titration of drugs with short half-lives.

For parenteral formulations, that most commonly used in medical care is the intravenous route, which avoids all natural barriers of the body for absorption, and therapeutic levels are reached almost instantaneously. Other advantages of this route include the greater predictability of the peak plasma concentration, as well as the generally smaller doses required. The principal adverse effect can be a depression of cardiovascular function, often called drug shock (see also Chapter 8).

Intramuscular injection is a very convenient, more practicable route for routine administration, and inherently safer for the patient.

Subcutaneous injections are administered into the loose connective and adipose tissue immediately beneath the skin. This route is particularly useful in the case of drugs that are not effective after oral administration and it permits self-medication by the patient on a regular basis as, for example, in the case of insulin for diabetics.

Local routes of injection can also be used for specific purposes and conditions. In this context, intrathecal (specialised route for anaesthetics) and intra-arterial (directly into an artery for local effect in a particular tissue or organ) injections can be mentioned. Disadvantages include: the need for qualified medical staff; not very good patient compliance; difficulty in counteracting the effects of the drug substance in the case of overdose; continuous care to avoid the injection of air or particulate matter into the body; severe allergic reactions; haematoma formation can occur, especially after fibrinolytic therapy.

From the above presentation of the main routes of drug administration, some conclusions can be drawn concerning oral versus parenteral administration: oral ingestion is more common, safer, convenient and economical, but its main disadvantages include: limited absorption of some drugs (because of their physicochemical characteristics), irritation to the GI mucosa, destruction of some drugs by low gastric pH or digestive enzymes, irregularities in absorption in the presence of food or other drugs, extensive deactivation of many drugs as a result of the 'first-pass effect', and necessity for cooperation on the part of the patient.

Over oral administration, the parenteral injection of drugs has certain distinct advantages: availability is usually more rapid, extensive, and predictable; the effective dose can be more accurately delivered; in emergency therapy and when a patient is unconscious or unable to retain anything given by mouth; avoidance of preliminary metabolism in the GI tract or liver; by use of electric pumps, facilitation for controlled intravenous infusion. Disadvantages include: asepsis must be maintained; pain may accompany the injection; difficulty for patients to perform the injections themselves if self-medication is necessary; high cost.

Bioequivalence
The chemical substance which is a pharmacologically active ingredient synthesised by the medicinal chemist is not per se the medicine which is administered to the patient. That is, drugs are not administered as such, but formulated into drug dosage forms. Typically only ~10% of modern dosage forms comprises the active ingredient (drug), which is mixed with a variety of pharmacologically inert ingredients or excipients that perform a number of functions (as bulking agents, colourants, antioxidants, preservers, binders, enhancers). It is important to emphasise that the manufacturing process, or changes in the excipients contained in a medicine, may have a profound effect on the bioavailability of a drug substance, and it is important to bear in mind that it is never the isolated drug substance, but a dosage form which is administered to the patient. In this context, a new concept appeared, namely that of bioequivalence [30]. It is assumed that drug products are pharmaceutically equivalent if they contain the same active ingredients and are identical in strength or concentration, dosage form and route of administration. Furthermore, such pharmaceutically equivalent drug products are considered to be bioequivalent when the rates and extents of bioavailability of the active ingredient in the respective products are not significantly different under suitable test conditions.

As noted in a recent report [31], the regulatory bioequivalence requirements of drug products have undergone major changes. The biopharmaceutics drug classification system (BCS) has been introduced into the guidelines of the FDA. The BCS is based on mechanistic approaches to drug absorption and dissolution, simplifying the drug approval process by regulatory bodies. This system is also useful for the formulation scientist who may now base development of optimised dosage forms on mechanistic rather than empirical approaches.

Factors affecting absorption
These may be subdivided into:

- factors depending on the physicochemical properties of the drug molecule and characteristics of dosage formulations, and
- biological factors (usually specific for the route of administration, surface area at the site, blood flow to the site, acid-base properties surrounding the absorbing surface), including genetically determined inter-individual variability.

Absorption from the GI tract is governed by factors such as surface area for absorption, blood flow to the site of absorption, the GI transit time (of major importance since the extent of absorption is very dependent on time spent in the small intestine with its very large surface area), the presence of food or liquid in the stomach (affecting especially the emptying time of the stomach), as well as co-administration of, for example, two drug substances, which may influence the absorption rate and extension of one of them, thus accelerating or delaying gastric emptying. For example, if salicylate (a weak acid) is administered with propantheline (which slows gastric emptying) its absorption will be retarded, whereas, co-administered with metoclopramide (which speeds up gastric emptying) its absorption is accelerated [6].

The physical properties of the formulation may also have a dramatic effect on the absorption of a selected drug substance. Poor absorption characteristics may be improved by the use of lipid adjuvants which form oil/water emulsions to achieve higher concentrations of lipophilic drugs that would not otherwise be possible. Usually, unsaturated fatty acids enhance absorption more than the saturated analogues.

In this context, we should also mention current interest in the use of absorption enhancers (for particular routes of administration). A recent review on the development of intestinal absorption enhancers [32] describes appropriate research methodology, the effects of the drug delivery system and physiological factors on absorption enhancing performance, the classification of enhancers, and issues of safety.

Absorption enhancers include e.g. L-lysophosphatidylcholine, N-trimethyl chitosan, chitosan chloride, and cyclodextrins (CDs). An example of the use of CD technology is the recent achievement of improved solubility and skin permeation of bupranolol in the form of its CD complex by the transdermal route [33]. Commercial preparations based on CD inclusion complexes are available for oral, sublingual, intranasal and intracavernosal administration. Some examples are piroxicam betadex (based on the inclusion complex between β-CD and the drug and displaying more rapid absorption than uncomplexed piroxicam), benexate betadex and nimesulide betadex (based on their respective β-CD complexes), itraconazole in an oral liquid formulation with hydroxypropyl-β-CD (displaying good bioavailability, with absorption independent of local acidity), clonazepam contained in dimethyl-β-CD as a nasal formulation (a useful alternative to buccal administration for patients with serial seizures), alprostadil alfadex (based on the α-CD complex of the drug, for intracavernosal delivery) and nicotine (as the β-CD inclusion complex in a sublingual tablet) [34] (See also Chapter 9).

On the other hand, it should be noted that diffusion of some drug substances may be reduced by their association with a cyclodextrin molecule [35], as it represents an average of ~20-fold increase in the molecular weight. This point is mentioned because the influence of CDs on diffusion through a semi-permeable membrane is very important, the absorption of biologically active molecules always occurring through such a membrane. The diffusion rate of a complex in homogenous solutions is always lower than that of the free guest. Also, the partition coefficient of lipophilic drugs in an octanol/water system is considerably reduced when CD is dissolved in the aqueous phase. Therefore, CDs can be used as reverse phase-transfer catalysts: the poorly soluble guest can be transferred to the aqueous phase, where its nucleophilic reactions, for example, can be accelerated.

Whilst rapid bioavailability of the drug substance from the dosage form is usually required, suitable sustained-release forms can be formulated to deliver effective levels of a medicine over long periods when this is appropriate.

For drugs given in solid form, the rate of dissolution may be the limiting factor in their absorption, especially if they have low water solubility. Since most drug absorption from the GI tract occurs via passive processes, absorption will be favoured when the drug is in the non-ionised and more lipophilic form. On the other hand, the particle size of the drug substance is a major consideration in virtually all formulation for oral and, notably, aerosol administration. The surface area per unit weight is increased by size-reduction, which aids both dissolution and the potential systemic bioavailability. The existence of polymorphism, or the ability of a compound to exist in more than one crystalline state with different internal structures, will also have significance for the development of a suitable dosage form. Metastable polymorphs will tend to have an increased solubility, and consequently faster dissolution than a stable polymorph. This property may become important for a drug substance with an inherently poor initial dissolution rate profile, provided that the metastable form does not convert to the stable modification during storage or in the GI tract. An example is provided by ampicillin, where the anhydrous and trihydrated forms result in significantly different serum levels in human subjects after oral administration, the more soluble anhydrous polymorph producing higher and earlier blood levels (see also Chapter 9).

Another important factor that will be discussed in the following subchapter is the pH of the drug substance administered.

Other factors that influence absorption from the GI tract include: disease of the GI tract, surgical interference with gastric function, and drug metabolism by intestinal flora [5-8].

In the case of buccal and sublingual administration, absorption is dependent on the fraction of unionised material available at the buccal membranes and on the partition coefficient of the drug. A careful balance between these properties is necessary, because buccal absorption is more dependent on lipid solubility than is absorption across the mucosa of the GI tract.

In the case of the rectal route, the mucosal membrane of the rectum being well supplied with blood and lymph vessels, drug absorption is usually high. However, it can be significantly increased by using enhancers such as chelating agents (e.g. EDTA), non-steroidal anti-inflammatory agents (NSAIDs), and surfactants [6]. Also, the particle size of the drug substance, as well as the base used in the suppository will play a significant role in drug absorption.

Factors affecting percutaneous drug absorption include: skin condition (inflammation and other conditions that increase cutaneous blood flow may enhance absorption), age, region, hydration of the stratum corneum, surface area to which the drug is applied, physical properties of the drug, and vehicle [3-8].

The most important physicochemical properties of a drug affecting its transdermal permeability are its partition coefficient and molecular weight. To reach the systemic circulation the drug substance must cross both the lipophilic stratum corneum and the hydrophilic viable epidermis. Although no direct correlation between percutaneous absorption and molecular weight of the drug substance can be demonstrated, it is obvious that macromolecules will penetrate the skin very slowly, if at all (peptides and proteins are not effectively absorbed through the skin). Increasing drug concentration in the dosage form generally increases absorption via the skin until the vehicle is saturated. The pH of the formulation will also affect its penetration, the drug molecule ideally being in its unionised form.

The skin presents a major barrier to the absorption of drugs. Challenges and progress in transdermal drug delivery have recently been reviewed [36], as have the clinical aspects [37]. The physicochemical constraints severely limit the number of molecules that can be considered as realistic candidates for transdermal delivery. Nonetheless, absorption through the skin can be enhanced by suspending the drug in an oily vehicle and rubbing the resulting preparation into the skin. Hydration of the skin for absorption is extremely important and many topical formulations simply increase hydration of the stratum corneum by reducing water loss with an impermeable layer of a paraffin or wax base, or with a high water content in the formulations. Thus, the dosage form may be modified or an occlusive dressing may be used to facilitate absorption. The use of a surface-active agent frequently enhances penetration of drug substances. Use of penetration enhancers such as DMSO (mentioned earlier) and urea may also dramatically increase transdermal drug absorption [26].

To increase the range of drugs available for transdermal delivery, several chemical and physical enhancement techniques have been developed. One such procedure aimed at enhancing penetration of hydrophilic and charged molecules across the skin is iontophoresis, which is an electrical stimulation modality primarily noted for affording control and the ability to individualise therapy [38]. The latter issue may achieve more significance as knowledge about inter-individual variations in protein expression and their effect on drug metabolism and drug efficacy accumulates. The components of an iontophoretic device include a power source and two electrode compartments. The drug formulation that contains the ionised molecule is placed in the electrode compartment having the same charge while the indifferent electrode is placed at a distal side on the skin. This technique not only has applications in pain management (for local pain relief or local anaesthesia), but significantly improves transdermal delivery of certain classes of drugs such as NSAIDs (e.g. piroxicam, diclofenac), opioids, local anaesthetics, anti-emetics, antivirals, cardiovascular agents, steroids, various peptides and proteins (e.g. insulin, human parathyroid hormone, luteinising hormone-releasing hormone (LHRH) and its analogues).

A related technique is phonophoresis (or sonophoresis), which employs ultrasound to increase percutaneous absorption of a drug. The method has been used extensively in sports medicine for the last forty years. With the typical parameters used (frequency 1-3 MHz, intensity 1-2 W/cm^2, duration 5-10 min, continuous or pulse mode), controlled human in vivo studies have shown either insignificant or only mild effects of this procedure. There has been renewed interest in the technique during the last decade, owing to the finding that administration of macromolecules with conserved biological activity was possible with low frequency ultrasound in animals. The status of the technique has been reviewed [39]. Recent use of both low and high frequency ultrasound is discussed, as are the roles of thermal, cavitational and non-cavitational effects on the reduction of the skin barrier.

As regards pulmonary absorption, it is obvious that the particle size of the drug substance is critical for optimum delivery in inhalation devices. If the particles are larger than 10 μm they impact the walls of the respiratory tract and never reach the alveolar sacs. If they are smaller than 1 μm then they are likely to be exhaled from the lungs before impact. Only 10-20% of the administered drug substance will reach the alveolar sacs owing to these particle size constraints, which are therefore critical for obtaining the desired, expected therapeutic effects.

As previously indicated, in the case of the parenteral route, especially from subcutaneous and intramuscular injection, absorption occurs by simple diffusion along the gradient from drug depot to plasma.

The rate of absorption is governed by the total surface area available for diffusion (area of the absorbing capillary membranes) and by the solubility of the drug substance in the interstitial fluid (lipid-soluble drugs generally diffusing freely through capillary walls). Transport away from the injection site is governed by muscle blood flow, and this varies from site to site (deltoid > vastus lateralis > gluteus maximus); blood flow to muscle can be increased by exercise (or massage) and thus absorption rates can be increased as well. Conversely, shock, heart failure or other conditions that decrease muscular blood flow reduce absorption. The incorporation of a vasoconstrictor agent in a solution of a drug to be injected subcutaneously also retards absorption.

As mentioned earlier, by intravenous injection of drugs in aqueous solution, bioavailability being complete and rapid, the factors relevant to absorption are circumvented.

1.2.1 Basic mechanisms of transport through membranes

The absorption of a drug (as well as the rest of the processes involved in the fate of a drug formulation in the body, namely distribution, biotransformation and excretion) involves its passage across cell membranes [40].

When a drug permeates a cell, it must obviously traverse cellular plasma membrane. Important in the process of transfer through membranes are both the mechanisms by which drugs cross membranes as well as the physicochemical properties of both the drug molecules and membranes.

Cell membranes consist of a bilayer of amphipathic lipids, with their hydrocarbon chains oriented inward to form a continuous hydrophobic phase and their polar groups oriented outward. Individual lipid molecules in the bilayer can move laterally, conferring on the membrane fluidity, flexibility, high electrical resistance, and relative impermeability to highly polar molecules. The membrane proteins embedded in the bilayer serve as receptors, ion channels, or carriers and provide selective targets for drug action.

There are two main mechanisms of drug absorption: passive diffusion and active transport. Considered to be the more important mechanism, passive diffusion ensures good absorption of non-polar, lipid-soluble agents from the gut (mainly from the small intestine) because of its enormous absorptive area. The drug molecule usually penetrates by passive diffusion along a concentration gradient by virtue of its solubility in the lipid bilayer. This transfer is directly proportional to the magnitude of the concentration gradient across the membrane, the lipid: water partition coefficient of the drug, and the cell surface area. The concentration gradient across the membrane becomes the driving force that establishes the rate of diffusion, with the direction from high towards lower drug concentration. In the case of weak acids or weak bases, the non-ionised form of the drug is relatively fat-soluble and thus diffuses easily. Thus, the greater the partition coefficient, the higher the concentration of the drug in the membrane and the faster is its diffusion. Many drugs, because of their chemical structures (which determine their physicochemical properties), behave as acids or bases in that they can take up or release a hydrogen ion. Within certain ranges of pH, these drugs will carry an electrical charge, whereas in other pH ranges the compounds will be uncharged. It is this uncharged form of a drug that is lipid-soluble and therefore crosses biological membranes readily. Absorption is therefore influenced by the pK_a of the drug and the pH at the absorption site. For example, consider the distribution of a drug substance acting like a weak acid (pK_a = 4.4), and its partitioning between plasma (pH = 7.4) and gastric juice (pH = 1.4). It is assumed that the gastric mucosal membrane behaves as a simple lipid barrier, permeable only to the lipid-soluble, non-ionised form of the acid (Figure 1.2).

The ratio of non-ionised to ionised drug at each pH can be calculated from the Henderson-Hasselbalch equation. In the case of weak acid, the total concentration ratio between the plasma and the gastric juice calculated by the above equation is 1000:1 (if the system comes to a steady state). In the case of a drug substance acting like a weak base with the same pK_a, the ratio would be reversed.

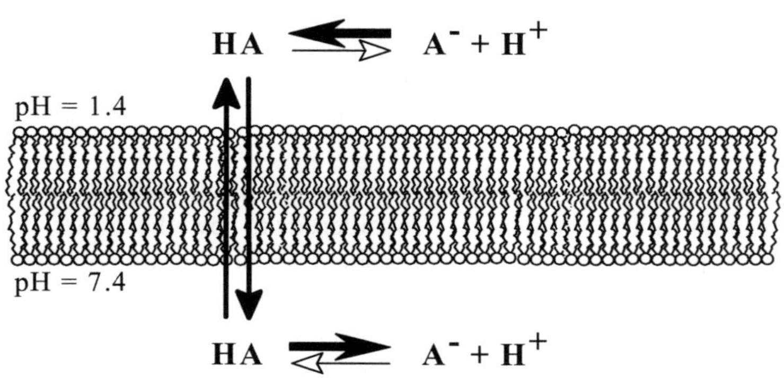

Fig.1.2 Equilibrium distribution from one side to the other of the cell membrane (Reproduced from ref. 4 (Fig. 1-2) with permission of The McGraw-Hill Companies)

Thus, it is assumed that, at steady state, an acidic drug will accumulate on the more basic site of the membrane, and a basic drug on the more acidic site. This phenomenon is termed ion trapping [4].

Based on the pH-partition concept presented, it would be predicted that drugs that are weak acids would be better absorbed from the stomach (pH 1 to 2) than from the upper intestine (pH 3 to 6), and vice versa for weak bases. On the other hand, it should be noted that since the drug is ionised to a very small extent in the stomach but appreciably so in blood, the drug substance should cross readily in the stomach-to-plasma direction but hardly at all in the reverse direction [41].

A summary of the effect of pH on degree of ionization of several acidic and basic drugs is presented in Figure 1.3.

While passive diffusion through the bilayer dominates in the disposition of most drugs, carrier-mediated mechanisms can also play an important role. Active transport requires energy, movement against an electrochemical gradient, saturability, selectivity, and competitive inhibition by co-transported compounds [42]. The specific carriers, transporter proteins (Figure 1.4), are often expressed within the cell membranes in a domain-specific fashion.

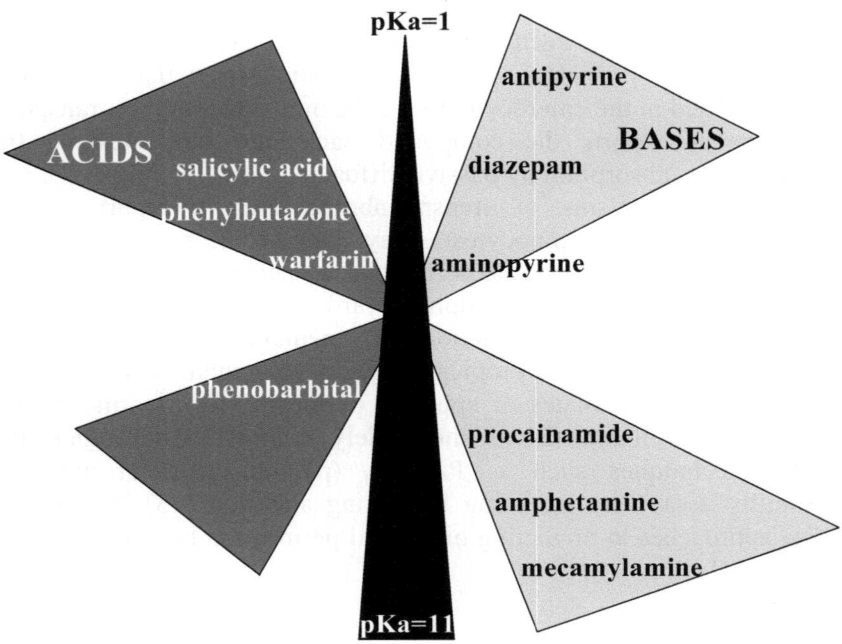

Fig.1.3 Prediction where drugs with certain pK_a values will be absorbed (Data from ref. [41])

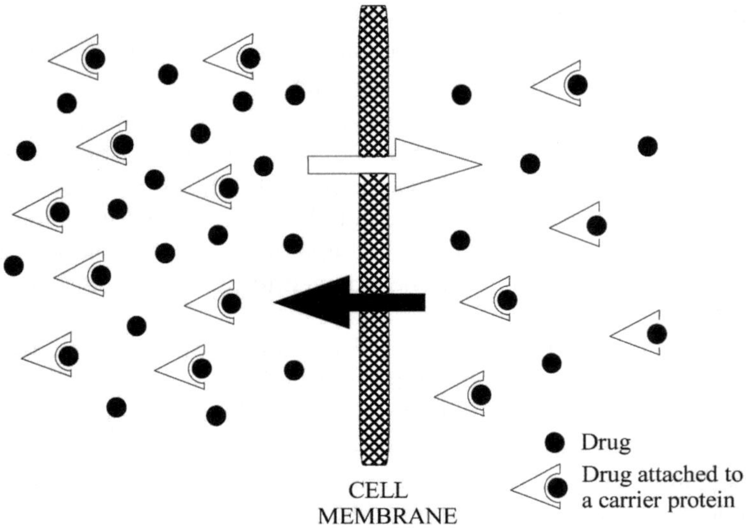

Fig.1.4 Comparative drug absorption via active (carrier) transport (the black arrow) and via passive diffusion (the white arrow) across a cellular membrane [41]

An example of such an important efflux transporter is the P-glycoprotein. Nevertheless, it should be emphasised that this special protein, localised in the enterocyte, limits the oral absorption of transported drugs since it exports the compound back into the intestinal tract subsequent to its absorption by passive diffusion.

Other mechanisms of transmembrane drug transport such as facilitated diffusion, or pinocytosis may also occur, but attempting to predict the type of transport expected for a specific drug and across a specific membrane type is not straightforward.

Measurement and prediction of membrane permeabilities of candidate drugs are important topics in drug development. The Caco-2 cell model that allows estimates of apparent permeabilities of drugs through membranes is well-established and widely employed, but other, non-biological techniques such as PAMPA (parallel artificial membrane permeability assay) are receiving increasing attention [43]. *In vitro* and *in silico* approaches to predicting biological permeation have recently been reviewed [44].

1.3 DRUG DISTRIBUTION

In section 1.2, the various routes of drug administration, as well as the factors that affect drug absorption in each case, were described. The present section addresses the subsequent phase, namely distribution of the drug to the various 'compartments' or 'volumes of distribution' that may be considered to comprise the human body. Following absorption or injection of the drug into the bloodstream, it is distributed into interstitial and cellular fluids, and interacts with macromolecules present in the various body fluids and tissues. The processes of drug diffusion into the fluids and drug binding to macromolecules affect both drug pharmacodynamics and pharmacokinetics. Distribution is thus effected by interaction of the drug with body components and the pattern of distribution depends on both the physicochemical properties of the drug in question (e.g. its lipid solubility, degree of ionisation, pK_a, molecular weight) and physiological parameters (e.g. pH, extent of plasma protein binding, permeability of membranes, blood flow, nature of the tissue) [45]. Following a qualitative overview of drug distribution, kinetic aspects are described that serve to introduce a basic pharmacokinetic parameter, the apparent volume of distribution, which is of crucial importance in optimising the dosage regimen.

1.3.1 Qualitative aspects

Following absorption of the drug into the general circulation, it is transported via the bloodstream and diffusion to the various tissues of the body (e.g. adipose tissue, muscle, brain) and body fluids. This distribution is aided by the rapidity of blood flow, the average circulation time being of the order of one minute. Drug distribution continues during blood recirculation.

It is noted here that certain disease states can alter drug distribution. For example, in patients with congestive cardiac failure, the apparent volume of distribution (defined below) of certain drugs may be approximately only one-third that of the normal, so that regular doses give rise to elevated plasma concentrations, with concomitant toxicity. Renal impairment can result in accumulation of several acidic compounds that compete with drugs for binding sites on plasma proteins. In liver disease, the lower than normal plasma albumin concentration will lead to reduced drug plasma protein binding. In addition, bilirubin and other endogenous compounds that accumulate in liver disease may also displace drugs from binding sites, thus altering drug distribution.

All drugs are bound to some extent to either plasma proteins (serum albumin primarily), tissue proteins, or both, and it is the unbound fraction that initially undergoes distribution. In the case of drugs such as propranolol, verapamil and aspirin, since more than 90% of the absorbed drug is bound in plasma, drug available to reach the site of action is limited. Owing to the equilibrium that is set up between bound drug and free drug, as distribution proceeds, the reduced concentration of drug in the bloodstream results in the blood-proteins releasing more bound drug.

More polar drugs, such as atenolol, tend to remain within the blood and interstitial fluids, whereas apolar drugs, such as the anaesthetic halothane, primarily concentrate in fatty tissues. The nature and extent of tissue distribution depends on numerous factors including e.g. the blood flow to specific tissues and the lipid-solubility of the drug. The anaesthetic thiopental, for example, is highly lipid-soluble, rapidly entering brain tissue, whereas penicillin is generally unable to do so due to its relatively high aqueous solubility. However, tissue distribution may be altered by disease state and increased penetration of penicillin into brain tissue in patients affected by pneumococcal and meningococcal meningitis occurs due to increased permeability of the inflamed meninges [46].

Just as the bound drug in the bloodstream acts as a reservoir, replenishing distributed drug, so too do many tissues in which drugs concentrate act as storage sites, slowly releasing the drug, thus maintaining high concentrations and prolonging drug efficacy.

From the above, it should be evident that the overall process of drug distribution is dynamic and very complex, involving release of the drug from the drug-plasma protein complex and its movement to major organs such as the liver, lungs, and kidneys, as well as to peripheral tissues. Pharmacokinetic modelling attempts to quantify not only this distribution phase but also simultaneous clearance of the drug (by metabolism and excretion). Depending on the drug and the level of accuracy required, models of different degrees of sophistication are employed to formulate mathematical equations describing these processes.

1.3.2 Kinetic aspects

For the purposes of modelling drug distribution, it is convenient to consider the body as being divided into discrete 'compartments', separated by boundaries. Superficially, this resembles a multiphase system in which a chemical component partitions itself among the distinct, non-miscible phases to an extent depending on its affinity for each, as dictated by relative solubilities. However, whereas in such a system an equilibrium distribution is eventually attained, the nature of drug distribution is much more complicated because the human body is not a simple receptacle and it is not always possible to associate a specific pharmacokinetic compartment with an actual tissue or organ, Furthermore, there is a dynamic distribution of the drug into and out of many peripheral tissue compartments while drug elimination proceeds simultaneously. Because drug distribution and elimination can overlap in time, they are usually treated together in any mathematical model that seeks to map the complete drug concentration-time profile in the phases following drug absorption.

In practice, the essential experimental parameter that is available to mirror the distribution and subsequent elimination of the drug is its concentration in whatever biological fluid is chosen for sampling. Most commonly, this is the blood plasma, since its composition resembles that of the extracellular fluid, which in turn is in contact with tissue cells containing the drug receptor sites. The blood plasma level of the drug is therefore taken as a measure of the drug concentration that reflects therapeutic efficacy. The discussion that follows is based mainly on the use of drug plasma concentration as the available experimental parameter.

The simplest of the pharmacokinetic models is the 'one-compartment model', for which it is assumed that the initially administered drug dose, after entering a central compartment (the bloodstream in the case of administration as an intravenous bolus), rapidly equilibrates with the peripheral compartments, leading to a constant drug concentration throughout. It is stressed that such a simplified model implies that the entire body is a single compartment and that distribution is instantaneous and uniform. In this case, the volume in which the drug dose is distributed is referred to as the 'apparent volume of distribution', denoted V_d. This represents the apparent volume of body fluid which yields the measured concentration of drug in plasma for a given drug dosage and it may be calculated from eqn. 1.1:

$$V_d = A / C \qquad 1.1$$

where A is the amount of drug in the body (measured in e.g. mg) and C is the measured drug concentration in the blood or plasma (measured in e.g. mg L^{-1}). Thus, if a 10 mg dose of a drug that is 100% bioavailable results in a measured plasma concentration of 4 mg L^{-1}, the apparent volume of distribution is 2.5 L. The value of V_d is generally regarded as a constant for a given drug and effectively represents the volume of body fluid that would be required to dissolve all of the drug present in the body at the same concentration as that found in the plasma.

That the apparent volume of distribution is actually a hypothetical volume is easily seen from the following illustration. If we consider three drugs, X, Y and Z, present in equal total amounts in the body, but e.g. appearing in plasma to different extents, then the apparent volumes of distribution may be calculated using equation 1.1, with C being the measured concentration in the plasma (assumed volume 3 L). The results are shown in the Table below.

Table 1.1 V_d calculated for drugs with varying fractions in plasma

	Mass in the body (mg)	Fraction in plasma	Measured drug concentration in plasma (mg L^{-1})	V_d (L)
Drug X	15	0.95	4.75	3.2
Drug Y	15	0.50	2.50	6.0
Drug Z	15	0.05	0.25	60.0

Although constant amounts of drugs X, Y, and Z in the body are involved, a wide range of apparent volumes of distribution is evident. For drug X, which is almost completely confined to the plasma, and hence has a very high concentration in this phase, the apparent volume of distribution is similar to that of the assumed plasma volume of 3 L. In contrast, for drug Y, only one-half of the total amount is present in the plasma, yielding a lower concentration in this phase compared with drug X; drug Y has a correspondingly higher V_d. In the case of drug Z, most of it is evidently distributed in the peripheral tissues (its relative concentration in the plasma being very low) and V_d is ten times that for drug Y. It follows that in general, higher apparent volumes of distribution must be associated with drugs that are either distributed extensively to tissue constituents or are dissolved in lipids, or both. In the case of the tricyclic antidepressant amitryptiline, V_d calculated for a 70 kg male is 1400 L, which exceeds the total body-water by a factor of ~30 [46]. Extensive distribution into tissues leaves a low measured concentration of drug in plasma, hence yielding a large V_d.

The illustrations above emphasise that V_d is not a real volume. It is, nevertheless, an important pharmacokinetic parameter for a drug. One of its primary uses in drug therapy is in the estimation of the loading dose. Knowing the desired plasma concentration C, and the apparent volume of distribution V_d, the required dose D (mg kg^{-1}) may be calculated from equation 1.2:

$$D = V_d C / f \qquad 1.2$$

where f is the bioavailability factor (the fraction of the drug dosage reaching the systemic circulation) and D is expressed in units of drug mass per unit of body mass [46].

While the one-compartment model is satisfactory for describing the dynamic behaviour of a drug that does, in practice, equilibrate rapidly between the central compartment (the bloodstream) and peripheral tissues (as in the case of e.g. aminoglycosides, with a distribution time of less than 30 min), the behaviour of many drugs requires simulation using a multi-compartmental or a physiological model. The two-compartmental model, for example, assumes that the drug displays a slow equilibration with the peripheral tissues. Here there is a clear distinction between the central compartment and the peripheral compartment in the sense that the mathematical treatment must take into account the finite values of the rate constants for transfer of the drug from the first to the second, and the reverse process. The use of simple pharmacokinetic models is resumed in section 1.5.1 where their relevance in analysing the process of drug elimination is considered.

1.4 DYNAMICS OF DRUG ACTION

1.4.1 Drug-receptor interaction

After absorption and distribution, a drug reaches its site of action to produce an effect. The means by which a drug elicits such an effect is known as the mechanism of action [3-8, 47-50].

The effect of a drug results from its interaction with its site of action in the biological system; this takes place during the pharmacodynamic phase (pharmaco – referring to drugs and dynamics – referring to what happens when two things meet and interact). This interaction, usually with macromolecular components of the body, alters the function of the relevant component and thereby initiates the biochemical and physiological changes that are characteristic of the response to the drug.

Those specific macromolecular components are referred to as receptive substances or drug receptors and denote the components of the body with which the chemical agent is presumed to interact. Besides drug-receptor interaction (stimulation or blockade), drugs may also produce effects via drug-enzyme interactions, or non-specific drug interactions.

Receptors are specific biological sites located on a cell surface or within a cell; they can be thought of as keyholes into which specific keys (drugs) may fit. Identification of the two functions of a receptor, ligand binding and message propagation, correctly suggests the existence of functional domains within the receptor, namely a ligand-binding domain and an effector domain. Certainly from a numerical viewpoint, proteins comprise the most important class of drug receptors. Being proteic in nature, an important property of physiological receptors that renders them excellent targets for drugs, is that they act catalytically and hence function as biochemical signal amplifiers. The largest group of receptors with intrinsic enzymatic activity are cell surface protein kinases, which exert their regulatory effects by phosphorylating various effector proteins at the inner face of the plasma membrane. Another large family of receptors uses distinct heterotrimeric GTP-binding regulatory proteins, known as G proteins, as transducers to convey signals to their effector proteins. However, although often regarded as drug receptors, they are in fact receptors for endogenous substances that mediate normal biological and physiological regulatory processes. A special group of receptors – acting as dimers with homologous cellular proteins – forms part of a larger family of transcription factors. These are soluble DNA-binding proteins that regulate transcription of specific genes and include receptors for steroid hormones, thyroid hormone, vitamin D and the retinoids.

The types of chemical bonds by which drugs bind to their receptors are (in decreasing order of strength): covalent, ionic, hydrogen, hydrophobic and van der Waals bonds. For the binding of a ligand (drug substance) to a receptor to be a genuine physiological phenomenon (and not just non-specific binding), the ligand binding should:

- be saturable (in which case a plot of amount of drug bound against drug concentration will level off and reach a plateau);
- be characterised by high affinity (binding constants less than 10^{-6}M);
- be linked to a pharmacological response characteristic of the particular ligand.

The so-called occupation theory defines that only when the receptor is actually occupied by the drug molecule is its function transformed in such a way as to elicit a response. In the classical occupation theory, two attributes of the drug are required: a) affinity, a measure of the equilibrium constant of the drug-receptor interaction, and b) intrinsic activity (or efficacy), a measure of the ability of the drug to induce a positive change in the function of the receptor. The probability that a molecule of drug will react with a receptor is a function of the concentrations of both drug and receptor.

Regulation of receptors

Receptors not only initiate regulation of physiological and biochemical function, but are also themselves subject to many regulatory and homeostatic controls. These controls include regulation of the synthesis and degradation of the receptor (by multiple mechanisms), covalent modification, association with other regulatory proteins, and/or re-localisation within the cell. Modulating inputs may come from other receptors, directly or indirectly, and receptors are almost always subject to feedback regulation by their own signalling outputs.

1.4.2 Mechanisms

Drugs have specific affinities for their specific receptors. Strong affinity for a receptor will allow a drug to elicit an agonist, antagonist, or mixed agonist/antagonist interaction. The organ on, or in which, the desired effect occurs is generally called the 'target organ'. The target organ can represent any organ or system in the body. Drugs that bind to physiological receptors and mimic the regulatory effects of the endogenous signalling compounds are termed agonists (Figure 1.5).

The physiological response is usually predictable: a drug agonist simply stimulates or enhances the body's natural response to stimulation. In contrast, drugs with antagonistic activity will block receptors for which they have affinity. Such compounds may, however, produce desired effects by inhibiting the action of an agonist (i.e. by competition for agonist binding sites).

Agents that do not elicit maximum response even at apparently maximum receptor occupancy are termed partial agonists, and those that stabilise the receptor in its inactive conformation are termed inverse agonists. Thus, antagonists are agents designed to inhibit or counteract effects produced by other drugs or undesired effects caused by cellular components during illness. Antagonists can be competitive or non-competitive.

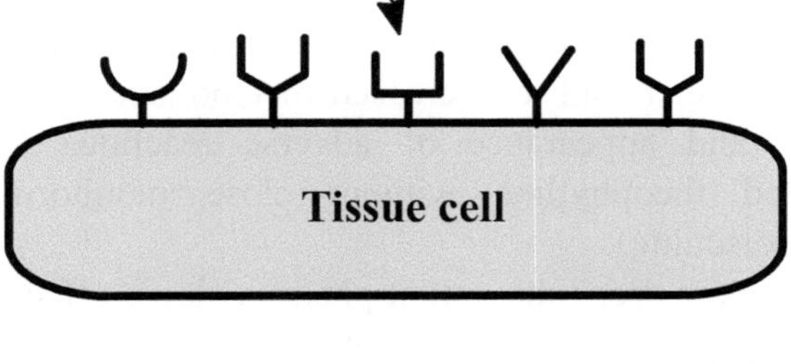

Fig.1.5 Schematic of drug-receptor interaction. The drug molecule (in black) has a high affinity for the receptor with which it makes the best fit

Competitive antagonists are agents with an affinity for the same receptor site as an agonist. Features of their action include the following: the competition with the agonist for the site inhibits the action of the agonist; increasing the concentration of the agonist tends to overcome the inhibition; competitive inhibition responses are usually reversible. In contrast, non-competitive antagonists are agents that combine with different parts of the receptor mechanism and inactivate the receptor so that the agonist cannot be effective regardless of its concentration; their effects are considered to be irreversible or nearly so. Antagonists often share some structural similarities with their agonists.

1.4.3 Further aspects

An important aspect to underline is that not all drugs work via receptors for endogenous mediators, and many drugs exert their effects by combining with an enzyme, transport protein or other cellular macromolecule (e.g. DNA) and interfering with its function.

Drug-enzyme interaction
Enzymes are generally considered as catalysts responsible for mediating biochemical reactions. Many enzymes begin working after becoming attached to a particular substrate; this is analogous to a drug attaching to a receptor. A drug/enzyme interaction occurs when a drug resembles the substrate that usually interacts with that enzyme. Stimulation or blockade of the enzyme will then be produced by the drug, and a pharmacodynamic reaction (effect) follows.

On the other hand, many very useful therapeutic drugs are enzyme inhibitors, which selectively inhibit the normal activity of only one type of enzyme, thereby reducing the ability of the enzyme to act on its normal biochemical substrate. Of particular relevance and importance, frequently seen in medicine, are the interactions between cytochrome P450 enzyme and various drug substances. As is well known, cytochrome P450 enzyme is responsible for metabolism of many drugs. Consequently, any interference with this enzyme can lead to decreased metabolism with concomitant drug accumulation (and appearance of adverse reactions – e.g. combining cimetidine and theophylline without close monitoring can lead to theophylline poisoning).

Finally, some drugs may elicit pharmacologic effects via non-specific drug interactions. For example, ointments and emollients may physically block underlying tissues from the outside environment. In other instances, drugs may penetrate cell membranes or accumulate within a cell or cavity so that interference with normal cell biochemical function occurs.

Drug-response relationships

Two other terms need to be introduced : efficacy and potency. Efficacy is the degree to which a drug is able to produce the desired effect. Potency is the relative concentration required to produce that effect.

1.5 DRUG CLEARANCE

1.5.1 Drug metabolism

At this point, we introduce the topic which is the focus of this book, namely drug metabolism. As implied in the title of this introductory chapter, the intention here is to describe its role in the context of the chain of events following ingestion of a drug or xenobiotic. As outlined below, all subsequent chapters will elaborate on the most important aspects and implications of drug metabolism in therapy and in the design of new medicinal agents.

The concept of clearance of a drug substance includes all elimination processes which act to remove it from the physiological areas; drugs may be eliminated from the body either unchanged (by the process of excretion – see following sub-chapter), or converted to metabolites with lower affinity characteristics (which obviously increase their elimination rate). The process of conversion is called biotransformation. Drug biotransformation reactions are classified as either phase I – functionalisation reactions, or phase II, – biosynthetic (conjugation) reactions [51, 52].

Phase I reactions introduce (or expose) a functional group on the parent compound, generally resulting in loss of pharmacological activity; however, active and chemically reactive intermediates may be also generated. Phase I reactions are especially important in the case of pro-drugs, which are rapidly converted to biologically active metabolites, often by hydrolysis of an ester or amide linkage (see Chapters 2 and 9). In rare instances, phase I metabolism is associated with an altered pharmacological activity.

Phase II conjugation reactions lead to the formation of a covalent linkage between a functional group on the parent compound (or on a phase I metabolite) with endogenously derived glucuronic acid, sulphate, glutathione, amino acids or acetate. These highly polar conjugates are generally inactive and are excreted rapidly in the urine and faeces.

Therefore, the usual net effect of biotransformation may be said to be one of inactivation or detoxification.

Biotransformations may be placed into four categories: oxidation, reduction, hydrolysis and conjugation. The first three comprise Phase I, whilst the last one comprises Phase II.

Oxidation – is the most common type of biotransformation; it includes side-chain hydroxylation, aromatic hydroxylation, deamination, N-, O-, and S-dealkylation, sulphoxide formation, dehydrogenations, and deamination of mono- and diamines.

Reduction – is relatively uncommon; it includes reduction of nitro, nitroso, and azo groups.

Hydrolysis – is a common biotransformation route for esters and amides.

Conjugation – represents the biosynthetic process of combining a chemical compound with a highly polar and water-soluble natural compound to yield a water-soluble, usually inactive and rapidly excreted product (details in Chapter 3).

Biotransformations take place principally in the liver, although the kidney, skeletal muscle, intestine, or even plasma may be important sites of metabolism. Within a given cell, most drug metabolising activity is found in the endoplasmic reticulum or cytosol, although drug biotransformations also occur in the mitochondria, nuclear envelope and plasma membrane.

It is emphasised that the metabolic conversion of drugs is generally enzymatic in nature. The most important group of drug metabolising enzymes is the Cytochrome P450 ('CYP450') Monooxygenase System represented by a superfamily of heme-thiolate proteins widely distributed across all living systems. These enzymes are involved in the metabolism of a very large range of diverse chemical structures, endo- and exogenous compounds including drugs, environmental chemicals and other xenobiotics (details in Chapter 4).

Hydrolytic enzymes include a number of non-specific esterases and amidases (identified in the endoplasmic reticulum of human liver, intestine and other tissues). We emphasise, as being of particular importance, the microsomal epoxide hydrolase, found in the endoplasmic reticulum of essentially all tissues and in close proximity to the cytochrome P450 enzymes; it is generally considered a detoxification enzyme, hydrolysing highly reactive arene oxides (generated from CYP450 oxidation reactions) to inactive, water-soluble trans-dihydrodiol metabolites (details in Chapter 4).

Of the conjugation enzymes the most important are considered to be the uridine diphosphate glucuronosyltransferases ('UGTs', microsomal enzymes), catalysing the transfer of glucuronic acid to aromatic and aliphatic compounds. Other important enzymes involved in this type of metabolic reaction include sulphotransferases and N-acetyltransferases. Details of these enzyme systems are also discussed in Chapter 4.

Some of the most important and common enzyme systems involved in drug biotransformation are presented in Figure 1.6. This figure is a significantly extended version of a similar representation in ref. 4.

The biotransformation of a drug may present large inter-individual variability that often results in significant differences in the extent of the process, and consequently in the rate of elimination of the drug, as well as in other characteristics of its concentration-time profile. The most important factors affecting drug metabolism include: genetic variation, environmental determinants and disease-state factors. It is crucial to know and if possible, to control these factors in optimising a dosage regimen for a particular individual.

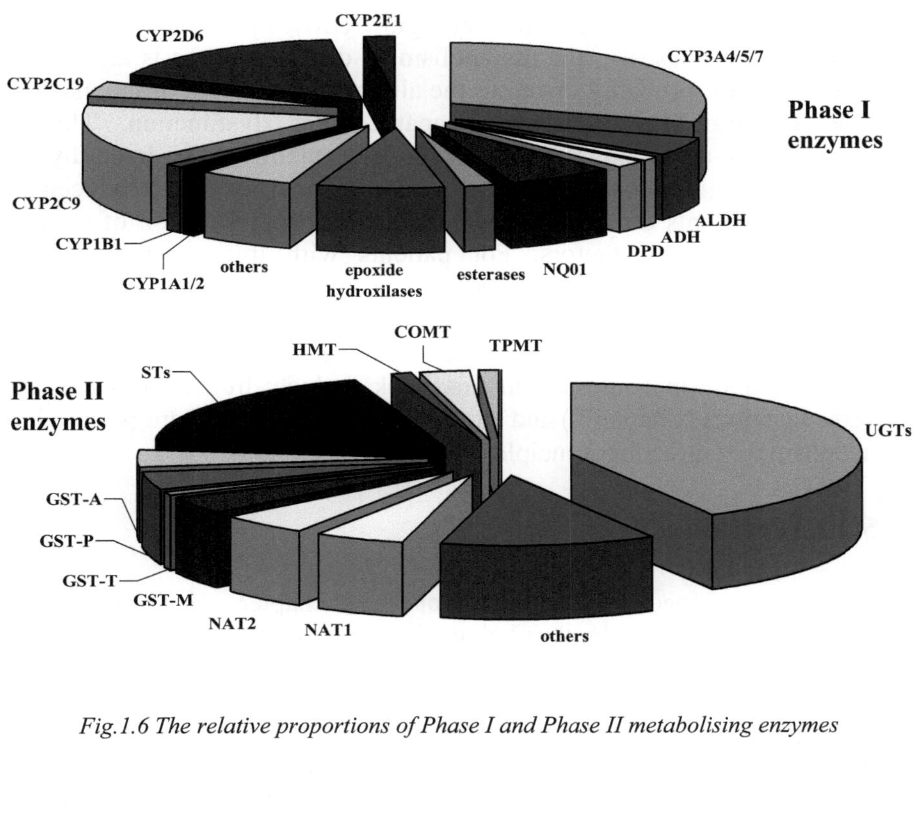

Fig.1.6 The relative proportions of Phase I and Phase II metabolising enzymes

Genetic variation. Existence of genetic polymorphisms leads to altered drug metabolising ability; differences involve a variety of molecular mechanisms leading to a complete lack of activity, a reduction in catalytic ability, or, in the case of gene duplication, enhanced activity (details in Chapter 4).

Environmental determinants can up- or down-regulate the enzymes; such modulation, termed induction and inhibition, respectively, is thought to be another major contributor to inter-individual variability in the metabolism of many drugs.

Disease factors. In renal failure, the metabolism of several drugs is reduced, but such effects are considered to be of relatively minor practical consequence.

Since the liver is the major location of drug-metabolising enzymes, any dysfunction in this organ can potentially lead to impaired drug biotransformation (in general, the severity of the liver damage determining the extent of reduced metabolism). In patients with very severe liver disease, cytochrome P450 levels are reduced, but moderate liver disease does not impair drug metabolism very significantly. In addition, in cases of very severe liver disease, the metabolism of different drugs is affected to different extents, probably owing to the altered composition of the multiple CYTP450 forms resulting from hepatocellular dysfunction. Thyroid dysfunction is also known to affect drug metabolism. In hyperthyroid patients, unusual prolongation of prothrombin time may be produced by oral anti-coagulants due to increased metabolic decomposition of vitamin K-dependent clotting factors. For patients with this condition, acute sensitivity to opioid analgesics can cause significant respiratory depression.

The above topics are treated in detail in Chapters 5-7. Further aspects of drug metabolism addressed in this book include drug interactions and adverse reactions (Chapter 8) and strategies for the design of drugs, based on metabolism as a directing principle (Chapter 9).

1.5.2 Excretion

As already mentioned at the beginning of the subchapter, some drugs are not biotransformed in the body, thus being eliminated from the body unchanged.

The most important organ of excretion is the kidney, although some substances are excreted in bile, sweat, saliva, and gastric juice or from the lungs. Renal excretion takes place principally by glomerular filtration; as the glomerular filtrate passes through the proximal tubule, some solute may be resorbed (tubular resorption) through the tubular epithelium and returned to the blood. Resorption occurs in part by passive diffusion and in part by active transport (especially with sodium and glucose). Also noteworthy here is the active transport of organic cations and anions into the lumen (tubular secretion), these active transport systems being extremely important in the excretion of a number of drugs.

Drugs also may be resorbed in the distal tubule, in which case the pH of the urine is extremely important in determining the rate of resorption (in accord with the principle of non-ionic diffusion and pH partition). It should be borne in mind that the urinary pH, and hence drug excretion, may fluctuate widely according to the diet, exercise level, drugs, time of day and other factors.

Biliary excretion and faecal elimination: Drugs that are secreted into the bile usually pass into the intestine; from here, they may be re-absorbed (and thus retained in the body) and this cycle is known as enterohepatic circulation (the system providing a reservoir for the drug). Examples of drugs that are enterohepatically circulated include morphine and the penicillins.

If a drug is not absorbed completely from the intestine, the unabsorbed fraction will be eliminated in the feces (such elimination being called fecal excretion).

Alveolar excretion: Due to the large alveolar area and high blood flow at this level, lungs are ideal for the excretion of appropriate substances such as gaseous and volatile anaesthetics.

Various disease states can alter drug excretion. Elimination of several drugs by the liver and/or kidneys is reduced in heart failure. Decreased hepatic perfusion attends reduced cardiac output and drug elimination is reduced. This increases the risk of toxicity from certain drugs or their metabolites (e.g. lignocaine). In patients with renal failure, glomerular filtration and tubular secretion of drugs usually fall at the same rate. The drop in glomerular filtration rate (GFR) is directly linked to the decline in drug excretion, which is why correct dosing relies on accurate GFR estimates for such patients. Thyroid dysfunction is another condition that may affect drug disposition, partly through its effects on drug metabolism (as mentioned earlier) and partly through changes in renal elimination. GFR is increased in thyrotoxicosis and decreased in myxoedema.

1.6 DYNAMICS OF DRUG CLEARANCE

It should be evident that the rate of elimination of the drug from the body is a crucial factor in its efficacy: if this is too rapid, frequent drug dosing is required to maintain the therapeutic efficacy, whereas too long a residence time in the body could lead to toxic effects.

1.6.1 Basic pharmacokinetic parameters

As indicated above, drug elimination occurs primarily by excretion of the original drug, its metabolites, or a combination of these, via many routes. Kinetically, this composite and irreversible process is conveniently characterised by an elimination rate constant (k_e) which takes all contributing processes into account and which can be related to other parameters reflecting drug elimination. One of these is drug clearance (CL), which can now be defined quantitatively as the volume of plasma that is completely emptied of the drug in unit time, measured in units of e.g. L h^{-1} [4]. The rate of elimination of a drug (measured as mass of drug eliminated per unit time e.g. mg h^{-1}) can be related to the clearance through the drug concentration C_t (measured in e.g. mg L^{-1}) at any time t as follows:

$$\text{Elimination rate (mg h}^{-1}\text{)} = \text{CL (L h}^{-1}\text{)} \times C_t \text{ (mg L}^{-1}\text{)} \qquad 1.3$$

where the total clearance CL may be considered to represent the sum of the clearances effected by metabolism and excretion.

As CL is constant for most drugs, elimination rate is proportional to concentration i.e. the higher the plasma drug concentration C_t at a particular time t, the faster the rate of drug elimination. Kinetically, this represents first-order behaviour, according to

$$-dC_t / dt = k_e C_t \qquad 1.4$$

reflecting the linear relation between the rate of decrease in drug concentration and the instantaneous drug concentration. The significance of the elimination rate constant k_e is that it represents the constant fraction of the amount of drug that is eliminated in unit time. On integration of eqn. 1.4 over the lapsed time period from t = 0 (corresponding to initial concentration C_o) up to some arbitrary time (t), one obtains the expression

$$C_t = C_o \exp(-k_e t) \qquad 1.5$$

showing that C_t decreases exponentially with time, as in Figure 1.7 (left). The rate law 1.5 can be cast into a linear form by taking natural logarithms on both sides, which gives eqn. 1.6:

$$\ln C_t = - k_e t + \ln C_o \qquad 1.6$$

so that a plot of ln C_t versus time yields a straight line with slope $-k_e$, as shown in Figure 1.7 (right).

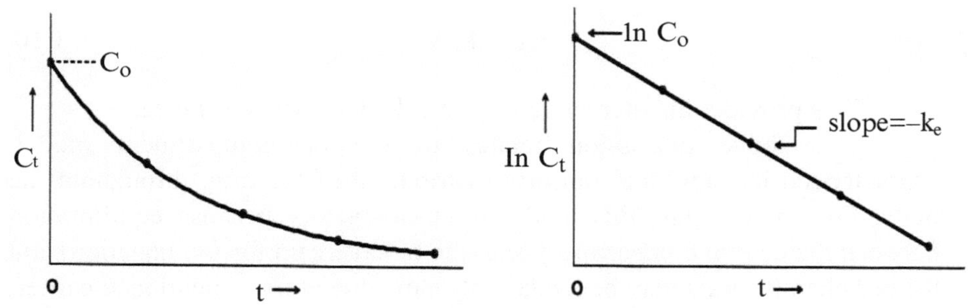

Fig.1.7 First-order kinetic behaviour showing (left) exponential decrease in concentration with time and (right) linear behaviour of the lnC_t versus time plot and determination of the elimination rate constant

Measurements of the plasma drug concentration at various times after drug administration are thus made and the data treated graphically as shown above. The value of the elimination rate constant k_e is then obtained from the slope of the linear graph.

The half-life ($t_{½}$) of the drug is defined as the time taken for concentration of the drug in the plasma to decrease to one-half of its initial value. Thus, at the time $t_{½}$, the value of C_t in equation 1.6 becomes $C_o/2$, and further manipulation leads to the relationship 1.7:

$$t_{½} = 0.693 / k_e \qquad 1.7$$

The inverse relationship between $t_{½}$ and k_e is expected and simply indicates that e.g. the longer the half-life of a drug, the smaller the rate constant for its elimination. Either the half-life or the elimination rate constant may thus be used to express the rate of clearance of the drug.

The rate of elimination of the drug was given in expression 1.3 above. An alternative way to express the rate of elimination is:

$$\text{elimination rate} = k_e \text{ x } A \qquad 1.8$$

where A is the amount of drug present. Equating expressions 1.3 and 1.8, we obtain

$$CL\ (L\ h^{-1}) \text{ x } C_t\ (mg\ L^{-1}) = k_e\ (h^{-1}) \text{ x } A\ (mg) \qquad 1.9$$

Finally, substitution of $A = V_d C_t$ (from eqn. 1.1) into the above expression and simplification yields 1.10:

$$CL = k_e V_d \qquad 1.10$$

This provides an alternative way to calculate drug clearance.

The above discussion relates to the one-compartment model, characterised by rapid and uniform distribution of the drug throughout the body. For some drugs, this model is unsatisfactory because equilibration between the central compartment (e.g. the bloodstream for i.v. injection) and the peripheral tissues may be a relatively slow process. As mentioned earlier, such a situation requires modelling by the two-compartment model shown in Figure 1.8, together with the corresponding profile for the drug concentration in the plasma.

Here, two distinct curves are evident, the one with the steeper initial slope representing drug distribution and elimination (the α-phase) while the second exponential curve, commencing after equilibrium is attained between the plasma and tissue, reflects the elimination of drug from the plasma (the β-phase).

As an example, in a recent study investigating the ADME of triethanolamine (TEA) in mice [53], it was found that the concentration-time profile of TEA in the blood following intravenous injection closely resembled that of Figure 1.8. The initial phase of the bi-exponential curve was characterised by a short half-life of only 0.3 h (corresponding to $k_\alpha = 2.3$ h^{-1} from eqn. 1.7) that was followed by a slower, terminal phase with a half-life of 10 h ($k_\beta = 0.07$ h^{-1}). Such biphasic elimination is consistent with a two-compartmental model.

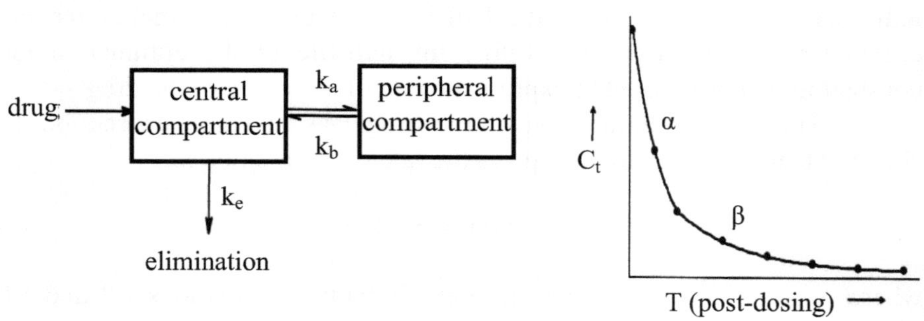

Fig.1.8 Two-compartment model (left) and the biphasic concentration-time profile (right)

With reference to the biphasic plot in Figure 1.8, we note that if distribution were instead to be complete in a very short period, the first part of the curve in Figure 1.8 (α-phase) would not be evident and the kinetics would reduce to that of a one-compartmental model.

In this chapter, only elementary aspects of pharmacokinetics were introduced, but these are adequate for following the remaining chapters. For more advanced treatments of pharmacokinetics, including clinical aspects, the reader is referred to the references above.

In the next chapter, the chemistry of Phase I and Phase II biotransformations outlined in section 1.5.1 is discussed in detail.

References

1. Thompson TN. 2001. Optimization of metabolic stability as a goal of modern drug design. Med Res Rev 21:412-449.
2. Pickett S. 2003. The biophore concept. Methods and Principles in Medicinal Chemistry 19:73-105.
3. Balant LP, Gex-Fabry M, Balant-Gorgia EA. 2000. Pharmacokinetics: from genes to therapeutic drug monitoring. In: Sirtori CR, Kuhlmann J, Tillement J-P, Vrhovac B, editors. Clinical Pharmacology. London: McGraw-Hill International Ltd., pp 9-24.
4. Wilkinson GR. 2001. Pharmacokinetics: The Dynamics of Drug Absorption, Distribution, and Elimination. In: Hardman JG, Limbird LE, Gilman GA, editors. Goodman&Gilman's The pharmacological Basis of Therapeutics, 10th ed. New York: McGraw-Hill International Ltd. (Medical Publishing Division), pp 3-29.
5. The Merck Manual – Second Home Edition. 2004. Section 2. Drugs, Chapter 11, Drug Administration and Kinetics, available at www.merckhomeedition.com/home.html.
6. Taylor JB, Kennewell PD. 1993. The Pharmaceutical phase. Routes of administration of medicines. In: Modern Medicinal Chemistry, New York: Ellis Horwood Ltd, pp 58-71.
7. Benet LZ, Perotti BYT. 1995. Drug Absorption, Distribution and Elimination. In: Woolf ME, editor. Burger's Medicinal Chemistry and Drug Discovery, 5th ed., Vol. I: Principles and Practice. New York: John Wiley & Sons, Inc.(A Wiley-Interscience Publication), pp 113-128.
8. Franklin MR, Franz DN. 2000. Drug Absorption, Action, and Disposition. In: Gennaro AR, editor. Remington: The Science and Practice of Pharmacy, 20th ed. Philadelphia: Lippincott Williams&Wilkins, pp 1098-1126.
9. Ying J, Zhiqiang J. 2003. Advances on oral mucosal drug delivery system. Zhongguo Shenghua Yaowu Zazhi 24:150-152.
10. Ward PD, Tippin TK, Thakker DR. 2000. Enhancing paracellular permeability by modulating epithelial tight junctions. Pharm Sci Technol To 3: 346-358.

11. Lennernas H. 2003. Intestinal drug absorption and bioavailability: Beyond involvement of single transport function. J Pharm Pharmacol 55:429-433.

12. Agoram B, Woltosz WS, Bolger MB. 2001. Predicting the impact of physiological and biochemical processes on oral drug bioavailability. Bulletin Technique Gattefosse 94: 41-65.

13. Levitt MD, Levitt DG. 2000. Appropriate use and misuse of blood concentration measurements to quantitate first-pass metabolism. J Lab Clin Med 136:275-280.

14. Kwan KC. 1997. Oral bioavailability and first-pass effects. Drug Metab Dispos 25:1329-1336.

15. Welling PG. 1996. Effects of food on drug absorption. Annu Rev Nutr 16:383-415.

16. Hiebert LM. 2002. Oral heparins. Clin Lab (Heidelberg, Germany) 48:111-116.

17. Miller AJ, Smith KJ. 2003. Controlled-release opioids. Pain 449-473.

18. The technology of the PHS (Positive Higher Structure), available at http://www.tupbyosystems.com/PHS/biotech.html.

19. Jeong DS, Shim CK, Lee MH, Kim SK. 1988. Manufacturing and evaluation of proliposomes. J Pharm Soc Kor 32:234-238.

20. Arbós P, Campanero MA, Arangoa MA, Irache JM. 2004. Nanoparticles with specific bioadhesive properties to circumvent the pre-systemic degradation of fluorinated pyrimidines. J Control Release 96:55-65.

21. Sinha VR, Singla AK, Wadhawan S, Kaushik R, Kumria R, Bansal K, Dhawan S. 2004. Chitosan microspheres as a potential carrier for drugs. Int J Pharm 274:1-33.

22. Gutiérrez Millán C, Sayalero Marinero ML, Castañeda AZ, Lanao JM. 2004. Drug, enzyme and peptide delivery using erythrocytes as carriers. J Control Release 95:27-49.

23. Fedorak RN, Bistritz L. 2005. Targeted delivery, safety, and efficacy of oral enteric-coated formulations of budesonide. Adv Drug Deliver Rev 57:303-316.

24. Ripamonti C, Bruera E. 1991. Rectal, bucal, and sublingual narcotics for the management of cancer pain. J Palliative Care 7:30-35.

25. Watanabe Y. 2000. A review of recent pharmaceutical preparations for rectal and vaginal administrations. Pharm Tech Japan 16:1767-1770, 1773-1776.

26. Casali F, Burgat V, Guerre P. 1999. Dimethyl sulfoxide (DMSO): properties and permitted uses. Rev Med Vet-Toulouse 150:207-220.

27. Vanbever R. 2003. Optimization of dry powder aerosols for systemic drug delivery. Optim Aerosol Drug Deliv 91-103.

28. Steckel H. 2003. Drug delivery systems for inhalable glucocorticoids. Pharmazie in Unserer Zeit 32:314-322.

29. Guofeng L, Jianhai C, Kang Z. 2003. Development of liposome for local administration and its efficacy against local inflammation. Zhongguo Yaoxue Zazhi (Beijing, China) 38:825-828.

30. Wilkinson GR. 2001. Pharmacokinetics: The Dynamics of Drug Absorption, Distribution, and Elimination. In: Hardman JG, Limbird LE, Gilman GA, editors.

Goodman & Gilman's The pharmacological Basis of Therapeutics, 10th ed. New York: McGraw-Hill International Ltd. (Medical Publishing Division), p 8.

31. Lodenberg R, Amidon GL. 2000. Modern bioavailability, bioequivalence and biopharmaceutics classification system. New scientific approaches to international regulatory standards. Eur J Pharm Biopharm 50:3-12.

32. Shen T, Huinan X. 2003. Development of intestinal absorption enhancers. Zhongguo Yiayao Gongye Zazhi 34:476-480.

33. Babu RJ, Pandit JK. 2004. Effect of cyclodextrins on the complexation and transdermal delivery of bupranolol through rat skin. Int J Pharm 271:155-165.

34. 2002. In: Sweetman SC, editor. Martindale: The Complete Drug Reference,33^{rd}ed. London: Pharmaceutical Press, electronic version (prepared by the editorial staff of the Royal Pharmaceutical Society of Great Britain).

35. Frömming K-H, Szejtli J. 1994. Pharmacokinetics and Biopharmaceutics. In: Cyclodextrins in Pharmacy. Dordrecht: Kluwer Academic Publishers pp 105-126.

36. Naik A, Kalia YN. Guy RH. 2000. Transdermal drug delivery: overcoming the skin's barrier function. Pharm Sci Technol To 3:318-326.

37. Kalia YN, Merino V, Guy RH. 1998. Transdermal drug delivery. Clinical aspects. Dermatol Clin 16:289-299.

38. Yogeshvar NK, Naik A, Garrison J, Guy RH. Iontophoretic drug delivery. 2004. Adv Drug Deliv Rev 56:619-658.

39. Machet L, Boucad A. 2002. Phonophoresis: efficiency, mechanisms and skin tolerance. Int J Pharm 243:1-15.

40. Bergelson LD. 1988, New Views on Lipid Dynamics: A Non-Equilibrium Model of Ligand-Receptor Interaction, Part I, Chapter 1, Basic Mechanisms of Medical Significance in Membrane Structure and Function. In: Benga Gh, Tager JM, editors. Biomembranes – Basic and Medical Research. Berlin: Springer-Verlag pp 1-12.

41. Jensen SC, Peppers MP. 1998. Biopharmaceutics and Pharmacokinetics. In: Rowland J, editor. Pharmacology and Drug Administration for Imaging Technologists. St. Louis (Missouri, USA): Mosby Inc., pp 30-38.

42. Winstanley P, Walley T. 1999. Basic principles. In: Simons B, editor. Churchill's Pharmacology, A clinical core text for integrated curricula with self-assessment. Edinburgh: Horne T Publisher, pp 4-15.

43. Kansy M, Avdeef A, Fischer H. 2004. Advances in screening for membrane permeability: high resolution PAMPA for medicinal chemists. Drug Discov Today Technol 1:349-355.

44. Mälkiä A, Murtomäki L, Urtti A, Konturri K. 2004. Drug permeation in biomembranes: In vitro and in silico prediction and influence of physicochemical properties. Eur J Pharm Sci 23:13-47.

45. Barre J, Urien S. 2000. Distribution of drugs. In: Sirtori CR, Kuhlmann J, Tillement J-P, Vrhovac B, editors. Clinical Pharmacology. London: McGraw-Hill International Ltd. pp 37-43.

46. Galinsky RE, Svensson CK. 2000. Basic Pharmacokinetics. In: Gennaro AR editor. Remington: The Science and Practice of Pharmacy, 20th ed. Philadelphia: Lippincott Williams&Wilkins, pp 1027-1144.

47. Rang HP, Dale MM, Ritter JM. 1999. Absorption, distribution and fate of drugs. In: Pharmacology, 4th ed. Edinburgh: Churchill Livingstone, pp 57-89.

48. Wingard LB Jr, Brody TM, Larner J, Schwartz A. 1991. Pharmacodynamics. In : Kist K, Steinborn E, Salway J, editors. Human Pharmacology, Molecular-to-Clinical. St. Louis, Missouri: Mosby Year Book Inc., pp 42-45.

49. Ritter JM, Lewis LD, Mant T.GK. 1999. Mechanisms of Drug Action (Pharmacodynamics). In: Radojicic R, Goodgame F, editors. A Textbook of Clinical Pharmacology, 4th ed. Oxford University Press Inc., pp 8-16.

50. Ross EM, Kenakin TP. 2001. Pharmacodynamics: Mechanism of Drug Action and the Relationship Between Drug Concentration and Effect. In: Hardman JG, Limbird LE, Gilman GA, editors. Goodman&Gilman's The pharmacological Basis of Therapeutics, 10th ed. New York: McGraw-Hill International Ltd. (Medical Publishing Division), pp 31-44.

51. Gibson GG, Skett P. 1994. Introduction to Drug Metabolism. London: Blackie Academic & Professional, An Imprint of Chapman & Hall, pp 1-2.

52. Correia MA. 2001. Drug Biotransformation. In: Katzung GB, editor. Basic & Clinical Pharmacology, 8th ed. New York: Lange Medical Books/Mc Graw Hill Medical Publishing Division, pp 51-63.

53. Stott WT, Waechter Jr JM, Rick DL, Mendrala AL. 2000. Absorption, distribution, metabolism and excretion of intravenously and dermally administered triethanolamine in mice. Food Chem Toxicol 38:1043-1051.

Chapter 2

PATHWAYS OF BIOTRANSFORMATION – PHASE I REACTIONS

2.1 INTRODUCTION

Drug metabolism is a complex and important part of biochemical pharmacology. The pharmacological activity of many drugs is reduced or nullified by enzymatic processes, and drug metabolism is one of the main mechanisms by which drugs may be inactivated.

Metabolism, or the biotransformation of a drug, is the process whereby living organisms effect chemical changes to a molecule [1-8]. The product of such a chemical change is called a "metabolite". In practice, all xenobiotics undergo transformations in living organisms.

In general, biotransformation converts a lipophilic xenobiotic to a polar compound, promoting a decline in its re-absorption by kidney tubules, thus allowing its excretion into the urine (the formation of polar metabolites from a non-polar drug facilitates efficient urinary excretion).

Implications for drug metabolism include drug interactions, carcinogenesis, toxication (bioactivation), substrate inhibition, enzyme induction, as well as termination of drug action.

Metabolism might convert an inactive agent (a prodrug) into the active agent that is responsible for producing the therapeutic effect. However, an important aspect to emphasise in this context is that delayed effects that manifest themselves many days after starting regular treatment with certain drugs can result from accumulation of long-lived metabolites that are at the same time the main cause of overdosing and the appearance of secondary or even adverse reactions.

It is convenient to divide drug metabolism into two phases (I and II) which sometimes, but not always, occur sequentially. Products of phase I reactions may be either pharmacologically active or inactive species and usually represent substrates for Phase II enzymes. Phase II reactions are synthetic conjugation reactions between a drug and an endogenous molecule (or between a phase I metabolite and an endogenous molecule). The resulting products have increased polarity compared to the parent drugs, being therefore more readily excreted in the urine (or, less often, in bile), and they are usually (but not always) pharmacologically inactive.

The principal organs of metabolism include the liver, kidneys and the GI tract, but drugs may be metabolised at other sites, including the lungs and the plasma. The microbial flora present in the gut play a role in the biotransformation of certain drugs (e.g. reduction of nitro- and azo-compounds).

Many of the enzymatic systems involved in drug metabolism are embedded in the membrane of smooth endoplasmic reticulum (sER), this being consequently the site of metabolism of many drugs.

An important factor that contributes to drug metabolism at the microsomal site is the lipophilicity of the drug. In contrast to a polar compound, a lipophilic compound will dissolve in the membrane of the sER, consequently serving as a substrate for the microsomal enzymes. Some endogenous compounds (steroids, thyroxine and bilirubin) are metabolised in the sER as well.

Genetic variation in drug metabolising enzymes is a factor that influences drug disposition; the implications of variations in the activity of an enzyme relate to blood levels of the drug, which in turn can result in either undesirable, unexpected toxic effects or expected therapeutic effects.

2.2 PHASE I AND PHASE II METABOLISM: GENERAL CONSIDERATIONS

Biotransformation reactions affecting drugs (as well as other xenobiotics) are traditionally separated (or, conveniently divided) into Phase I and Phase II reactions (see also Chapter 1, subchapter 1.5.1). The reactions of Phase I are thought to act as a preparation of the drug for phase II, i.e. phase I "functionalises" the parent drug molecule by producing or uncovering a chemically reactive group on which the phase II reactions can occur. For example, a $-CH_3$ moiety can be functionalized to become a $-CH_2OH$ or even a $-COOH$ group. Through introduction of oxygen into the molecule or following hydrolysis of esters or amides, the resulting metabolites are usually more polar (subsequently, less lipid-soluble) than the parent drug, therefore presenting reduced ability to penetrate tissues and less renal tubular resorption than the parent drug. These primary metabolites are then further converted to secondary metabolites, involving a process of conjugation of an endogenous molecule or fragment to the substrate, yielding a metabolite known as a conjugate. Conjugates are usually more hydrophilic than the parent compound, and subsequently much more easily excreted *via* the kidney. This is the concept of sequential metabolism (Figure 2.1.) [9].

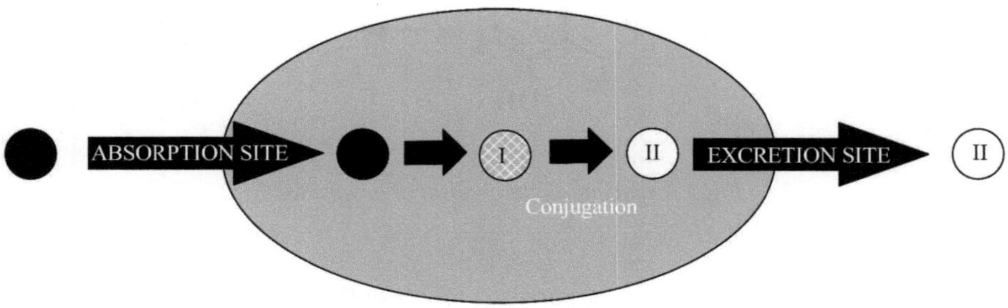

Fig.2.1 Scheme of sequential metabolism

where, ● represents the parent drug,
⊕ the primary metabolite, and
(II) the secondary metabolite.

There is a third class of metabolites, recognized as xenobiotic-macromolecule adducts (also called macromolecular conjugates), formed when a xenobiotic binds covalently to a biological macromolecule [9].

As a very recent example of sequential metabolism, we mention that of 2,3,7-trichlorodibenzo-p-dioxin (2,3,7-triCDD) by cytochrome P450 and UDP-glucuronosyltransferase in human liver microsomes [10]. This study investigated the glucuronidation of 2,3,7-triCDD by rat CYP1A1 and human UGT. The ability of ten human liver microsomes to metabolise this polychlorinated compound was assessed.

As another representative example of sequential metabolism, we present the biotransformation of propranolol, a process that leads to two metabolites, as shown in Figure 2.2. Propranolol is first oxidised to 4-hydroxypropranolol, which then undergoes sequential metabolism to 4-hydroxypropranolol glucuronide [11]. Other biotransformation reactions of propranolol will be presented later.

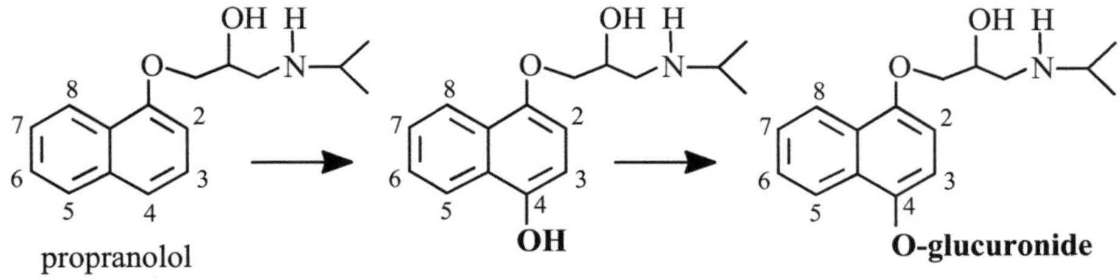

Fig.2.2 Sequential metabolism of propranolol

Another possibility, occurring frequently, is that of parallel metabolism leading to a common metabolite (Figure 2.3) [9].

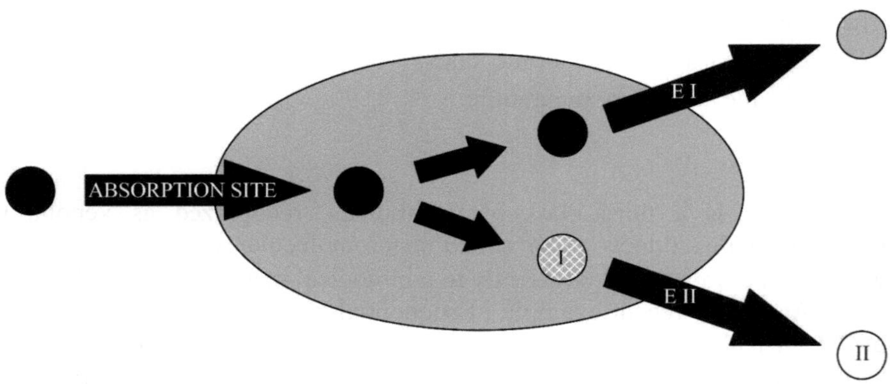

Fig.2.3 Scheme of parallel metabolism: the drug may undergo biotransformation to the primary metabolite, and be subsequently eliminated by excretion (as a Phase II metabolite), or it may follow other elimination (metabolic or excretion) routes

where, ● represents the parent drug,
the primary metabolite
the secondary metabolite, and
other metabolic or excretion form of elimination

We give as a representative example of parallel metabolism, the biotransformation of dextromethorphan [9], via two CYTP450 isoforms; both pathways involve N- and O-demethylation steps, but in reverse order, leading to a common metabolite (Figure 2.4).

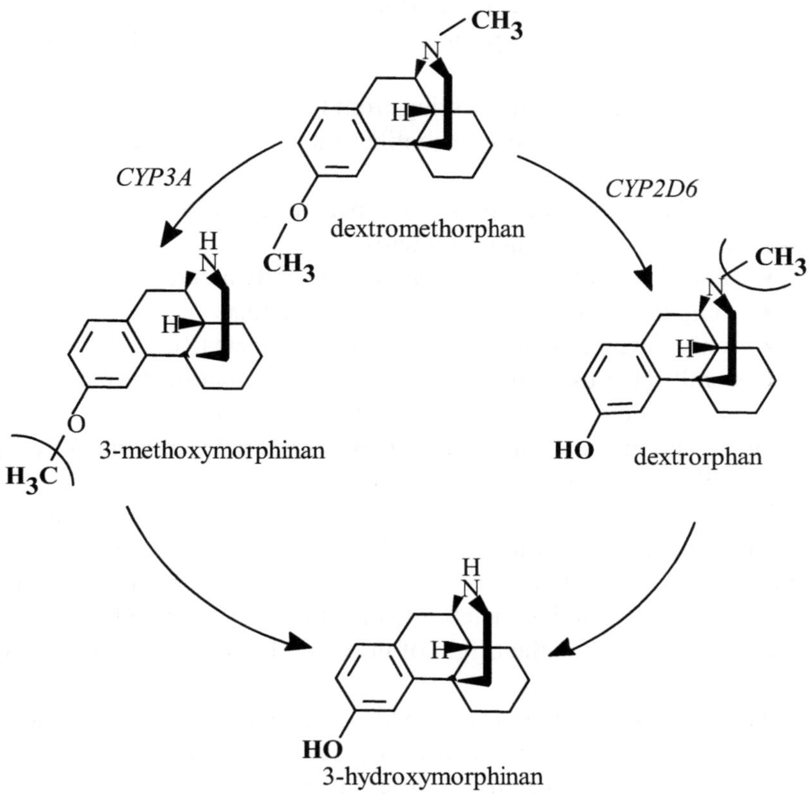

Fig.2.4 Parallel biotransformation pathways of dextromethorphan

Reversible metabolism may occur when a metabolite or biotransformation product and the parent drug undergo interconversion. Although reversible metabolism is less common, there are examples occurring across a variety of compounds, including phase I metabolic pathways (for some amines, corticosteroids, lactones and sulphides/sulphoxides), as well as phase II metabolic pathways (including reactions of glucuronidation, sulphation, acetylation etc.) [9].

A recently published, detailed account of the subject of reversible metabolism of drugs [12] highlights the complexity of the pharmacokinetic treatment of such processes as well as the fact that two compounds undergoing metabolic interconversion may have different activities. Thus, for example, in the well-known prednisone-prednisolone system, both compounds are active but in the case of the reversible metabolism involving haloperidol and it metabolite, reduced haloperidol, the latter compound is an inactive, and possibly toxic species. Clinical implications of this system are discussed in depth.

Thus, an aspect worth stressing from the outset is that in the case of a xenobiotic having a single metabolite, the following scenarios present themselves:

- neither the xenobiotic nor its metabolite exerts a biological effect (within the concentration range of interest)
- both of the above species are biologically active
- only the xenobiotic exerts biological effects
- only the metabolite exerts biological effects [8].

Generally, the usual net effect of biotransformation may be said to be one of inactivation or detoxication, the duration and intensity of a xenobiotic's actions being influenced (sometimes predominantly) by its rate and extent of metabolism.

There are, however, numerous examples in which biotransformation does not result in inactivation; many drugs generate active metabolites and moreover, in a few instances activity derives entirely from the metabolite. The production of an active metabolite may therefore be beneficial, or it may be detrimental when it is the origin of undesirable (adverse) effects (see also Chapter 8).

There are also examples in which the parent drug has little or no activity of its own but is instead converted to an active metabolite. A particular case of 'inactive' drugs that yield active metabolites is represented by the well-known prodrugs (See Chapter 9).

When a delayed or prolonged response to a drug is desired (or an unpleasant taste or local reaction is to be avoided), it is a common pharmaceutical practice to prepare an inactive (or non-offending) precursor, such that the active form may be generated in the body. This practice has been termed drug latentiation [9]. Examples of such precursors include chloramphenicol palmitate, dichlorphenazone and the estolates of various steroid hormones.

Transformation of a drug, or other xenobiotic, into a toxic metabolite, on the other hand, is effected by a toxication reaction. Toxic responses from such a metabolite may manifest at a number of levels, ranging from the molecular to that of an organ or organism, with the former not necessarily implying the latter. What can be stated is that metabolic toxication processes are always counterbalanced by competitive and/or sequential detoxication processes that may lead to inactivation of the toxic metabolite.

Factors affecting drug metabolism

A great number of physiological and pathological factors affecting drug metabolism have been characterized; these are of importance both in drug research and toxicology (details in Chapter 6).

Among the inter-individual factors we stress the species differences determined by genetic differences; the consequences of this genetic polymorphism are a greatly impaired metabolism of drugs (or prodrugs), and a marked risk of adverse reactions (see also Chapter 8).

Pharmacogenetics has thus become in recent years a major issue in clinical pharmacology and pharmacotherapy [13] (see also Chapter 7).

Intra-individual factors are related to physiological changes or pathological states (affecting for example the hormonal balance and immunological mechanisms of individuals). Biological rhythms (still not always duly recognized) are of the utmost importance and their study is the realm of chronopharmacology [14].

There are, however, factors from outside the body (intimately connected with the intra-individual factors) that can also have a profound influence on drug metabolism. Physical exposure to these factors can be either deliberate (e.g. alcohol, tobacco smoke, substances taken as food) or accidental (from air, water, different pollutants). Usually, the first group falls into the category of dietary factors while the second group comprises environmental factors.

Factors of even greater significance (as far as drug therapy and toxicology are concerned) are enzyme induction and enzyme inhibition [15]. Enzyme inducers act by increasing the concentration and subsequently, the activity of some enzymes (or isoenzymes), while inhibitors decrease the activity of some enzymes (or isoenzymes) by reversible or irreversible inactivation (for details see Chapter 5). A given drug may induce and/or inhibit its own metabolism, thus acting respectively as an auto-inducer and/or auto-inhibitor; still, the vast majority of available data document the influence of one drug on the biotransformation of another, pre- or, co-administered, this being one of the major causes of drug-drug interactions [16] (details in Chapter 8, subchapter 8.1, subsubchaper 8.1.3).

The influence of drug molecular configurational and conformational factors is a well-known and common phenomenon, resulting in substrate stereoselectivity and product stereoselectivity (enantio- and diastereo selectivity) [8].

In conceiving and preparing this monograph we have followed two approaches:

- the molecular level – which covers the biochemistry of drug metabolism (e.g. enzymes and their properties, catalytic reactions and their mechanisms, structure-metabolism relationships) and,

- the systemic level – which covers the physiology of drug metabolism (e.g. enzymes and their regulation, factors affecting drug metabolism, its pharmacological and toxicological consequences, and different related aspects).

In the following subsections, we describe the principal Phase I metabolic reactions including illustrative examples with their mechanisms.

2.3 OXIDATIONS INVOLVING THE MICROSOMAL MIXED-FUNCTION OXIDASE SYSTEM

2.3.1 Components of the enzyme system and selected miscellaneous oxidative reactions (mechanisms of action)

Phase I metabolism is dominated by the microsomal mixed-function oxidase (MMFO) system and this is known to be involved both in the metabolism of endogenous compounds (steroid hormones, thyroid hormones, fatty acids, prostaglandins and derivatives) as well as in the biotransformation of drugs (or other xenobiotics) [6,8,17].

The mixed-function oxidase system (found in microsomes of many cells – notably those of liver, kidney, lung and intestine) performs many different functionalisation reactions.

The most important Phase I reaction is that of oxidation – by incorporation of oxygen into the substrate; therefore, this reaction characterizes oxygenases. Most frequently, these reactions are mono-oxygenation reactions (incorporating only one of the two atoms of molecular oxygen – see reaction below), the corresponding enzymes thus being categorised as monooxygenases [6,17].

The presence of such enzymes in the kidney can result both in the formation of toxic compounds leading to nephrotoxicity, and to the detoxification of metabolites generated elsewhere e.g. in the liver. The potential roles of renal flavin-containing monooxygenases and cytochrome P450s in the metabolism and toxicity of the model industrial compounds 1, 3-butadiene, trichloroethylene, and tetrachloroethylene have been the subject of a recent publication [18]. A feature highlighted there is the strong dependence of particular metabolic reactions on factors such as species-, tissue-, and sex-related differences.

The phase I oxidative enzymes are almost exclusively localised in the endoplasmic reticulum (by contrast, most phase II enzymes being found predominantly in the cytoplasm). They differ markedly in structure and properties; among them, the most intensively studied both for drug and endogenous compound metabolism is the cytochrome P450.

Cytochrome P450-mediated oxidations are unique in their ability to introduce polar functionalities into systems that are as unreactive as saturated or aromatic hydrocarbons, being at the same time critical for the metabolism of lipophilic compounds without functional groups suitable for conjugation reactions. On the other hand, we have to stress that reactions catalysed by cytochrome P450 sometimes transform relatively innocuous substrates to chemically reactive toxic or carcinogenic species [8,19] (details in Ch.4).

The general cytochrome P450-catalysed reaction is:

$$\mathrm{NADPH + H^{+} + O_2 + \mathbf{RH}} \xrightarrow{\text{cytochrome P450}} \mathrm{NADP^{+} + H_2O + R\mathbf{OH}} \qquad 2.1$$

where RH represents an oxidisable drug substance and ROH, the hydroxylated metabolite. As can be seen from the above reaction, reducing equivalents (derived from NADPH + H^{+}) are consumed and only one atom of the molecular oxygen is incorporated into the substrate (generating the hydroxylated metabolite), whereas the other oxygen atom is reduced to water (the reaction is actually a hydroxylation rather than a genuine oxidation).

In addition to hydroxylation reactions, cytochrome P450 also catalyses the N-, O- and S-dealkylation of many drugs (details in further subsections); these types of reactions can be considered as a special form of hydroxylation reaction in that the initial event is a carbon hydroxylation (followed by heteroatom elimination).

Components of the M.F.O. system include [17,20]:

- the cytochrome P450
- the NADPH- cytochrome P450 reductase and
- lipids

Cytochrome P450 is the terminal oxidase component of an electron transfer system present in the endoplasmic reticulum responsible for many drug oxidation reactions. It is a haemoprotein having unusual properties (with iron protoporphyrin IX as the prosthetic group) and is found in almost all living organisms. Mammalian cytochromes P450 are found in almost all organs and tissues, located as already mentioned, mostly in the endoplasmic reticulum but also in mitochondria. It is important to note that in contrast to the porphyrin moiety, which is constant, the protein part of the enzyme varies markedly from one isoenzyme to the other (as a consequence of genetic polymorphism) [21], subsequently determining differences in their properties, substrate and product specificities, and sensitivity to inhibitors (details in Ch.4). This explains the great number (more than 500) of P450 isoenzymes identified and characterised and their resemblance in the so-called cytochrome P450 superfamily.

NADPH-cytochrome P450 reductase is a flavin-containing enzyme, consisting of one mole of FAD (flavin adenine dinucleotide) and one mole of FMN (flavin mononucleotide) per mole of apoprotein; this is quite unusual as most other flavoproteins contain only FAD or FMN as their prosthetic group. The enzyme exists in close association with cytochrome P450 in the endoplasmic reticulum membrane and represents an essential component of the M.F.O. system in that the flavoprotein transfers reducing equivalents from NADPH + H^+ to cytochrome P450 as shown in Eq. 2.2:

$$\text{NADPH} + H^+ \rightarrow (\text{FAD} \xrightarrow[\text{reductase}]{\text{NADPH-cyt.P450-}} \text{FMN}) \rightarrow \text{cytochrome P450} \qquad 2.2$$

According to Eq.2.2, NADPH-cytochrome P450 reductase is thought to act as "transducer" of reducing equivalents by accepting electrons from NADPH and transferring them sequentially to cytochrome P450.

The lipid component was originally identified as phosphatidylcholine and later studies showed that fatty acid composition of the phospholipids could be critical in determining functional reconstitution of M.F.O. activity. It has been suggested that lipid may be required for substrate building, facilitation of electron transfer or even providing a "template" for the interaction of cytochrome P450 and NADPH- cytochrome P450 reductase molecules. Nevertheless, it must be stressed that the precise mode of action of lipids is still unknown.

Among the most important non-P450 oxidative enzymes participating in phase I reactions, the following are noteworthy: microsomal flavin-containing monooxygenase (FMO), the xanthine-dehydrogenase and the aldehyde oxidase (details appear in subchapter 2.4 and Chapter 4).

Reactions catalysed specifically by the bacterial cytochromes P450 and the potential for applying the oxidising power of these enzymes have been discussed recently [22]. Oxidative reactions described include aliphatic and aromatic hydroxylation, alkene epoxidation, oxidative phenolic coupling, heteroatom oxidation and dealkylation, as well as multiple oxidations.

Some of the most common CYTP450-catalysed reactions are summarised in Figure 2.5:

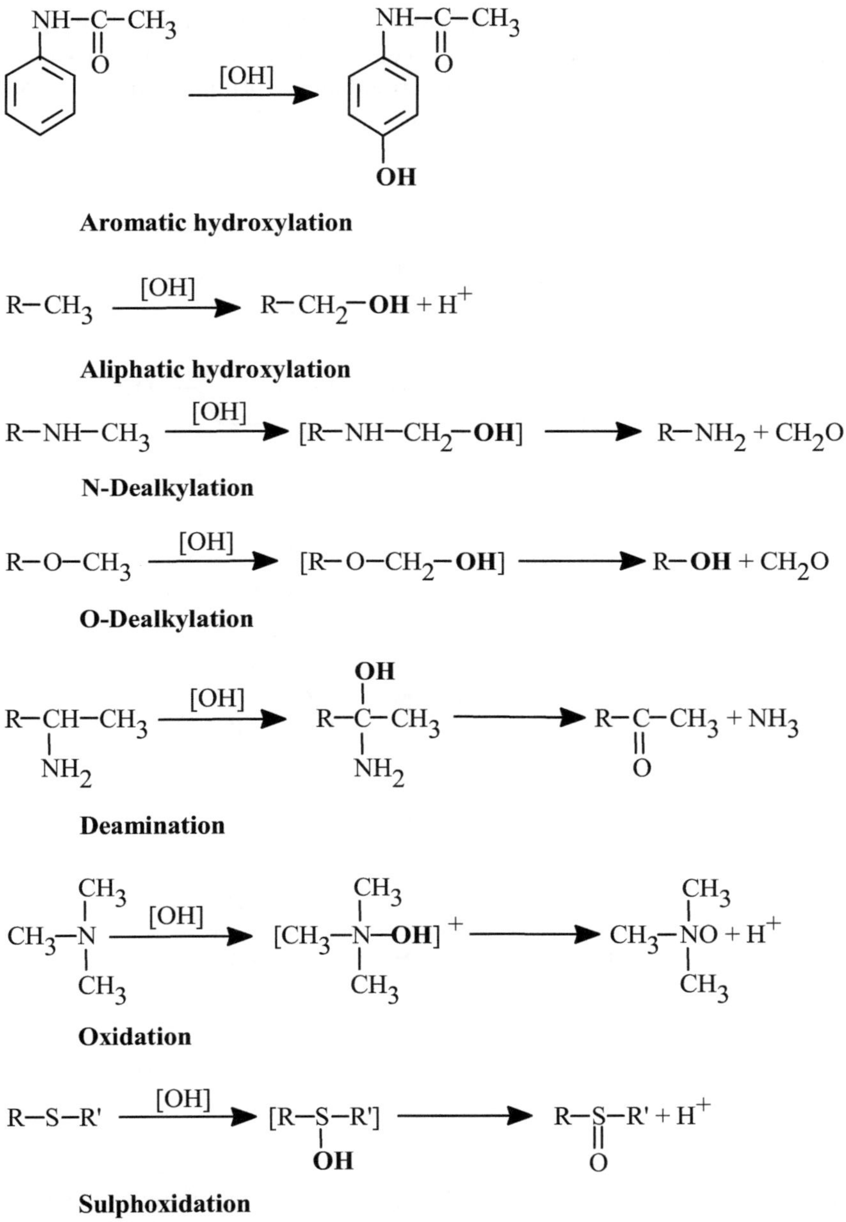

Fig.2.5 Common reactions catalysed by CYTP450; note the hydroxyl intermediates commonly occurring in these reactions

However, it ought to be stressed that the majority of oxidations are carbon oxidations, with carbon atoms in organic compounds greatly differing in their hybridisation and molecular environment, and consequently yielding a variety of oxidised intermediates (primary and secondary alcohols, phenols, epoxides) as presented in Figure 2.6:

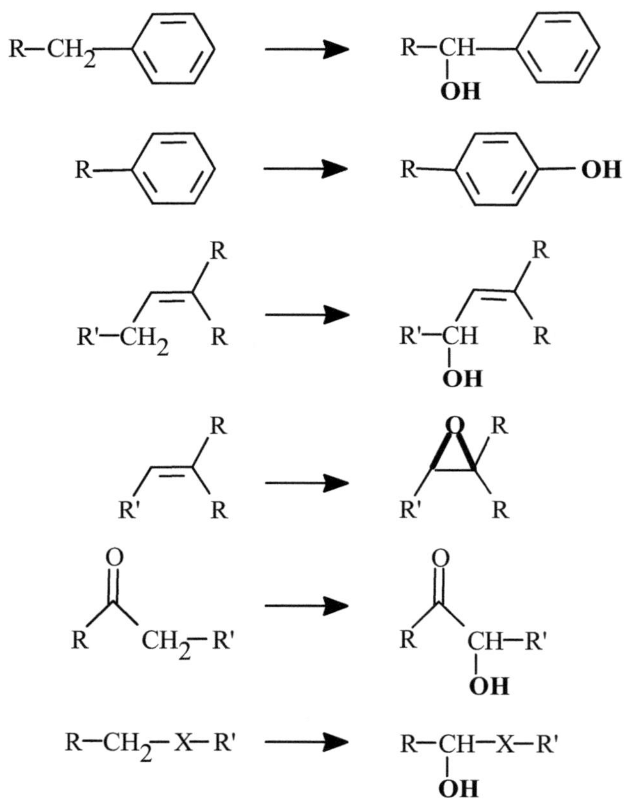

Fig.2.6 General P450-catalysed oxidation reactions at carbon centres

The general mechanism for aromatic hydroxylation involves an epoxide intermediate, illustrated in Figure 2.7, with naphthalene as substrate [23]. The formation of the epoxide involves the so-called NIH shift (NIH stands for U.S. National Institute of Health where the shift was discovered).

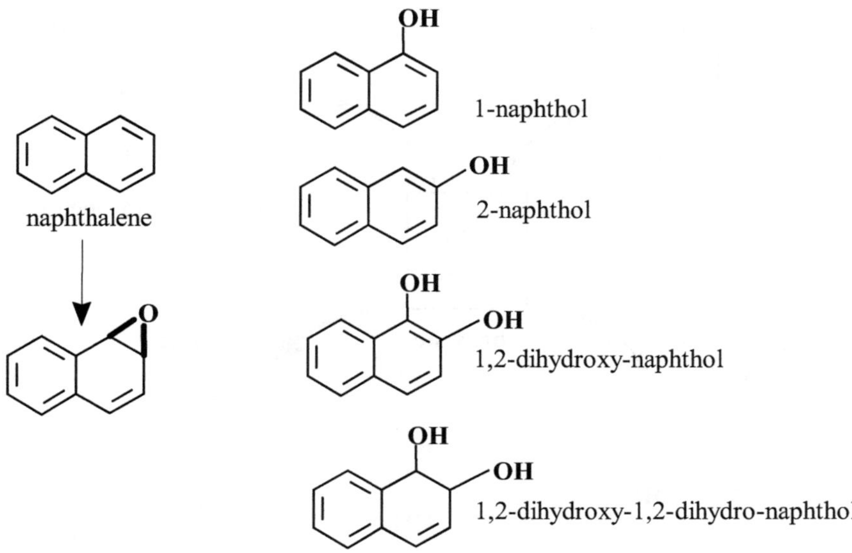

Fig.2.7 Epoxide intermediate in the biotransformation of naphthalene (Reproduced from ref. 25 with permission from R Paselk, Humboldt State University)

Degradation of naphthalene by specific Pseudomonas putida bacteria in soils has been reported [24] together with an assessment of the metabolites formed and their toxicities. The survival of the bacteria in non-sterile soil samples was measured in the presence and in the absence of naphthalene. The results of the study suggested that the metabolites catechol, related compounds and their condensation products may reach toxic levels in the stationary phase of the bacterial cells.

Oxidative metabolism of benzene gives a variety of products, with phenol as the major metabolite, as well as di- or even tri-hydroxylated metabolites [25], as indicated in Figure 2.8:

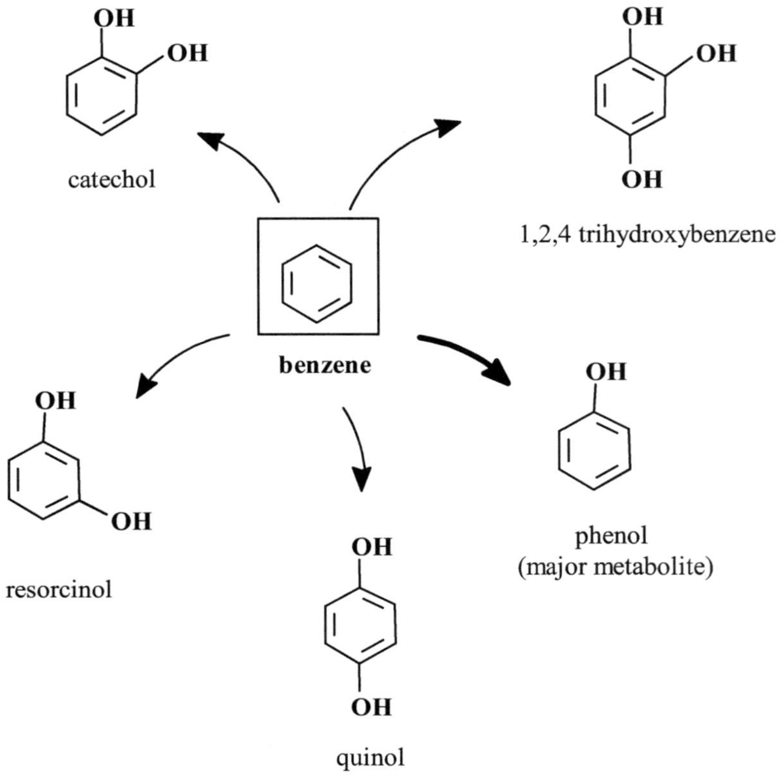

Fig.2.8 Benzene hydroxylation yielding mono-, di-, and tri-hydroxylated metabolites (Reproduced from ref. 25 with permission from R Paselk, Humboldt State University)

The carcinogenicity of benzene is related to the production of reactive oxygen species from its metabolites. A recent study of the mechanism of antiapoptotic effects (i.e. leading to prolonged cell survival) by benzene metabolites p-benzoquinone and hydroquinone in relation to carcinogenesis was reported [26]. Both metabolites were found to inhibit the apoptotic death of NIH3T3 cells induced by serum starvation as well as lack of an extracellular matrix. This inhibiting effect was reduced in the presence of an antioxidant, implicating the role of reactive oxygen species derived from the benzene metabolites. Further experiments suggested that the metabolites contribute to carcinogenesis by inducing dysregulation of apoptosis due to caspase-3 inhibition.

Reactions of the type shown in Figure 2.8 are important, because by a similar mechanism, aromatic hydroxylations can become metabolically activating, as seen in benzopyrenes, where the epoxide is a potent carcinogen [27] (Figure 2.9):

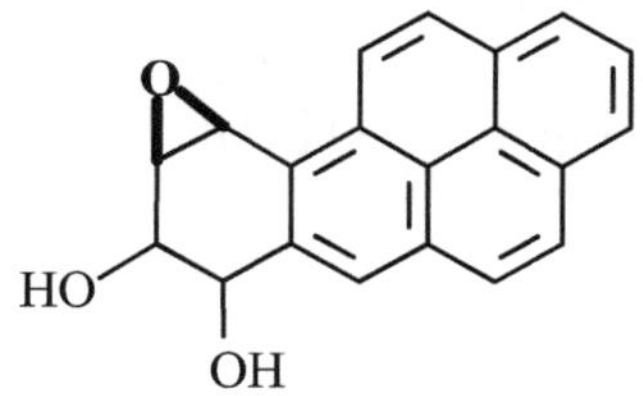

Fig.2.9 Example of toxic activation, yielding a potent carcinogen

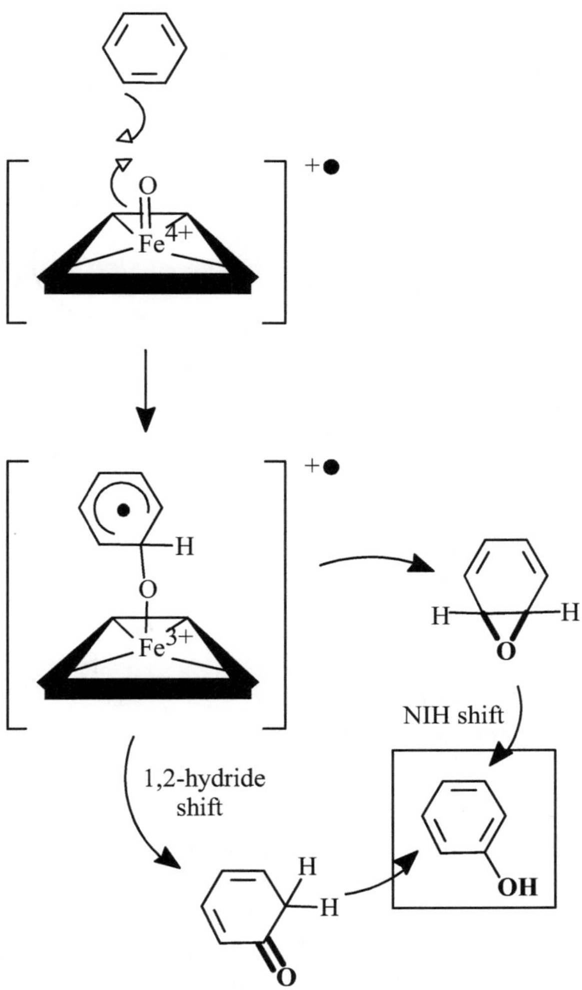

Fig.2.10 CYTP450-catalysed aromatic hydroxylations; radical iron-oxo species delivering oxygen. (Reproduced from ref. 23 with permission from Abby L. Parrill, University of Memphis)

As the aromatic hydroxylation mechanism involved, we present the radical iron-oxo species delivering oxygen (Figure 2.10), as well as details of the NIH shift mechanism (Figure 2.11) [23]:

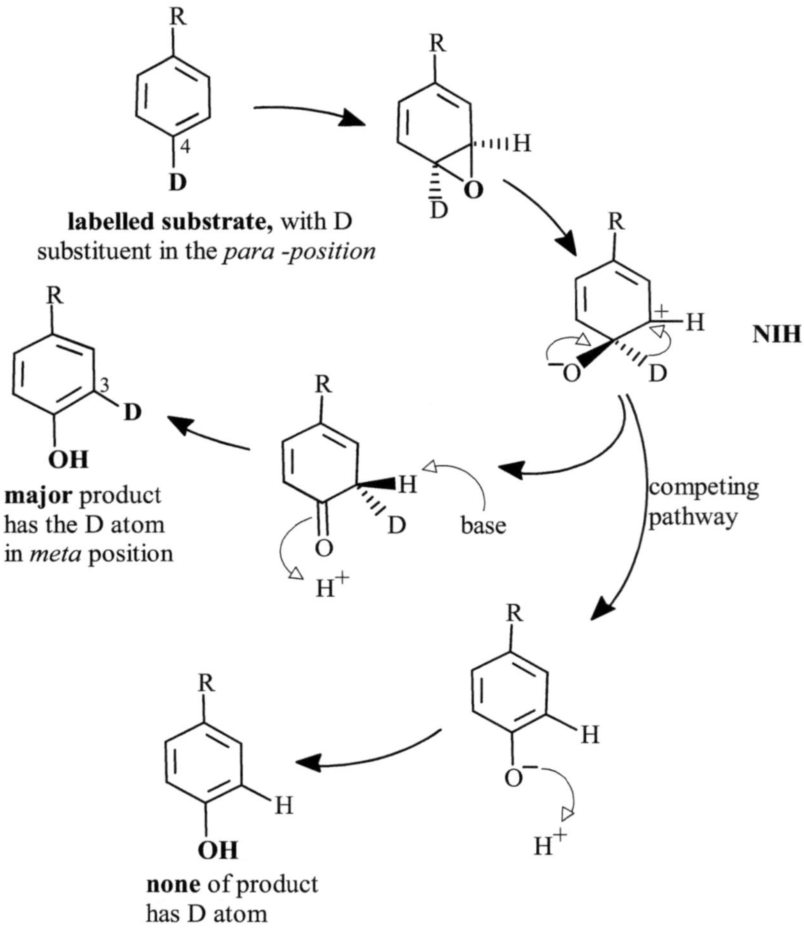

Fig.2.11 NIH shift mechanism
(Reproduced from ref. 23 with permission from Abby L. Parrill, University of Memphis)

The NIH shift is an intramolecular 1,2-hydride migration which can be observed in enzymatic and chemical hydroxylations of aromatic rings [23].

In enzymatic reactions the NIH shift is generally thought to derive from the rearrangement of arene oxide intermediates, but other pathways have been suggested.

The mechanism was first documented for a number of substrate molecules containing deuterium substituents ("D" in Figure 2.11) on their aromatic rings. Studies showed that hydroxylation at the labelled position will cause either migration (see the major intermediate in the figure), or loss, of the labelled substituent (the competing pathway, leading to the product without "D", in the figure presented). In addition, oxidative attack and hydroxylation *ortho-* to the label will also lead to both retention and loss of the label. Besides deuterium, the NIH shift may also affect halogenated substituents such as fluoro-, chloro- and bromo-.

Aliphatic compounds are not readily oxidised or metabolised unless there is an aromatic side chain; primary and secondary alcohols are formed (Figure 2.12).

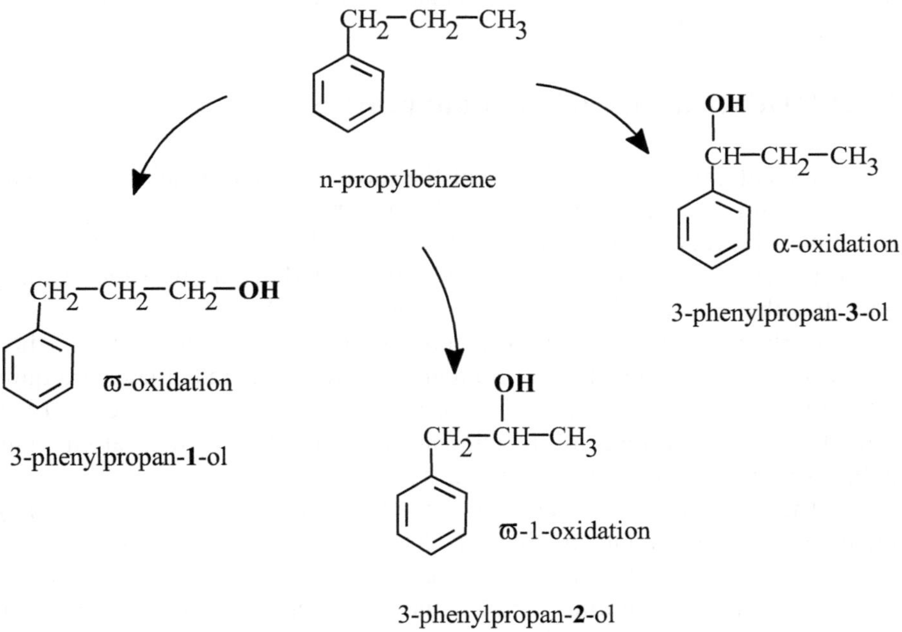

Fig.2.12 Aliphatic hydroxylations: preferred positions yielding primary or secondary alcohols (Reproduced from ref. 25 with permission from R Paselk, Humboldt State University)

Heterocyclic compounds are hydroxylated at the 3-position (Figure 2.13):

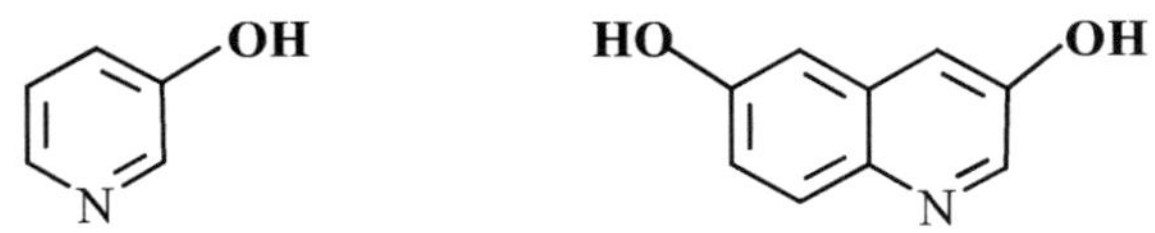

Fig.2.13 Preferred positions of hydroxylation for heterocyclic species (Reproduced from ref. 25 with permission from R Paselk, Humboldt State University)

Aromatic ring-hydroxylating dioxygenases (ARHD) are enzymes that effect reactions such as those shown above on aromatic hydrocarbons and heterocyclic molecules bearing various substituents. Aspects of their discovery, classification, enzymology, structure and properties have recently been reviewed [28].

2.3.2 Oxidations at carbon atom centres

Oxidations at carbon atoms represent the most common metabolic pathway for "attacking" the drug molecule. The major redox system catalyses the reductive cleavage of molecular oxygen, transferring one of the oxygen atoms to the substrate (resulting in the hydroxylated metabolite) and forming with the other one a molecule of water (see Eq. 2.1) [8,17,20].

Carbon atoms in organic compounds differ greatly in their hybridisation and molecular environment, and these characteristics are quite relevant as far as reactivity towards monooxygenases is concerned. Therefore, it is correct to distinguish the saturated carbon atoms (sp^3 hybridisation) from the unsaturated ones (sp^2 or sp).

Targets for such reactions are represented by methyl (CH_3-), methylene (-CH_2-), and methine (-CH=) groups respectively, and the resulting products are primary, secondary, and tertiary alcohols respectively. The resulting metabolites may undergo further biotransformations (dehydrogenations, oxygenations and/or conjugations) serving therefore as examples of sequential metabolism.

Primary alcohols are oxidised first to aldehydes. In aqueous solution, aldehydes being more easily oxidised than alcohols, oxidation usually continues until the carboxylic acid is formed (a metabolite that may further undergo conjugations).

Secondary alcohols are oxidised to ketones, which in alkaline solution can be oxidised further (the same situation as above).

Tertiary alcohols are not oxidised under alkaline conditions. In acidic solution, the tertiary alcohol undergoes dehydration and then the resulting alkene is oxidised.

Oxidations of sp^3-hybridised carbon atoms

Reaction mechanism of C-sp^3 hydroxylation

The general CytP450-mediated hydroxylation reaction of sp^3-hybridised carbons is described by Eq.2.3 and represents the overall substitution of a hydrogen atom by a hydroxyl group.

$$RR'R''C\text{-}\mathbf{H} + [O] \rightarrow RR'R''C\text{-}\mathbf{OH} \qquad 2.3$$

The mechanism of hydrogen radical abstraction is known as the oxygen rebound mechanism [29], and has been extensively studied and finally understood at the molecular level (Figure 2.14):

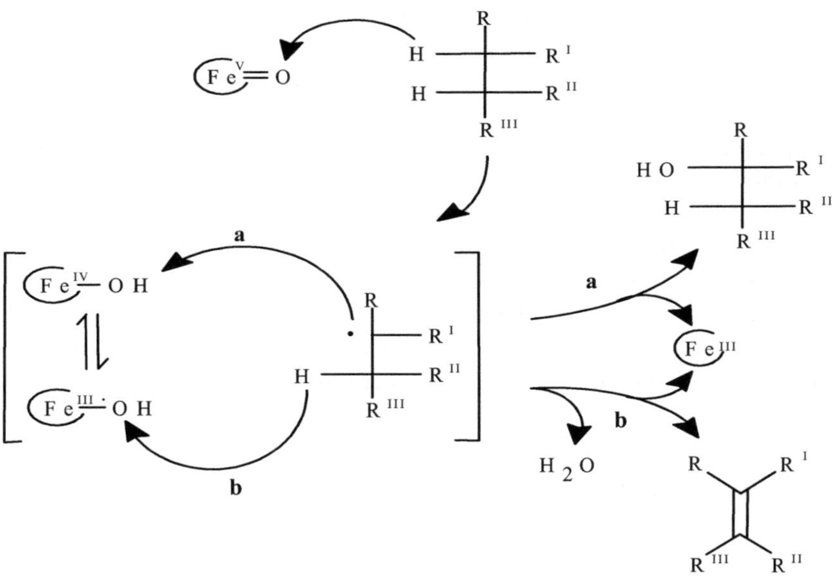

Fig.2.14 Mechanism of sp^3-carbon atom oxidation involving the CYTP450 system (Reprinted from ref. 29, p.125, with permission from Elsevier)

It is assumed that the perferryl-oxygen intermediate is responsible for homolytic cleavage of the C-H bond, the substrate consequently being transformed into a carbon-centred free radical. The enzyme thence becomes an iron-hydroxide intermediate to which the hydroxyl radical can be bound with variable strength. Following hydrogen abstraction, two outcomes are possible, namely oxygen rebound (reaction a, leading to the corresponding hydroxylated metabolite), or abstraction of a second hydrogen atom (reaction b), eliminating water and producing an olefin. Trapping of the carbon-radical intermediate by the iron-coordinated hydroxyl radical (before it can rearrange or break out of the solvent cage) is a point of crucial significance, since it explains why CytP450-mediated C-sp^3 hydroxylations are usually not toxication reactions liberating carbon-centred free radicals [29].

Reaction mechanism of C-sp^3 desaturation:
From Figure 2.14 an additional reaction of the carbon-centred radical intermediate is evident, namely desaturation, with consequent formation of an olefin (reaction b); the implicit condition is the presence in the substrate of two vicinal hydrogen atoms.

This type of reaction is of particular interest in the toxicological context of potent carcinogenic compounds/intermediates that undergo oxidative biotransformations to yield epoxides, as the latter can react with important biological macromolecules such as nucleic acids [30].

Methyl groups undergoing CytP450-mediated hydroxylation may present a variety of positions (e.g. in branched alkyl groups and on alicyclic compounds). In the case of the antihistaminic terfenadine the first biotransformation reaction is one of hydroxylation (Figure 2.15). The resulting metabolite may be further oxidised to the corresponding acid (documented also for finasteride and some other drugs). Moreover, in the case of terfenadine, the acid formed by oxidation of a methyl group has been identified as the major metabolite in human urine. As regards enzyme involvement, it has been proven that the first oxidation step is mediated by CYP3A isoenzymes [31]. Under certain circumstances, the resulting carboxylic acid may react further, undergoing decarboxylation.

In a study aimed at explaining and predicting adverse drug interactions associated with terfenadine, its extensive metabolism in a variety of intact hepatocytes from human and rat cultures was investigated recently [32], the rates and routes of metabolism being established by HPLC. Metabolites identified included products of C-oxidation and the N-dealkylation product azacyclonol. Various substrates and inhibitors of cytochrome P 4503A (CYP3A) were then tested for their ability to inhibit terfenadine metabolism, with a range of outcomes depending on the inhibitor and the type of hepatocyte. Human hepatocytes were suggested as having potential as a screening system for such inhibitors.

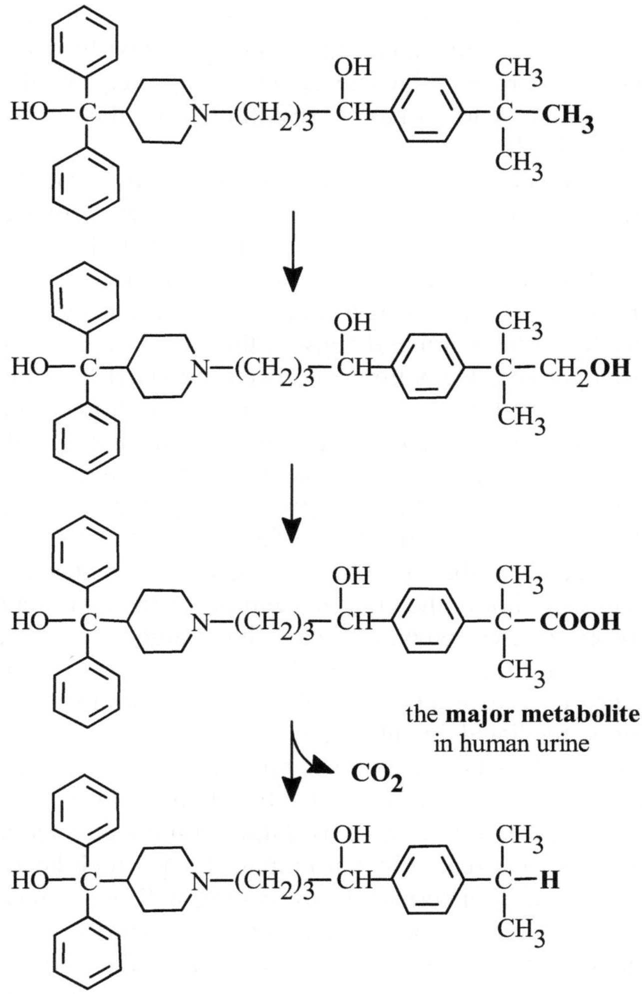

Fig.2.15 Steps in the biotransformation of terfenadine

A case of particular biochemical and physiological significance is offered by metabolism at the 10-methyl group in androgens. The methyl group being adjacent to a quaternary C-sp^3 centre in fact undergoes a reaction of C-demethylation, leading subsequently to ring A aromatization and estrogen formation. The enzymatic system involved is a CYTP450 aromatase (CYP19, also known as estrogen synthetase) [33] and the methyl group is oxidised to formic acid. Nevertheless, the reaction partly resembles the usual C-sp^3 hydroxylation, being initiated by a hydrogen abstraction and consequently yielding a carbon-centred radical. It is noteworthy that a similar mechanism might be involved in the oxidative breakdown of cardiac glycosides, involving sequential loss of two digitoxose units and finally resulting in the formation of a monodigitoxoside [34].

Several recent studies relating to androgen metabolising enzymes have appeared in the literature. One study relates aromatase activity to clinical effects associated with osteoporosis [35]. Evidence was found suggesting that in postmenopausal women, circulating adrenal androgen may be transformed into estrogen in peripheral tissues and may contribute to maintenance of bone mineral density, thus resulting in a protective effect against osteoporosis. Dehydroepiandrosterone (DHEA) can be converted sequentially into androstenedione and estrone in cultured human osteoblast apparently through aromatase activity. Localisation and function of androgen metabolising enzymes in the brain has been reviewed [36], as has the role of aromatase in the neuroprotective properties of estradiol [37]. In the latter work, neuroprotective effects of precursors of estradiol (e.g. testosterone) are described as being mediated by aromatase, suggesting that formation of estradiol in the brain is neuroprotective. Aromatase, described as a neuroprotective enzyme, was thus suggested as an important pharmacological target for therapies aimed at prevention of neurodegenerative disorders. Rapid changes in brain aromatase activity are evidently mediated by phosphorylation processes, as described in a recent review [37], where it was shown that such activity in hypothalamic homogenates is rapidly down-regulated by addition of Ca^{2+}, Mg^{2+} and ATP.

Another interesting case of oxidation of non-activated sp^3-hybridized carbon atoms is represented by barbiturates [38], which bearing alkyl side-chains attached to a quaternary C-sp^3 centre, will experience only limited activation. Barbiturates are metabolised primarily by the hepatic microsomal enzyme system and the metabolic products are excreted in the urine, and less commonly, in the faeces. Approximately 25 to 50% of a dose of aprobarbital or phenobarbital is eliminated unchanged in the urine, whereas the amount of other barbiturates excreted unchanged in the urine is negligible. The excretion of unmetabolised barbiturate is one feature that distinguishes the long-acting category from those belonging to other categories that are almost entirely metabolised. The inactive metabolites of the barbiturates are excreted as conjugates of glucuronic acid.

In the case of 3-carbon chains, hydroxylation occurs preferentially at the terminal carbon, while for side-chains of four or more carbon atoms, the antepenultimate carbon is preferred (position 3') [39] (Figure 2.16). In the case of pentobarbital, hydroxylation occurs on the pentyl side chain, the main metabolite (~40%) in human urine being the 3'-hydroxy derivative [40].

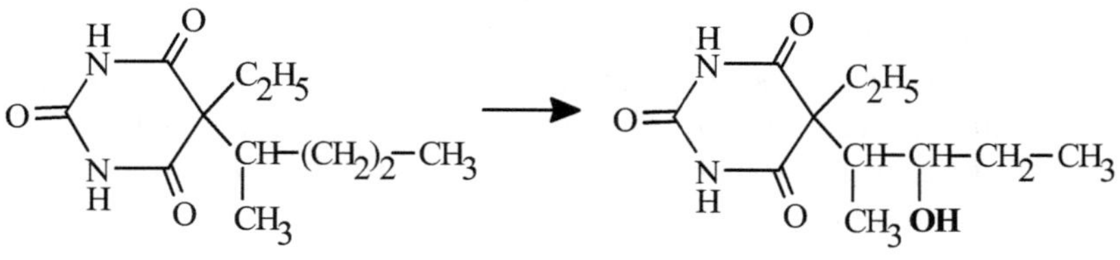

Fig.2.16 Hydroxylation of pentobarbital

Hydroxylations on ethyl groups may occur at either of the carbon atoms; a minor but very interesting metabolic reaction involves phenacetin [41] (Figure 2.17):

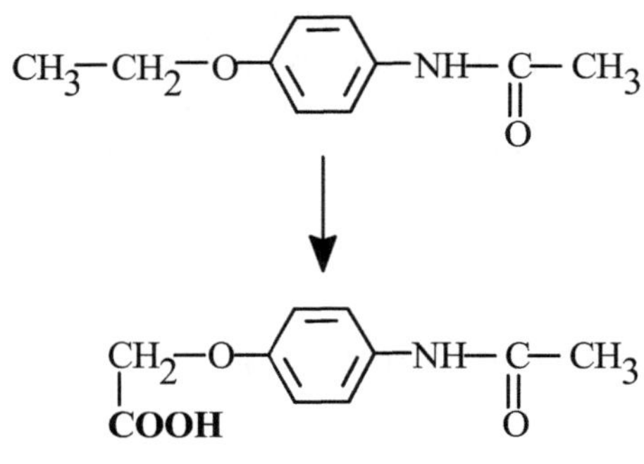

4-acetamidophenoxyacetic acid

Fig.2.17 β-hydroxylation of phenacetin

The resulting acid is the main urinary metabolite. A fact of particular relevance is that the reaction is strongly dependent on biological factors as well as pre-or co-administration of phenobarbital, which markedly increases the metabolite formation by enzyme induction. Also noteworthy is that phenacetin may undergo two other types of biotransformation: an N-hydroxylation (mediated by CYP1A enzymes), a reaction of great toxicological significance, yielding reactive electrophile intermediates which can form adducts with biological nucleophiles, and O-deethylation, mediated by two specific enzymes, CYP1A1 (aryl hydrocarbon hydroxylase) and CYP1A2, known as phenacetin O-deethylase. The O-deethylated metabolite resulted in the well-known acetaminophen (paracetamol). It is important to note that both enzymes are inducible by PAHs [42].

An *in vivo* and *in vitro* study of the metabolism of phenacetin in rats revealed its disappearance rate from blood and the activity of the enzyme phenacetin O-deethylase in liver to be at maximum in the morning and at a minimum in the evening [43]. However, these circadian variations are not completely responsible for previously observed rhythmical variations in the antipyretic action of phenacetin.

A study was undertaken to determine the *in vivo* role of the enzyme CYP1A2 in phenacetin-induced toxicity in mice [44], this enzyme being known to metabolise phenacetin *in vitro*. From experiments involving long-term feeding of phenacetin, the drug was found to be more toxic for mice lacking the enzyme than for controls, substantiating the conclusions that metabolism of phenacetin by this enzyme does alter *in vivo* toxicity and that alternate metabolic pathways contribute to its toxicity.

As in the case of phenacetin above, similar behaviour was demonstrated in the case of chlorpropamide undergoing hydroxylations at either of the carbon atoms on the *n*-propyl group (Figure 2.18), with the β-hydroxylation as the major metabolic route in humans [45]. We may stress here species differences in this drug biotransformation, the main pathway in rats being for instance α-hydroxylation.

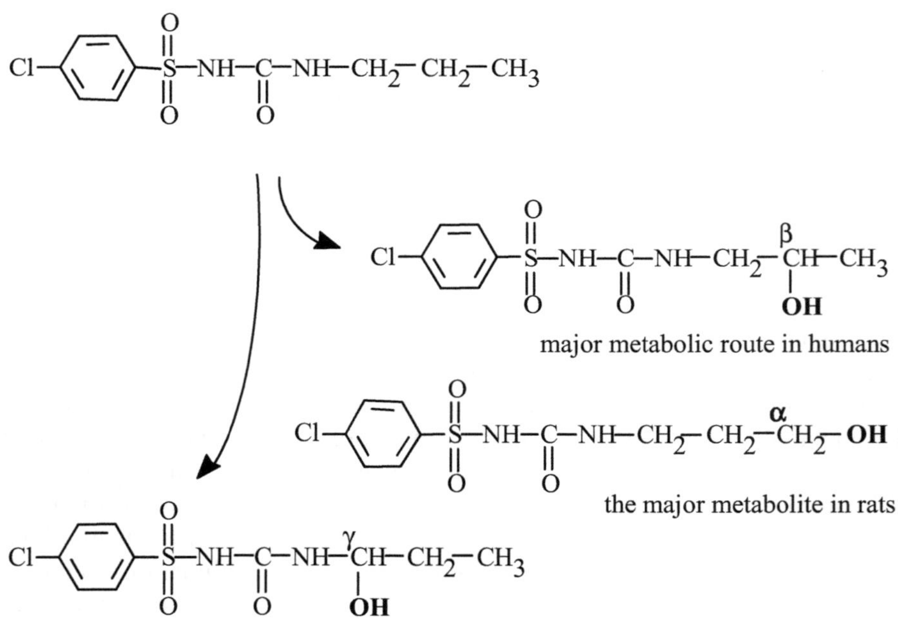

Fig.2.18 β-hydroxylation on the n-propyl group of chlorpropamide

As seen in Figure 2.18, we note that both α- and γ-hydroxylations occur in humans as well, but these represent only minor routes of biotransformation.

Non-activated C-sp^3 atoms in cycloalkyl groups are of particular interest because they appear as substituents in a number of drugs. At least two groups of cycloalkane derivatives warrant mention here, namely steroid hormones (Figure 2.19) and terpenes.

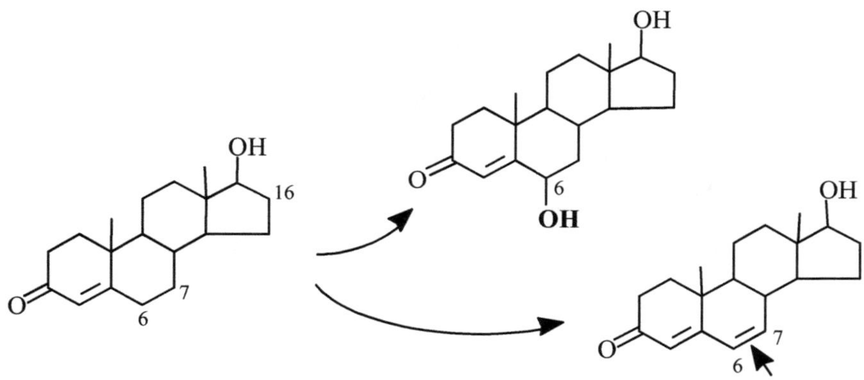

Fig.2.19 Hydroxylation and desaturation reactions on the molecule of testosterone (Reprinted from ref. 29, p.133, with permission from Elsevier)

They indeed deserve special mention due to the physiological and toxicological significance of their regio- and stereoselective CYTP450-catalysed hydroxylations [46].

Of relevance in the above example is the allylic oxidation at the **6**-position, although hydroxylated metabolites may occur with the –OH group at other positions too, in particular 2β, 15β, 16α and 16β. When incubated with liver microsomes (of dexamethasone-treated rats), the hormone yields two metabolites: the 6β-hydroxylated and the 6,7-desaturated. It is emphasised that it has been proven that desaturation in this case does not result from dehydration of corresponding 6β- or 7β-hydroxytestosterone, but occurs simultaneously with the 6β-hydroxylation, under the action of the same isozymes. Studies using CYP2A1 confirmed a double hydrogen abstraction mechanism, with the first abstraction involving the 6α-hydrogen [47].

Of clinical relevance, a report postulating the induction of testosterone metabolism by esomeprazole recently appeared [48]. This was based on an unusual episode in which a female patient gradually developed loss of libido while being treated with esomeprazole; testosterone supplementation or discontinuation of esomeprazole treatment reversed the effect.

Another interesting compound in the present context – due to its cyclohexyl and piperidyl groups, is phencyclidine (PCP) [49]. The drug presents two monohydroxylated metabolites with the –OH group in the 4-and 4'-positions, both appearing as major urinary metabolites; the ratio of the two products is species dependent, being 2:1 in humans and 4:1 in dogs (Figure 2.20).

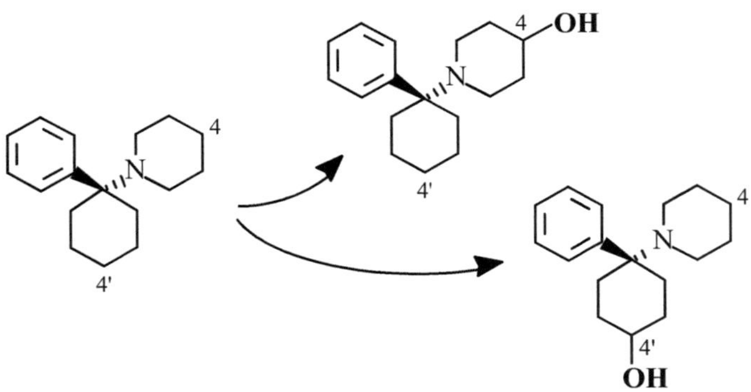

Fig.2.20 Major urinary metabolites of phencyclidine

A study of the *in vitro* metabolism of PCP using rabbit liver preparations [50] revealed the formation of four metabolites originating from oxidative metabolism of the piperidine ring, namely 5-(1-phenylcyclohexylamino)valeraldehyde,N-(1-phenylcyclohexyl)-1,2,3,4-tetrahydropyridine, 5-(1-phenylcyclohexylamino)valeric acid, and 1-phenylcyclohexylamine. The second of these was proposed as a work-up elimination product of the carbinolamine α-hydroxy-N-(1-phenylcyclohexyl)piperidine. Microsomal enzymes were necessary for the formation of all observed metabolites. Another study of PCP metabolism revealed that the cytochrome P450 3A plays a major role in the biotransformation of this drug [51]. Concurrent administration of potential inhibitors of cytochrome P450 3A could reduce the PCP elimination rate whereas potential inducers were able to accelerate it. Quantitative analysis of PCP and its main metabolites and analogues has been the subject of a recent review [52]. The report discusses (*inter alia*) analytical methodology and presents a scheme of PCP metabolism.

Several more complex saturated cyclic systems have also been investigated. Among them a well-studied example is that of tolazamide (with the methyl group in a toluyl moiety) which undergoes only one specific hydroxylation in humans [53], the 4'-hydroxylated derivative (Figure 2.21) being demonstrated to be the major urinary metabolite; no other ring-hydroxylated products were identified.

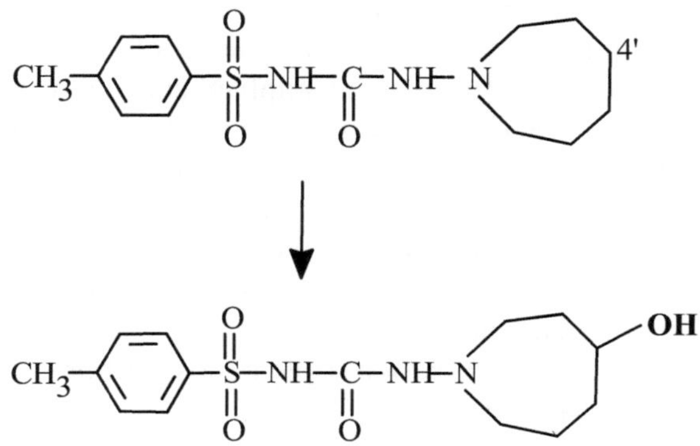

Fig.2.21 Specific hydroxylation of tolazamide in humans

The study of tolazamide metabolism in humans and rats employed tritium-labelled drug to identify metabolites [54]. Following administration of tritiated tolazamide to human subjects, 85% of the radioactivity was excreted in urine after several days in the form of both unchanged tolazamide and as many as six of its metabolites. The structure of one of these [1-(4-hydroxyhexahydroazepin-1-yl)-3-p-tolylsulfonylurea)] (Figure 2.21) was established by X-ray analysis. The other metabolites identified included 1-(hexahydroazepin-1-yl)-3-p-carboxyphenyl sulphonylurea, p-toluenesulphonamide,1-(hexahydroazepin-1-yl)-3-p-ydroxymethylphenyl) sulphonylurea, 1 - (4 - hydroxyhexa - hydroazepin - 1- yl) - 3 - p - tolylsulphonylurea, as well as a labile, unidentified metabolite. Relative amounts of these species present in urine of humans and rats were also reported [54].

Among other drugs undergoing similar specific hydroxylations we may mention also gliclazide and zolpidem. Regarding enzyme involvement, participation of the CYTP450 isoforms CYP2C8, CYP2C9 and CYP2C10 has been proven [55].

In an account of the metabolism of benzodiazepines [56] the extensive metabolism that the hypnotic zolpidem undergoes is mentioned. This includes oxidation of methyl groups and hydroxylation at a site on the imidazolepyridine ring. The CYP3A4 isoform is also known to be involved in the metabolism of zolpidem, indicating that interactions could occur between this drug and those that may be inhibitors or substrates of this isoform.

In another study of the metabolism of zolpidem [57], the kinetics of biotransformation to its three hydroxylated metabolites was determined *in vitro* using human liver microsomes. Microsomes that contained the human cytochrome P450 isoforms CYP1A2, 2C9, 2C19, 2D6, and 3A4 mediated zolpidem biotransformation. Inhibition of zolpidem metabolism in liver microsomes by ketoconazole and sulphaphenazole was established.

An interesting reaction is given by compounds containing unsaturated functional groups; it was observed that these groups direct hydroxylation to adjacent sp^3 carbons. Moreover, depending on a number of chemical and biological factors, it was noticed that the resulting regioselectivity may be either high or low. We may mention the following unsaturated systems as having been found to activate adjacent carbon atoms: aromatic rings, carbon-carbon double bonds, carbon-carbon triple bonds, carbonyl groups in ketones and amides, as well as cyano groups.

A shared characteristic displayed by the α-positions in such compounds is a common, larger electron density in the C-H bonds, and smaller electron densities on the C atoms. These results (from molecular orbital calculations) appear as important electronic indices for prediction of regioselective aliphatic hydroxylations. Midazolam is an example of therapeutic relevance, containing methyl groups adjacent to several aromatic heterocycles. The methyl group undergoes hydroxylation yielding 1'-hydroxymidazolam, found as the major metabolite in human plasma (Figure 2.22) [58]. The principal isoenzyme involved has been proven to be the CYTP3A4.

In another investigation, midazolam was used as a substance probe to determine the ability of hepatocytes from a whole adult human liver to serve as a model for studying xenobiotic metabolism [59].

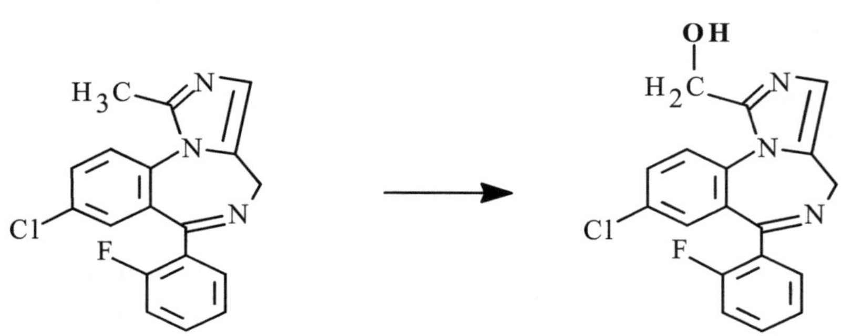

Fig.2.22 CYTP3A4-mediated hydroxylation of midazolam

Both Phase I and Phase II reactions were investigated. The hepatocytes resulted in the metabolism of midazolam to various hydroxylated metabolites, mainly 1-hydroxymidazolam, as such and as its glucuronide conjugate. The metabolism of midazolam in microsomal fractions obtained from human livers was found to be extensive and mediated primarily by a single cytochrome P450 enzyme. A very recent study aimed at quantitative prediction of *in vivo* drug interactions with macrolide antibiotics in humans [60] centred around the metabolism of midazolam. Using human liver microsomes, α- and 4-hydroxylation of midazolam were evaluated as CYP3A-mediated reactions. This metabolism was found to be inhibited following pre-incubation with macrolides such as erythromycin and azithromycin, in the presence of NADPH. (Kinetic data for enzyme inactivation were subsequently used in a simulation of *in vivo* interactions based on a physiological flow model, yielding results that were consistent with experimental *in vivo* data).

In the case of reactions of side-chains longer than a methyl group, we may mention benzylic hydroxylation, which is also a significant reaction in the biotransformation of a number of drugs. We illustrate an interesting example, namely the hydroxylation of metoprolol (Figure 2.23); it undergoes 1'-hydroxylation (in rats), with the diastereomeric benzylic metabolites predominantly in the 1'-(R-) configuration [61]:

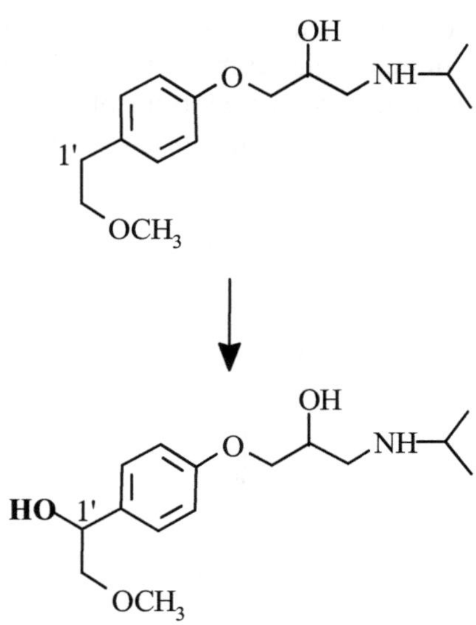

Fig.2.23 Stereospecific 1'-hydroxylation of metoprolol yielding mainly the 1'-(R-) product

A report on the frequency distribution of the 8h urinary ratio metoprolol/hydroxymetoprolol in a specific population has appeared [62] showing that age may be a factor.

Interesting oxidations of C-sp^3 atoms adjacent to other unsaturated systems involve hydroxylation of allylic positions (in side-chains) or cycloalkenyl groups. An example of medical relevance is hexobarbital; its major metabolic route involves 3'-hydroxylation (Figure 2.24), followed by dehydrogenation yielding the corresponding 3'-keto metabolite:

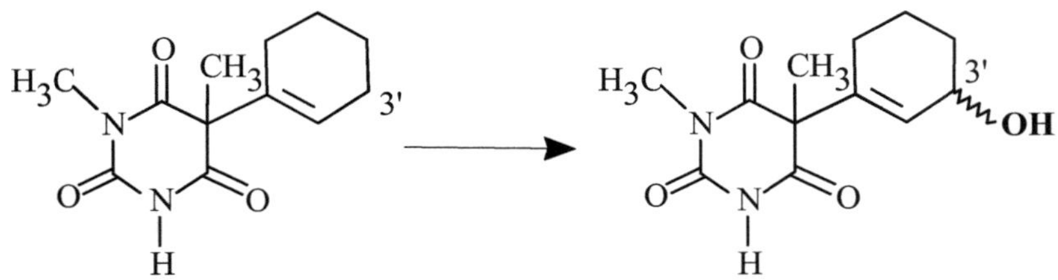

Fig. 2.24 Hydroxylation of hexobarbital

An important point to stress is that this hydroxylation reaction displays a complex array of substrate and product stereoselectivities, since the molecule is chiral and the 3'-carbon is prochiral. The phenomenon has been proven to be markedly influenced by biological factors [63].

With an even higher selectivity with respect to hydroxylation, glutethimide deserves mention as an important pharmacological example in view of the properties of the resulting metabolite (Figure 2.25):

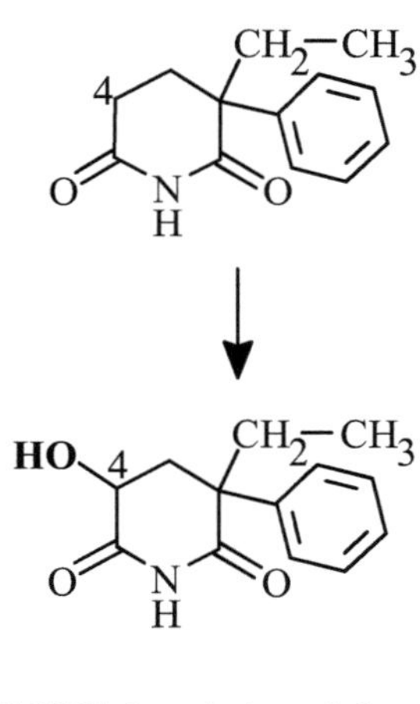

Fig.2.25 Hydroxylation of glutethimide

It is an interesting example illustrating the case of carbons adjacent to carbonyl groups, with the α-position having high selectivity. From a range of resulting hydroxylated intermediates, the 4-hydroxy derivative (either free or in conjugated form) has been shown to be the major plasmatic and urinary metabolite [64]. The reason for highlighting this case is that this particular metabolite is a more active sedative-hypnotic agent than the parent drug, while, on the other hand it is believed to be responsible for most of the severe symptoms displayed by intoxicated patients [64]. As an aside, the fine structural line dividing e.g. convulsant/anticonvulsant or sedative/stimulant properties was some years ago indicated as potentially exploitable for generating drugs of abuse. Glutethimide was specifically mentioned in this context [65].

Hydroxylations of carbon atoms adjacent to acetylenic or cyano groups are not very specific, nor relevant for drug metabolism. Such reactions have been studied for e.g. acetonitrile (or higher nitriles), which clearly are not drug substances; nevertheless, the reaction mechanism, the enzymatic systems involved, as well as the α-hydroxylation regioselectivity are well known and explained at the molecular level [66].

As previously mentioned in this subchapter, carbon atoms undergoing oxidations may be unsaturated as well, thus presenting either sp^2- or, sp-hybridisation.

We present briefly the CYTP450-mediated oxidation of C-sp^2 atoms in aromatic rings, as a highly complex metabolic route leading to a variety of products; these can be either unstable intermediates or stable metabolites. A noteworthy feature here is the heavy dependence of the chemical reactivity of the intermediates (e.g. epoxides) on the chemical nature of the target group and molecular properties of the substrate. This reactivity also determines the nature and the relative amounts of stable metabolites produced.

A well-documented example concerns the mechanism of ring oxygenation (Figure 2.26) [67]. The reaction results in loss of aromaticity, due to the formation of tetrahedral transition states following the activation of the CYTP450-oxygen complex (reaction a). Three oxygenated intermediates may be formed, through alternative reactions c, d and/or e and f. The essence of the entire pathway is the formation of the cation-radical (reaction b) that will bind the activated oxygen (reaction a). The products of this phase are a biradical and a cationic oxygenated intermediate. The second phase subsequently provides two possible rearrangement pathways, leading to the stable phenolic metabolites. An interesting aspect to emphasise is that while the biradical pathways show little (or even no) substituent effects, quite the opposite applies to the cationic pathways.

The arene oxide intermediates (reaction g) are usually highly unstable, easily undergoing ring opening by a mechanism of general acid catalysis, leading ultimately to the stable phenols [68]. More stable epoxides are those of polycyclic aromatic hydrocarbons and olefins.

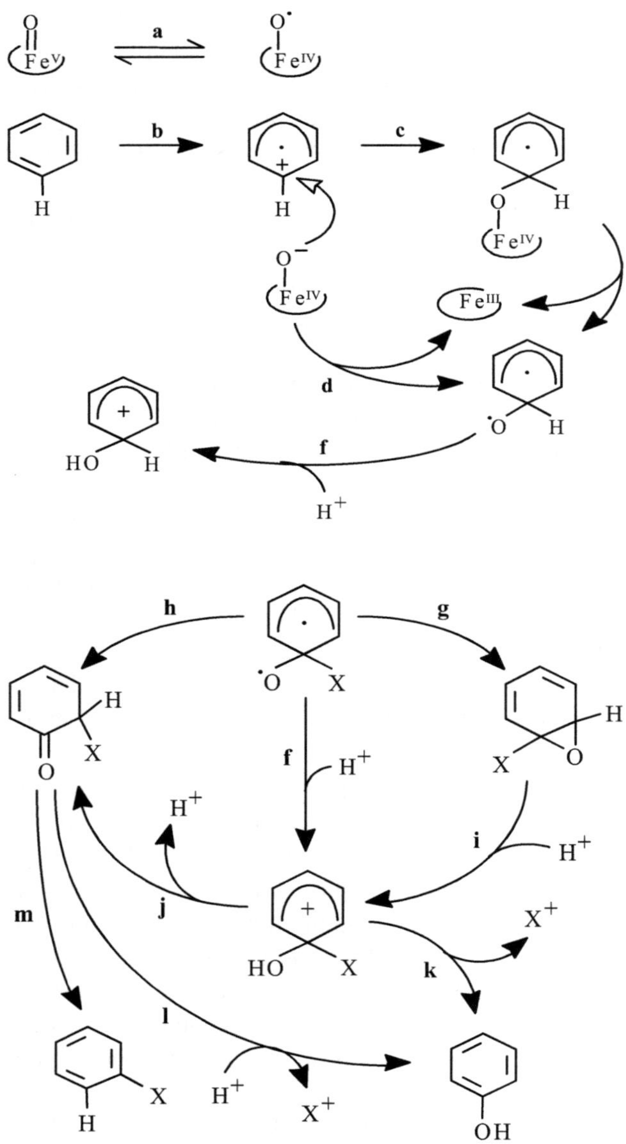

Fig.2.26 Mechanism of ring oxygenation, leading ultimately to the stable phenols (Reprinted from ref. 67 with permission from Elsevier)

Other interesting examples in this context refer to NIH shift (displacement involving migration of the geminal hydrogen atom), regio- and stereoselectivity in aryl oxidation of certain drugs.

Regioselectivity is well-documented for diclofenac (only three of the seven possible positions being hydroxylated [69]), while in the case of (S)-mephenytoin, the substrate regioselectivity as well as enantioselectivity are evident. Cytochromes P450 from the CYP2C subfamily catalyse the para-hydroxylation of mephenytoin with high efficacy and a marked preference for the (*S*-)-enantiomer [70].

Biotransformation of mephenytoin to its two major metabolites, 4-hydroxymephenytoin and 5-phenyl-5-ethylhydantoin in human liver microsomes has been investigated [71]. Metabolism was found to be stereoselective, (*S*-)-mephenytoin being preferentially converted to the 4-hydroxy derivative at low substrate concentrations while the (*R-)*-enantiomer was demethylated to 5-phenyl-5-ethylhydantoin over a wide concentration range. Mediation by cytochrome P450-type monooxygenases was established.

A more complex situation, combining product regioselectivity with substrate enantioselectivity, is encountered in the metabolism of propranolol [72]. Figure 2.27 illustrates the regioselective aspects.

In mammals, oxidative metabolic pathways include hydroxylations of the naphthalene ring at the 4-, 5-, and 7-positions as well as side-chain N-desisopropylation [73]. Cytochrome P450 isozymes are involved in propranolol metabolism in human liver microsomes, where the 4-OH, 5-OH and N-desisopropyl derivatives occur as primary metabolites and the 7-OH species is present in trace quantities.

The main route in humans is the CYP2D6-mediated 4-hydroxylation; alternatively, 5-, 2-, and 7-hydroxylations may also occur. Among the four monohydroxylated metabolites, the 4- and 5-hydroxy species are poor substrates for a second hydroxylation. In contrast, the 2- and the 7-hydroxylated metabolites may subsequently be easily hydroxylated at the preferred 3-position, yielding the corresponding dihydroxylated metabolites, shown in the central part of the figure. Three other dihydroxylated metabolites can be also found in human urine, though in very small amounts, namely the 4,6-, 4,8- and 3,4-derivatives.

While 4-hydroxylation showed no apparent substrate stereoselectivity, the 7-hydroxylation has been proven to be selective for (+)-(*R*-)-propranolol (in a ratio of about 20:1), while the 5-hydroxylation was selective for (-)-(*S*-)-propranolol, in a 3:1 ratio [72].

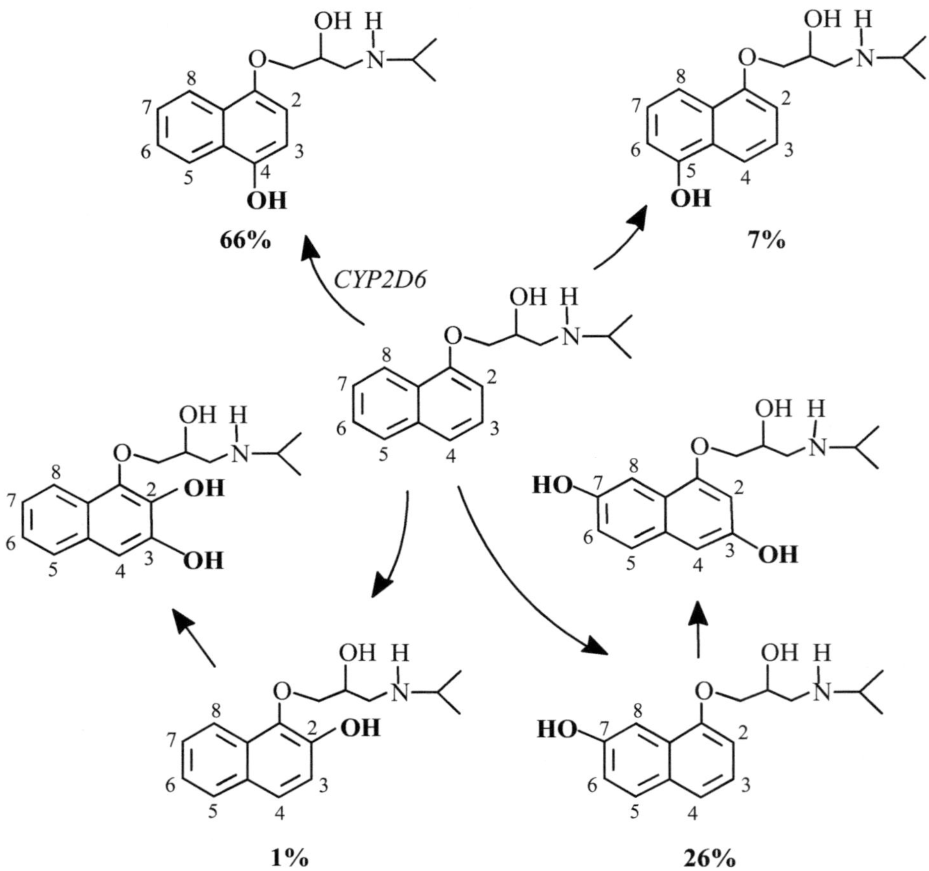

Fig.2.27 Complex regioselective biotransformation reactions of propranolol and the corresponding relative amounts of the monophenolic metabolites

Concerning product enantioselectivity in aryl oxidation, the traditional example for illustrating this is phenytoin (Figure 2.28).

In humans this phenyl hydroxylation is mediated by the isoform CYTP450C and occurs almost exclusively at the para-position, with the ratio of the two enantiomeric metabolites (S-/R-) about 10:1 [73]. Again, it is important to stress the species differences: in contrast to humans, the meta-phenol is formed preferentially in dogs and is the pure (R-)-enantiomer.

Another interesting aspect to note is that both meta- and para-phenols are formed from the same intermediate – the 3,4-epoxide [74].

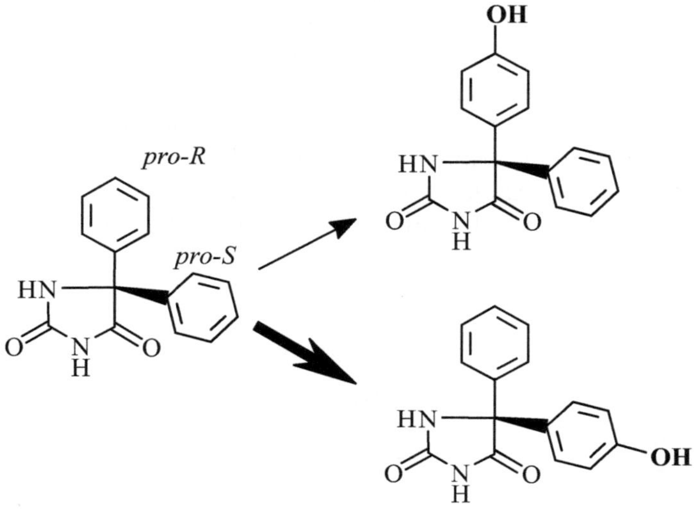

Fig.2.28 Enantioselectivity in the biotransformation of phenytoin

The 5-(p-hydroxyphenyl)-5-phenylhydantoin may be further metabolised to a catechol. Spontaneous oxidation of the catechol then leads to semiquinone and quinone species that modify proteins by forming covalent linkages [75].

The hydroxylation of phenols is of particular interest. As a rule, it has been demonstrated that when the position para- to the first hydroxyl group is free, it will generally be hydroxylated more rapidly than the ortho-position. An important example of a drug following this pattern is given by salicylamide (Figure 2.29) [76]; about 50% of an oral dose of salicylamide administered to mice was recovered as the 5-hydroxylated metabolite, while only about 20% of the dose underwent 3-hydroxylation.

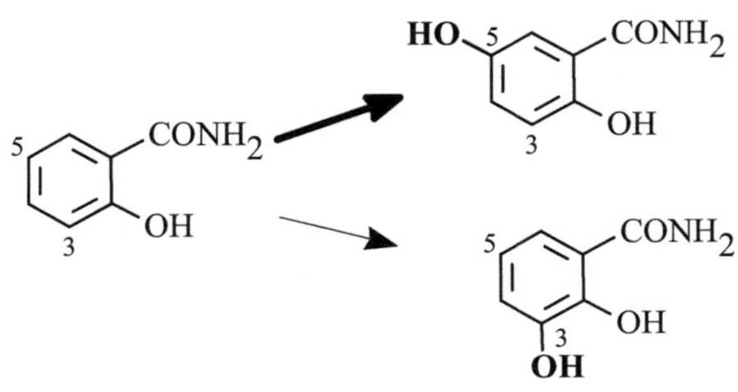

Fig.2.29 The two preferred positions of hydroxylation of salicylamide

Polycyclic aromatic hydrocarbons (PAHs) have been the subject of intensive study due to their toxicological significance. Representatives of these compounds display high carcinogenic potencies following their toxication to reactive metabolites - ultimately called carcinogens [77].

One of the most carcinogenic PAHs is benzo[α]pyrene, present in tobacco smoke. Figure 2.30 presents the three major epoxide metabolites, their hydration to dihydrodiols (by epoxide hydrolase), as well as the epoxidation of the M-region (which is the most electron-rich region in the molecule) dihydrodiol to a dihydrodiolepoxide, considered to be the ultimate carcinogen:

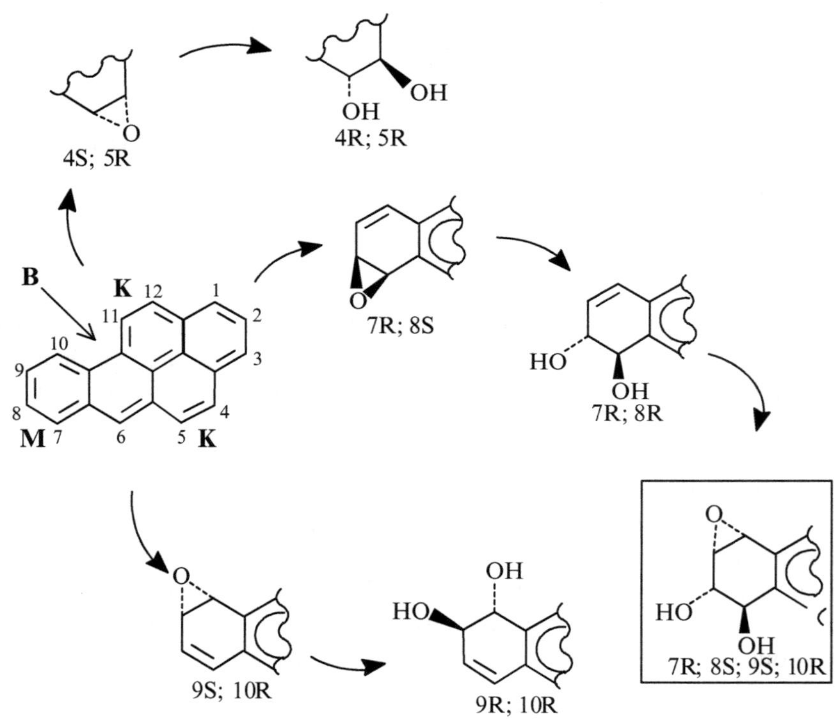

Fig.2.30 Major epoxide metabolites of PAHs
(Reprinted from ref. 27 with permission from Elsevier)

Diol-epoxides rearrange to a triol carbonium ion (Figure 2.31) which will then react covalently with e.g. nucleophilic sites in nucleic acids [78].

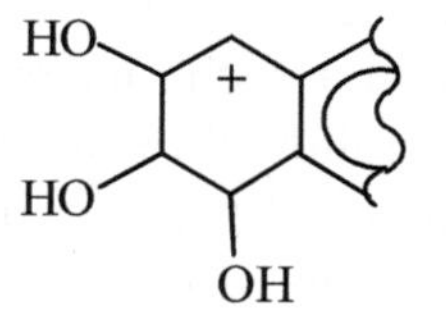

Fig.2.31 The triol carbonium ion

Oxidation of sp^2-hybridised carbon atoms

Apart from their presence in carbon-carbon bonds of aromatic systems, sp^2-hybridised carbon atoms occur, either isolated or conjugated, in olefinic bonds. Bonds of this type are found in a large variety of xenobiotics, as well as in various endogenous substrates (e.g. arachidonic acid) [79]. They usually undergo CYTP450-catalysed oxidation to epoxides and a few other products.

The mechanism of olefin oxidation involves two distinct formations of C-O bonds (pathways a and b) shown in Figure 2.32.

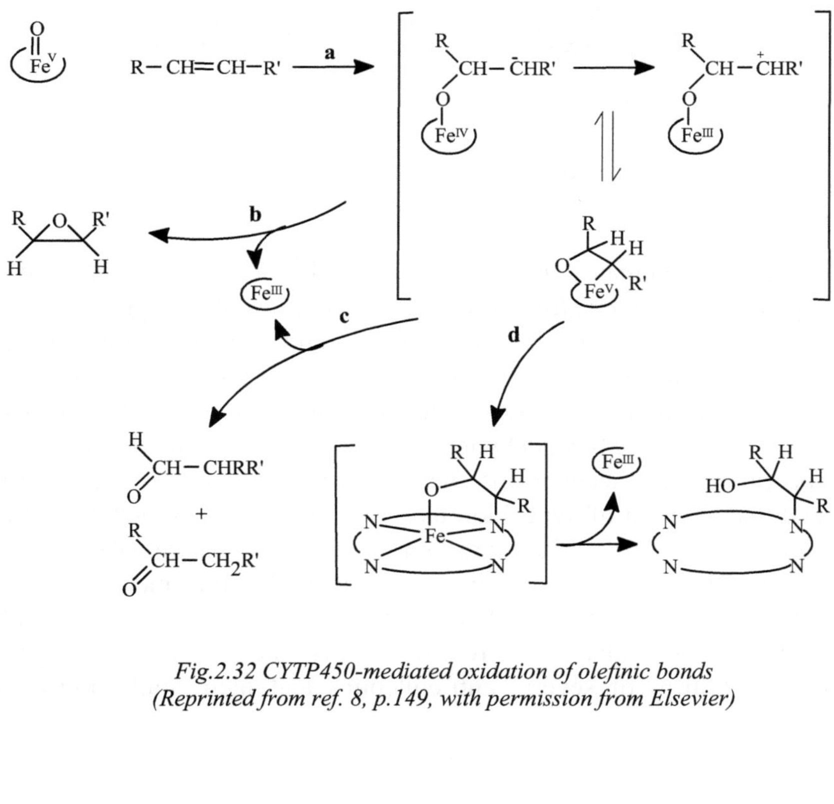

Fig.2.32 CYTP450-mediated oxidation of olefinic bonds (Reprinted from ref. 8, p.149, with permission from Elsevier)

After the first C-O bond forms (reaction a), at least three intermediates arise: a radical, a carbocation and a cyclic intermediate.These highly reactive intermediates can subsequently follow three alternative pathways: formation of the second C-O bond (reaction b), generation of carbonyl derivatives (reaction c) and covalent binding to heme (reaction d) with the subsequent formation of abnormal N-alkylporphyrins ("green pigments"). It is this last reaction by which some compounds (called "suicide-substrates") can act as irreversible mechanism-based enzyme inhibitors [80]. Further information appears in Chapter 5, section 5.2.

A number of drugs (or metabolites) form olefinic epoxides which can either be stable or rearrange intramolecularly. An example is given by carbamazepine (Figure 2.33). This tricyclic drug yields more then 30 metabolites, among which appears the 10-11-epoxide, not as a predominant one, but a nevertheless pharmacologically active species [81].

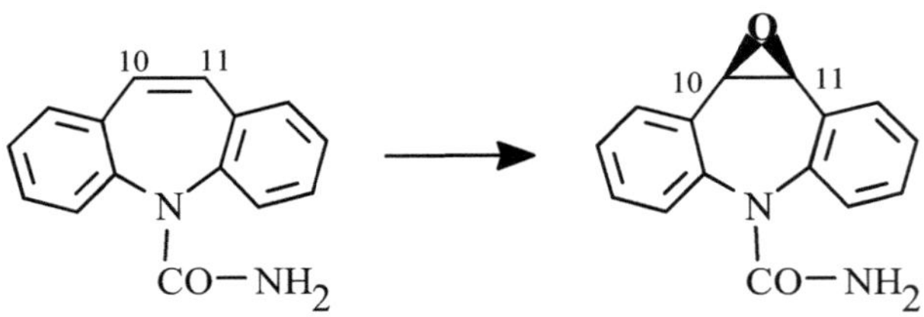

Fig.2.33 Olefinic type epoxidation of carbamazepine

Actually, in humans, epoxidation followed by enzymatic hydration is a major pathway of biotransformation of tricyclic drugs of this type.

From a study of the metabolism and covalent binding of carbamazepine with the MPO/H_2O_2/Cl^- system and neutrophils, a common pathway was identified [82]. Metabolites detected included an intermediate aldehyde, 9-acridine carboxaldehyde, acridine, acridone, chloracridone and dichloroacridone. To account for the observed ring contraction, it was suggested that reaction of hypochlorous acid with the 10,11-double bond of carbamazepine yields a carbonium ion as the first intermediate in its metabolism. This pathway is similar to that for the metabolism of iminostilbene, a metabolite of carbamazepine, but differs in rate and details of mechanism. The reader is also referred to a recent review describing the variable metabolism of several anti-epileptics and their implications for therapy [83].

Epoxide rearrangement reactions generally include formation of a lactone. Such an intramolecular nucleophilic reaction occurs during the metabolism of hexobarbital (see Figure 2.24 above). The major metabolic route involves 3'-hydroxylation followed by dehydrogenation to the corresponding 3'-keto metabolite. Through an alternative pathway (species dependent) the epoxide intermediate may arise, followed by cyclisation, involving an intramolecular rearrangement. In contrast, olefinic epoxidation in allylic chains, such as occurs in alclofenac, was found to represent only a very minor biotransformation, the resulting metabolite accounting for 0.01% or even less of a dose in humans [84] (Figure 2.34).

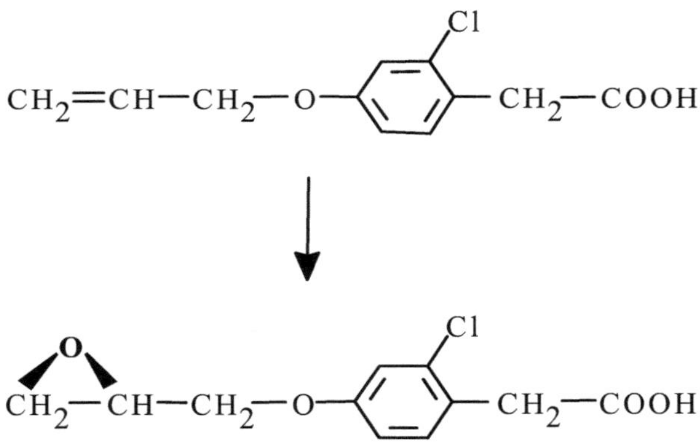

Fig.2.34 Epoxidation of a double bond in an allylic chain illustrated for alclofenac

Alclofenac underwent no metabolism in control mouse hepatic microsomes, but in microsomes induced by phenobarbitone or 3-methylcholanthrene, it was found to biotransform to its dihydroxy and phenolic derivatives [85]. These metabolites did not destroy cytochrome P450 *in vitro* but formation of the reactive epoxide intermediate was cited as partly mediating destruction of the enzyme.

Oxidation of sp-hybridised carbon atoms

sp-hybridised carbon atoms, found in carbon-carbon triple bonds (e.g. in alkynes) undergoCYTP450-mediated oxidation to a number of products. The mechanism is very similar to that applying to oxidation of olefinic bonds and it is postulated that several reactive intermediates are generated (Figure 2.35).

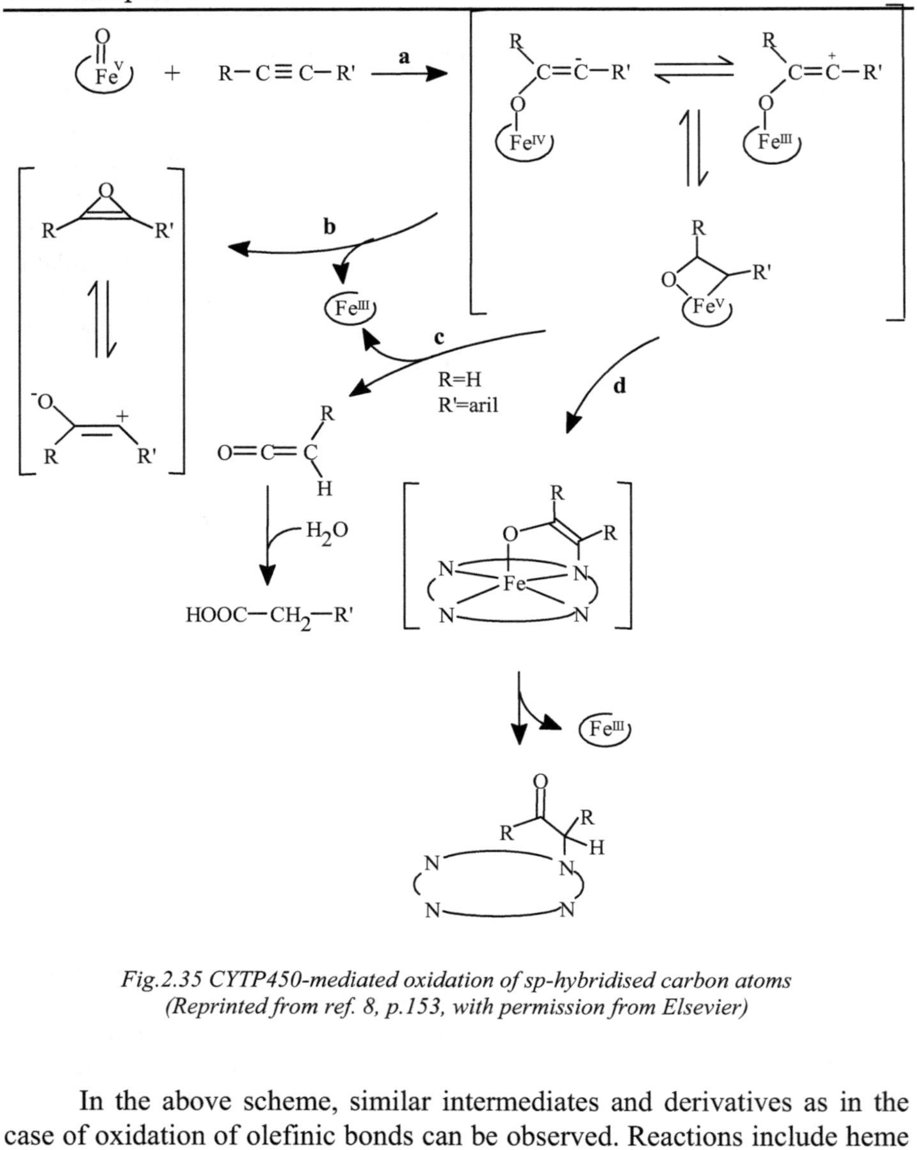

Fig.2.35 CYTP450-mediated oxidation of sp-hybridised carbon atoms (Reprinted from ref. 8, p.153, with permission from Elsevier)

In the above scheme, similar intermediates and derivatives as in the case of oxidation of olefinic bonds can be observed. Reactions include heme alkylation and consecutive destruction of CYTP450 with subsequent appearance of abnormal "green pigments" [86].

An interesting example of acetylenic oxidation yielding D-homosteroids (steroids with a six-membered D-ring) is afforded by the 17α-ethynyl steroids (Figure 2.36). The reactions are illustrated for norethindrone, derived from hydrolysis of its acetate in most tissues including skin and blood. The site of primary metabolism of norethindrone is the liver, but the first-pass effect may be significantly reduced by administering the drug transdermally.

The main reactive intermediate, the epoxide, following successive oxidations and a decarboxylation, will finally yield the rearranged, six-membered D-ring [87].

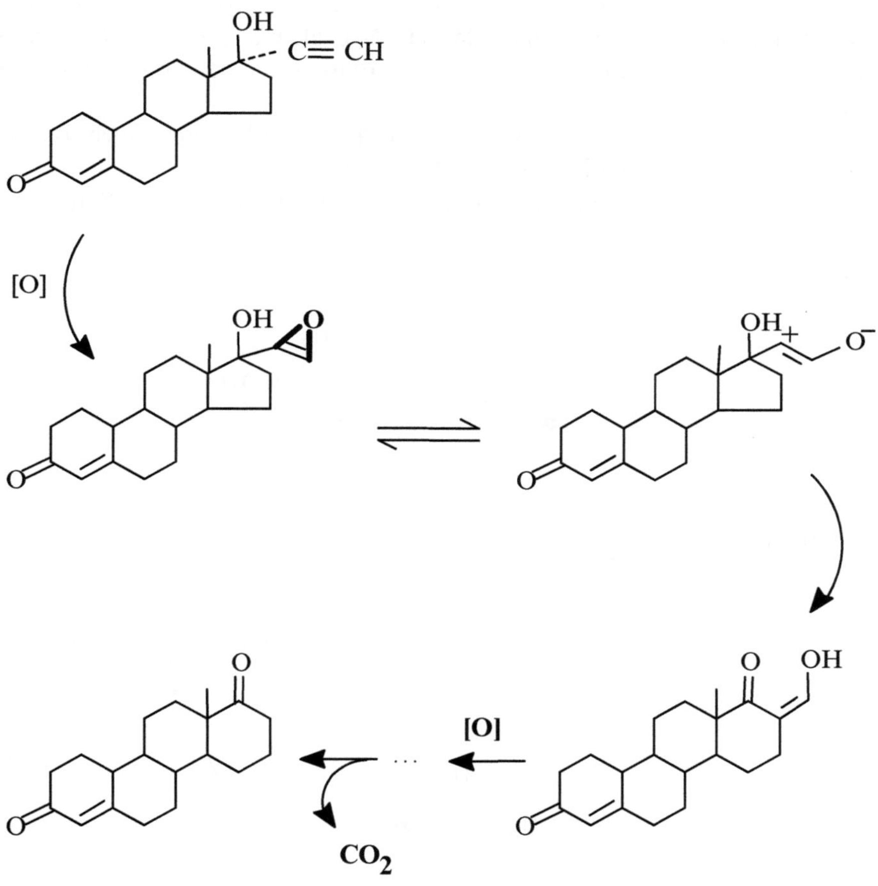

Fig.2.36 Mechanism of D-homoannulation of norethindrone

2.3.3 Oxidations at hetero-atoms

A large variety of drugs known to contain hetero-atoms such as O, N, S or P are substrates for reactions of oxidation, reduction and hydrolysis.

We may mention from the outset the diversity of these reactions, including principally nitrogen oxidations, N-C cleavage, oxidation of oxygen and sulphur-containing compounds, oxidative dehalogenations and dealkylations.

From the numerous examples of drugs undergoing such types of biotransformations we give some representative cases:

- the group of primary amines: e.g. phentermine [88] (Figure 2.37) (Ar = Ph):

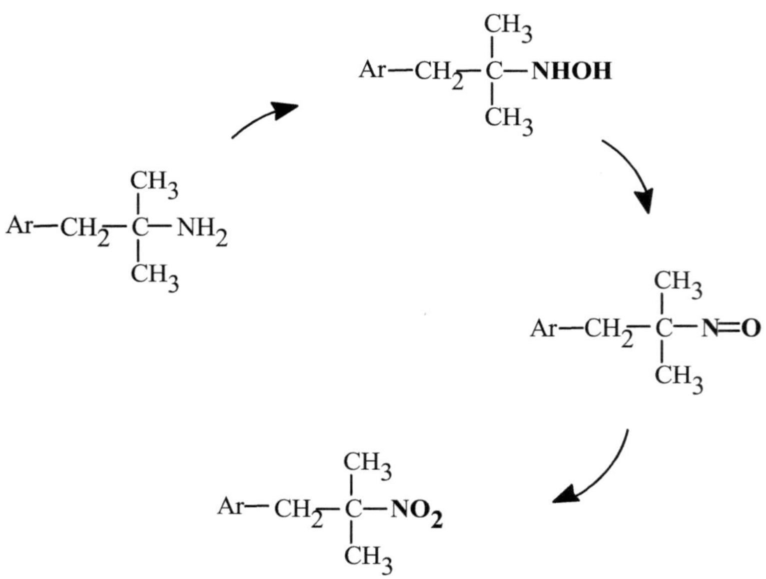

Fig.2.37 N-oxidation of primary amines, partially MFO- and CYTP450-catalysed

Oral and intraperitoneal dosing of phentermine yielded a p-hydroxy phentermine conjugate as the major metabolite in urine; N-hydroxy phentermine yielded a p-hydroxyphentermine conjugate [89].

- the group of primary arylamines : procainamide [90] (Figure 2.38):

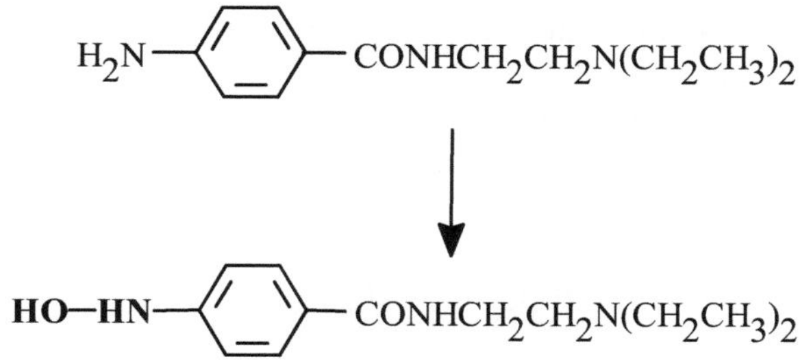

Fig.2.38 Procainamide hydroxylation (common pathway of metabolism in rat and human liver microsomes)

The resulting hydroxylamine is very reactive, undergoing non-enzymatic oxidation (autoxidation) to the corresponding nitroso-compound, which may covalently react with glutathione and thiol groups in proteins yielding sulphinamide adducts. It is assumed that this reaction may be responsible for procainamide-induced lupus [91]. Yet another noteworthy possibility is the coupling of the hydroxylamine and nitroso intermediates to form an azoxy derivative. Alternatively, either of the intermediates can react with the "parent" primary amine to yield an azo derivative (C-N=N-C).

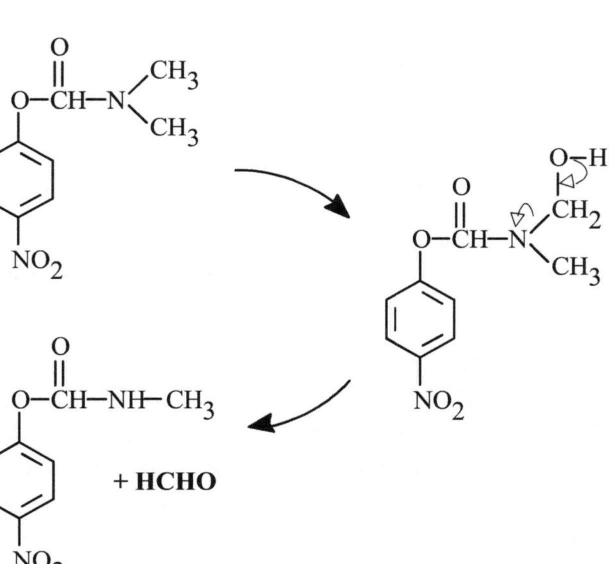

Fig.2.39 The dealkylation of N,N-dimethyl-p-nitrophenylcarbamate (Reproduced from ref. 25 with permission from R Paselk, Humboldt State University)

• the group of N,N-dimethylamino derivatives: methyl groups attached to a nitrogen atom are hydroxylated and rearrange to release an aldehyde, as shown in Figure 2.39 for the dealkylation of N,N-dimethyl-p-nitrophenylcarbamate.

An interesting alternative is that of di-dealkylation, as presented in Figure 2.40, yielding a primary amine:

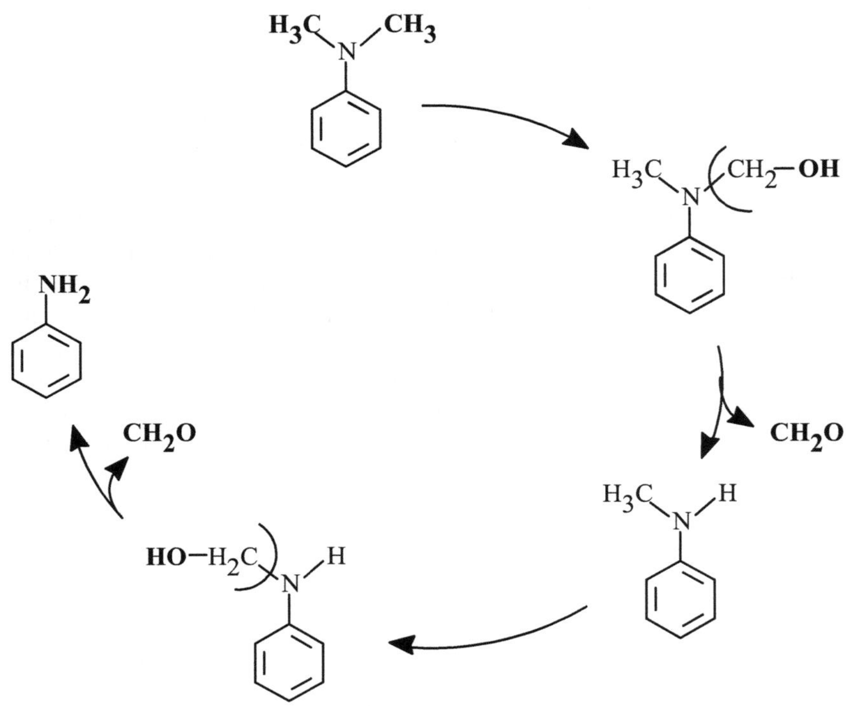

Fig.2.40 The di-dealkylation of N,N-dimethylaniline
(Reproduced from ref. 25 with permission from R Paselk, Humboldt State University)

• the group of N,N-diethylamino derivatives. For a long time it was believed that N,N-diethylamino derivatives, could not yield N-oxide metabolites due to steric hindrance [92]. Generally, N,N-dimethylamino derivatives are better substrates for N-oxygenation than their N,N-diethylamino homologues, although some xenobiotics belonging to the latter class are known to yield small amounts of N-oxides. Examples include clomiphene and lidocaine [93] (Figure 2.41):

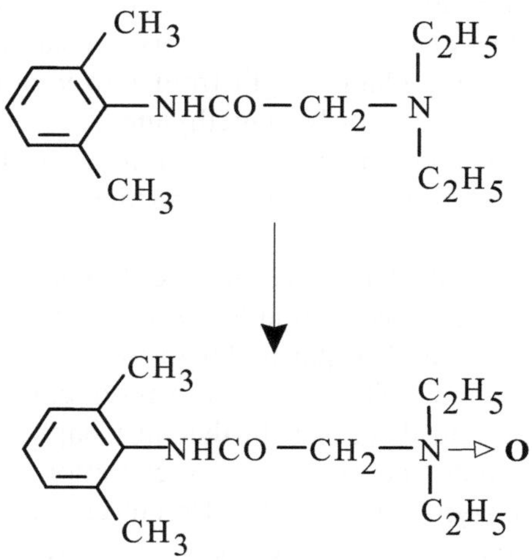

Fig.2.41 Formation of lidocaine N-oxide

The metabolism of lidocaine to its major metabolite monoethylglycinexylidide (MEGX) has been studied in human liver microsomes [94]. At least two distinct enzymatic activities were identified.

A review describing the advantages and disadvantages of using MEGX as a probe of hepatic function in liver transplantation has appeared [95]. Transformation of lidocaine to MEGX in the liver is the basis of a flow-dependent test of liver function. This test, though still subject to limitations, is significant in the context of assessing risk in liver transplantation as it reflects 'real-time' hepatic metabolising activity.

- the group of tertiary alicyclic amines: morphine [96] (Figure 2.42):

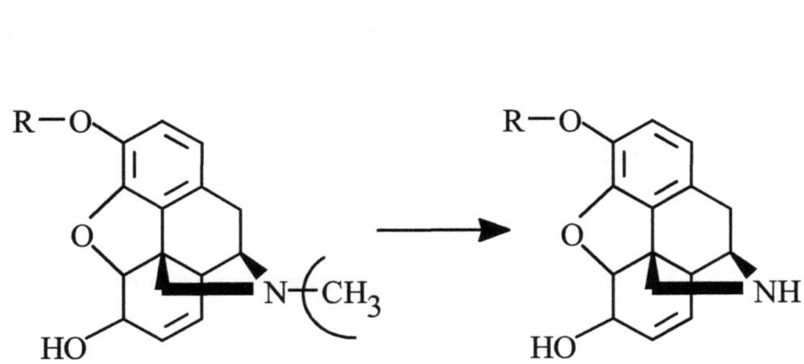

Fig.2.42 N-demethylation of morphine (R = H)

This N-demethylation is well established both in animals and humans. It was shown that the value of the Michaelis constant, K_m, for the reaction decreased with increasing chain length from 1 to 9 carbon atoms; the decyl and dodecyl analogues were not N-demethylated.

However, morphine, as a good example of a complex molecule enclosing a piperidine ring, may also be oxidized, yielding an N-oxide of particular relevance.

A detailed account of features of opioid pharmacology, including the metabolism of morphine, is available [97]. Some emphasis is given to morphine glucuronides, and in particular to morphine 6-glucuronide owing to its clinical importance. Though this is a product of Phase II metabolism, it is introduced here and discussed further in Chapter 3. These compounds are formed in the liver and their fate is excretion in the bile and urine. Depending on the enzymes involved, different conjugates may form. In the case of morphine, the process is stereospecific and dependent on the body region. The biotransformation of morphine-3-glucuronide to the active morphine-6-glucuronide is well known. A review of this topic also describes the discovery of a unique opioid receptor for morphine-6-glucuronide [98]. More recently, a review on the clinical implications of this metabolite appeared [99].

- the group of 1,4-dihydropyridines: their aromatization to the corresponding pyridine metabolites has been extensively studied, both *in vitro* and *in vivo*. The most common example is given by felodipine, which undergoes biotransformation in a CYP3A4-mediated reaction yielding a metabolite that contains the pyridine moiety (Figure 2.43):

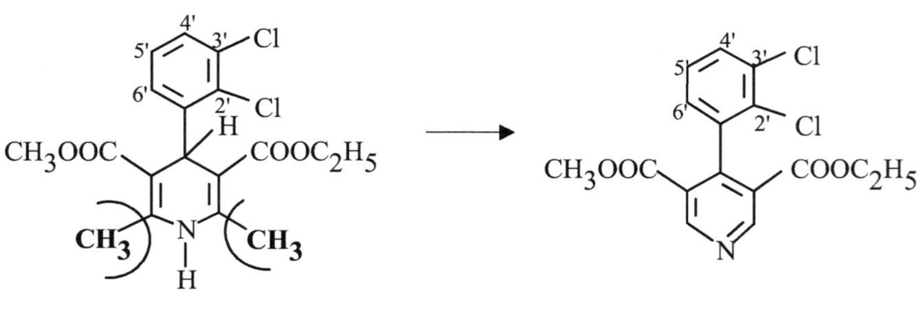

Fig.2.43 Structure of felodipine undergoing aromatization

The fastest rate of aromatization was observed for the 2',6'-disubstituted derivatives, variations being correlated with electronic properties of the substrates. The slowest rates were associated with the 2',3'-,2',4'-,3',4'- and 3',5'-disubstituted derivatives [100].

This biotransformation was studied in rat-liver microsomes [101], kinetic data indicating it as a major metabolic pathway and cytochrome P450 was implicated in the aromatization of felodipine.

- the group of amino azaheterocyclic compounds: trimethoprim is not N-hydroxylated, but forms two isomeric N-oxides, oxidation occurring at the 1- and 3- positions [102] (Figure 2.44). In the specific case of this antibacterial drug, hydroxylamine formation (considered as a route of toxication) is limited by the amine-imine tautomerism, which controls the metabolic processes, preventing the N-hydroxylation, and instead favouring the appearance of the isomeric N-oxides shown below.

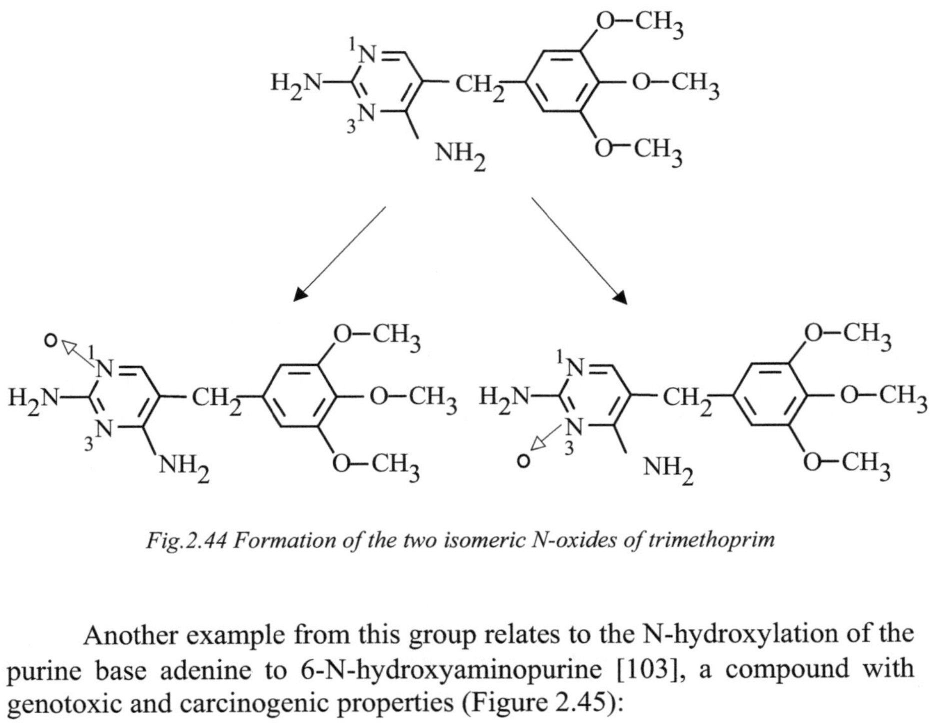

Fig.2.44 Formation of the two isomeric N-oxides of trimethoprim

Another example from this group relates to the N-hydroxylation of the purine base adenine to 6-N-hydroxyaminopurine [103], a compound with genotoxic and carcinogenic properties (Figure 2.45):

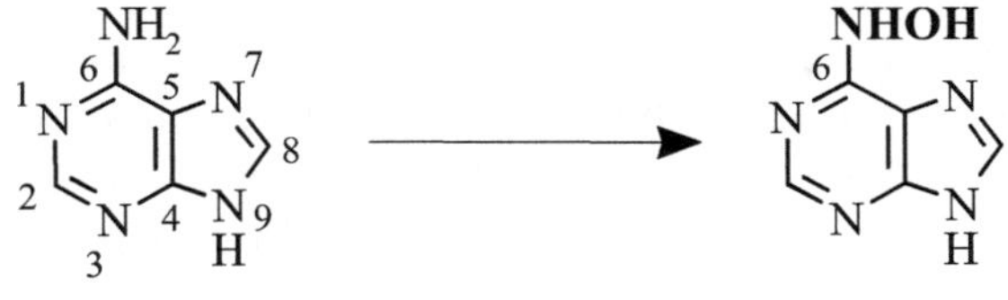

Fig.2.45 6-N-hydroxylation of adenine

The 6-substituent was also found to play a role in influencing N-oxide formation [104]. For such compounds (heterocyclic hydroxylamines), it is assumed that the formation of a nitrenium ion is the step leading to the ultimate carcinogen or mutagen [105]. In a study of the effect of oxygen on adenine hydroxylation by the hydroxyl radical in aqueous solution, the 8-hydroxyadenine derivative was isolated [106].

- the group of hydrazines (1-substituted, 1,1-disubstituted, 1,2-di-substituted and azo derivatives): hydralazine (a 1-substituted hydrazine) [107] (Figure 2.46). The reaction of biotransformation in this case proceeds *via* radical pathways, with loss of the hydrazine moiety to yield phthalazine.

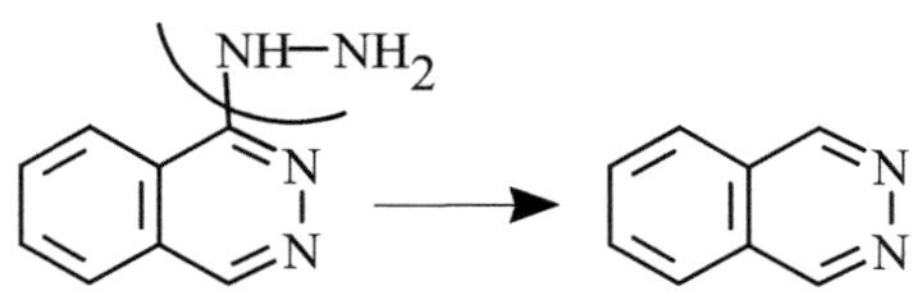

Fig.2.46 Metabolism of hydralazine

From the same group, we refer to procarbazine (a 1,1-disubstituted hydrazine, Figure 2.47) for which azo formation and subsequent N-C cleavage reactions are well documented. However, the mechanism of N-dealkylation may involve α-carbon hydroxylation rather than hydrazone hydrolysis [108].

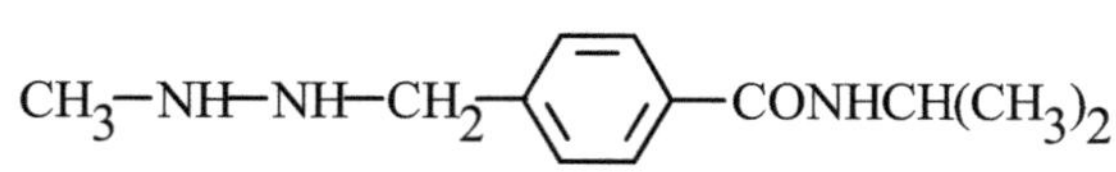

Fig.2.47 Structure of procarbazine

The intermediate derived from N-oxidation may be either a diazene or a nitrene resonance form, existing in tautomeric equilibrium with an azomethinimine. The nitrene intermediate may form an iron-nitrene complex with CYTP450, while the azomethinimine can rearrange to a hydrazone.

Particular examples are the **N**-dealkylations. They represent the simplest case of N-C cleavage; see the example of morphine above (Figure 2.42).

Another interesting example is given by a seven-membered azaheterocycle, belonging to the group of benzodiazepines, namely diazepam (Figure 2.48). In humans, its N-demethylation [109] to the long-acting desmethyldiazepam is a major route of metabolism. Diazepam displays competition with the structurally related pinazepam, for the same metabolic route. N-dealkylation of these drugs occurs at markedly different rates; in rat liver microsomes for example, the N-depropargylation of pinazepam is eightfold faster than the N-demethylation of diazepam [110,111].

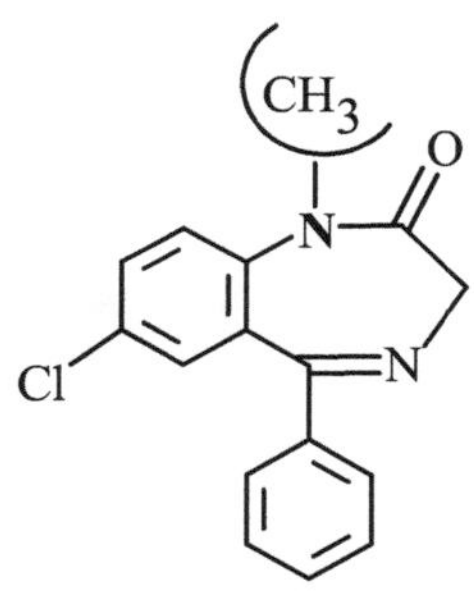

Fig.2.48 N-demethylation of diazepam

Environmental and genetic factors may influence interindividual metabolism of diazepam and these have been discussed [112].

The benzodiazepine pinazepam contains an unsaturated bond (the propargyl group) at the N1-position and its metabolism involves N1-dealkylation and C3-hydroxylation. N-desmethyldiazepam is the main metabolite in dogs. Both pinazepam and N-desmethyldiazepam are converted to the inactive oxazepam [113].

N-demethylation of caffeine is another well-studied case [114] (Figure 2.49). As can be observed from the figure a preferred position for dealkylation is the N(3) atom, the reaction yielding para-xanthine; indeed, this metabolite was shown to predominate markedly over N(1)-, and N(7)-demethylated metabolites (theobromine and theophylline respectively). For the N(3)-demethylation of caffeine in human liver, the enzyme found to be primarily responsible is the isoform CYP1A2 (while other P450 enzymes, at least in part, are involved in the formation of the N(1)-, and N(7)-demethylated metabolites).

An alternative biotransformation pathway for caffeine involves a non-P450 enzyme system, namely the xanthine oxidase; in particular, this enzymatic system catalyses the 8-hydroxylation of certain N-demethylated metabolites of caffeine, such as theophylline and 1-methylxanthine.

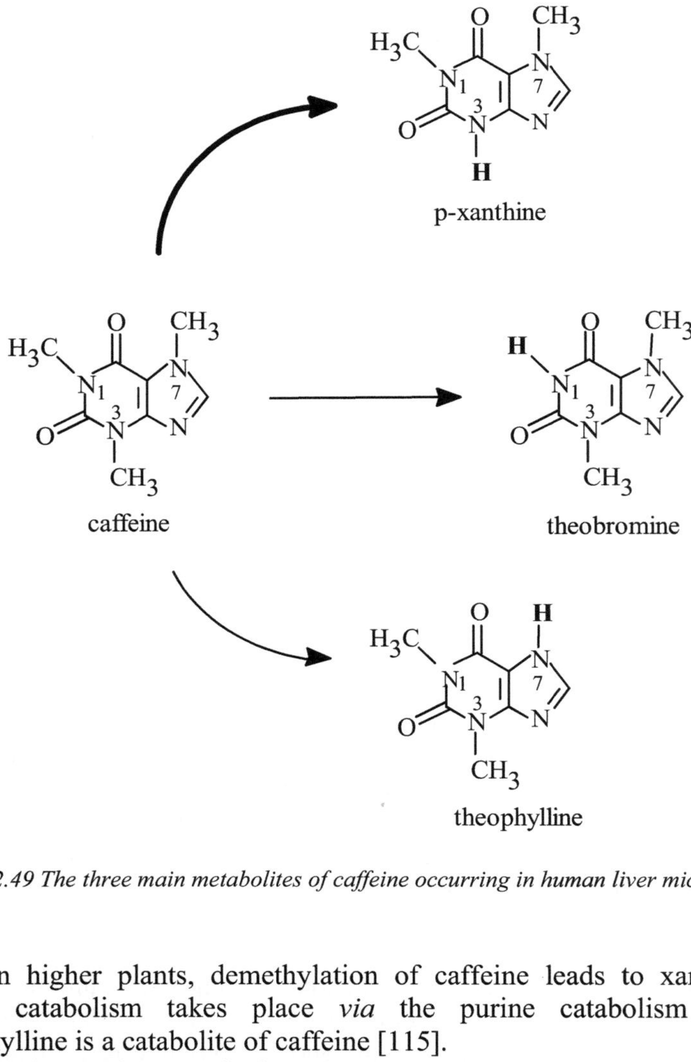

Fig.2.49 The three main metabolites of caffeine occurring in human liver microsomes

In higher plants, demethylation of caffeine leads to xanthine and further catabolism takes place *via* the purine catabolism pathway. Theophylline is a catabolite of caffeine [115].

An interesting case is that of propranolol, which can undergo not only hydroxylations (Figure 2.27), but also N-dealkylation and deamination, these in fact being its major metabolic routes [116] (Figure 2.50):

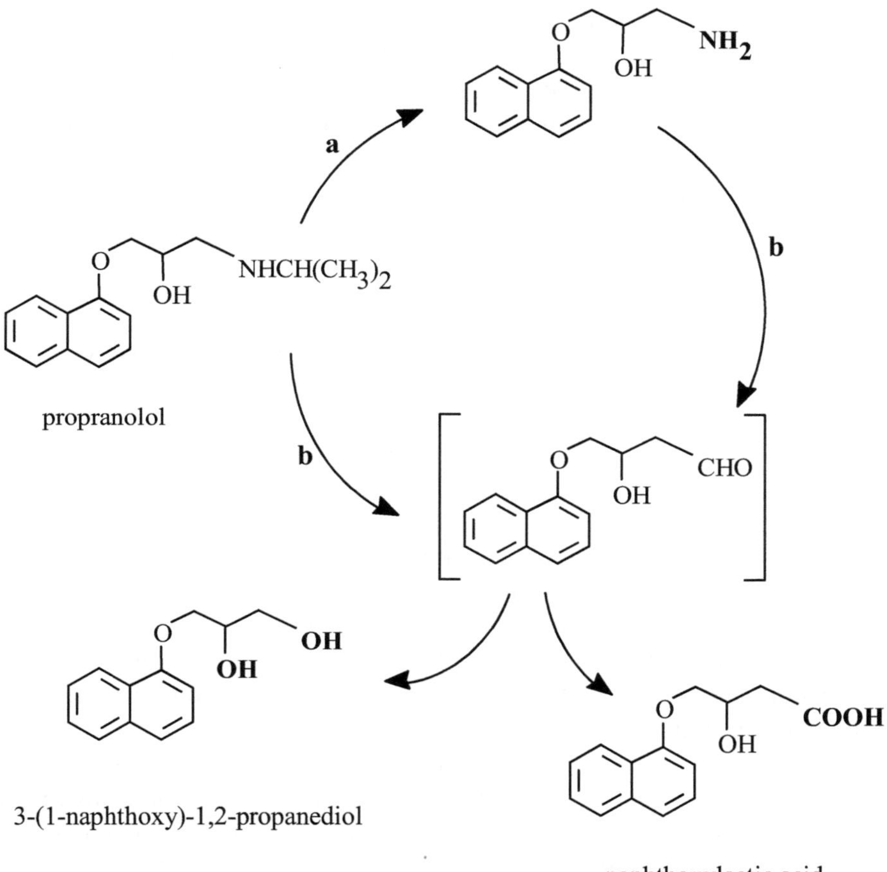

Fig.2.50 Alternative biotransformation reactions of propranolol yielding a diol and an acid metabolite (Reprinted from ref. 8, p.213, with permission from Elsevier)

Most of the administered dose is dealkylated (path a) and then deaminated (path b), the dashed arrow indicating that the deamination of the "parent" drug is minor compared to that of deisopropylpropranolol.

The aldehyde produced by deamination is either rapidly reduced – yielding the diol, or oxidised to the corresponding acid. This oxidative degradation of the side-chain of propranolol is assumed to be an important process in humans, accounting for some 15-30% of a dose on chronic administration of the drug [117].

Deamination is an important pathway of metabolism for certain other drugs with a basic side-chain such as β-blockers, antihistamines and antipsychotics. These drugs usually being arylalkylamines with a secondary or tertiary amino group, deamination will involve either the parent drug and/or its N-dealkylated metabolite(s).

Oxidative dehalogenation is another particular case of CYTP450-catalysed oxidation.
One of the most important examples involves halothane [118] (Figure 2.51):

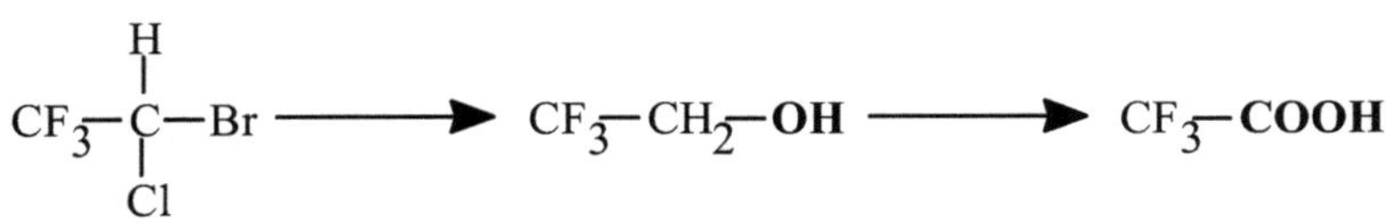

Fig.2.51 Oxidative dehalogenation of halothane

The compound can induce post-anaesthetic jaundice or hepatitis, its reductive metabolism partly and perhaps mainly accounting for such unwanted effects (see also subchapter 8.3). It undergoes CYTP450-catalysed dehalogenation by both oxidative and reductive routes. Its oxidative biotransformation occurs at the –CHClBr group, leading eventually to trifluoroacetic acid.

The metabolism of polyhalogenated compounds used as anaesthetics is a subject with important toxicological implications. Metabolism of the compounds occurs mainly in the liver and hepatotoxicity is not unusual. Molecular processes underlying such adverse reactions have been reviewed [119].

Finally, we present another example of a specific oxidation (albeit not involving a drug), namely the CYTP450-catalysed oxidative ester cleavage. Not many years ago, a few isolated observations of such a reaction were published; they referred to the oxidative de-esterification of flampropisopropyl (a herbicide) to the corresponding acid, the proposed mechanism involving the formation of a hydroxylated intermediate, followed by its post-enzymatic breakdown to the acid and acetone [120] (Figure 2.52):

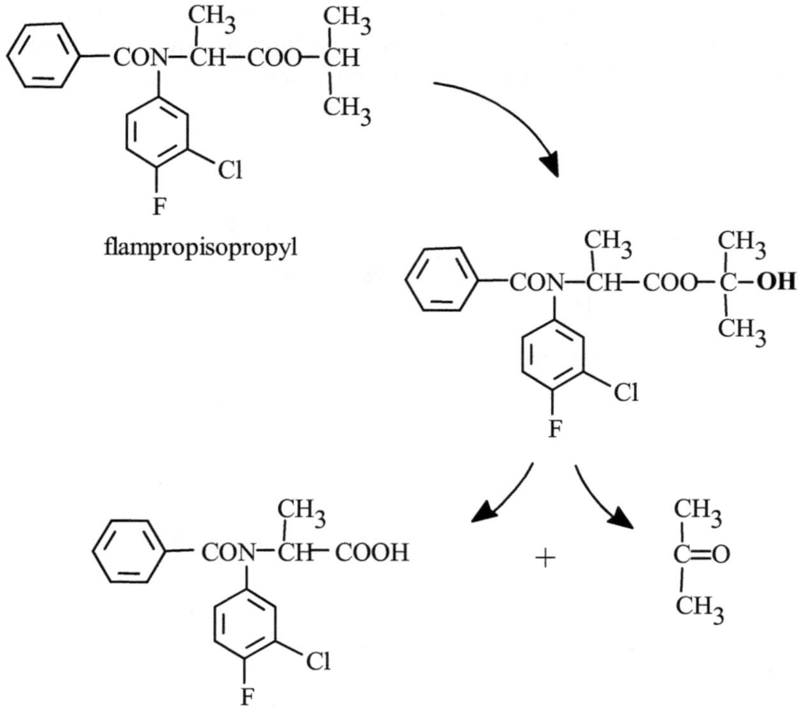

Fig.2.52 Oxidative ester cleavage of flampropisopropyl

Another interesting group of dealkylations is that of O-alkylated compounds: alkyl groups are hydroxylated adjacent to oxygen and rearrange to release an aldehyde, as shown for p-nitroanisole (Figure 2.53):

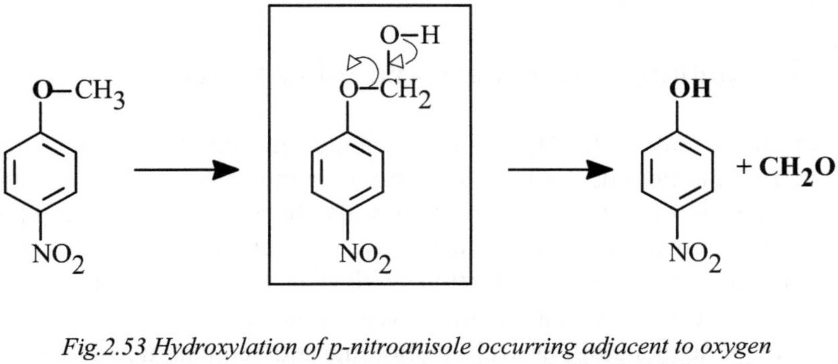

Fig.2.53 Hydroxylation of p-nitroanisole occurring adjacent to oxygen (Reproduced from ref. 25 with permission from R Paselk, Humboldt State University)

S-dealkylations, analogous to O-dealkylations, may also occur. They are likewise CYTP450-catalysed reactions, the intermediate undergoing S-C cleavage, yielding a thiol and a carbonyl compound (Figure 2.54):

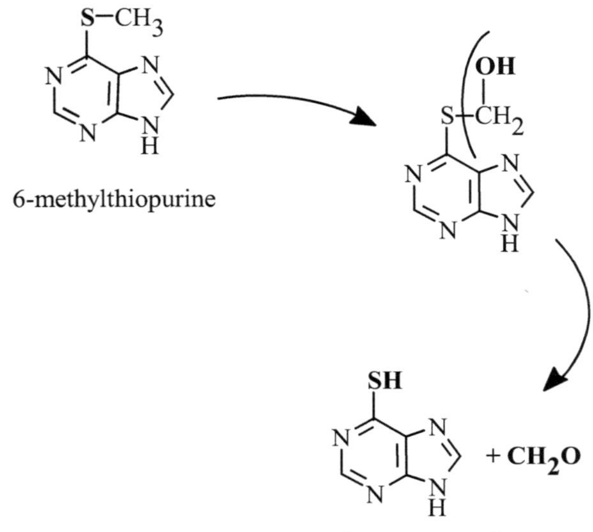

Fig.2.54 Dealkylation of 6-methylthiopurine, with a hydroxylated intermediate and final demethylation and formation of an aldehyde (Reproduced from ref. 25 with permission from R Paselk, Humboldt State University)

2.4 OXIDATIONS INVOLVING OTHER ENZYMATIC SYSTEMS

2.4.1 The monoamine oxidase and other systems

Monoamine oxidase (MAO), widely distributed in most tissues of mammals is a membrane-bound, FAD-containing enzyme, mainly located in the mitochondria. However, some activity has also been detected in microsomes, cytosol and even in the extracellular space. Its presence in the brain is of particular importance in connection with the therapeutic profile of its inhibitors and its role as an activator of xenobiotics [121-126].

Protein sequencing, as well as cloning and sequencing cDNA coding for humans, have proven the existence of two different forms of the enzyme, conventionally designated as MAO-A and MAO-B.

Physiological substrates of MAO are predominantly primary amines; they are oxidatively deaminated according to the following reaction:

$$RCH_2NH_2 + O_2 + H_2O \rightarrow RCHO + NH_3 + H_2O_2 \qquad 2.4$$

Normally, this is a two-step reaction, producing first the aldehyde, the amine and the enzyme in the reduced form; subsequently, the reduced enzyme is re-oxidised by molecular oxygen, with concomitant production of hydrogen peroxide (Eq. 2.5 and 2.6):

$$[FAD] + RCH_2NH_2 + H_2O \rightarrow [FAD\mathbf{H_2}] + RCHO + NH_3 \qquad 2.5$$

$$[FAD\mathbf{H_2}] + O_2 \rightarrow [FAD] + H_2O_2 \qquad 2.6$$

If the resulting hydrogen peroxide is not quickly decomposed by peroxidases, it may activate some neurotoxins, which is of potential toxicological significance.

The MAO catalytic mechanism is understood at the molecular level [124]. Usually, reaction begins with a single-electron oxidation (of the nitrogen atom), yielding an amine radical cation and thus facilitating the next step, namely abstraction of a hydrogen atom followed by fast electronic loss. Then, a second oxidation step occurs, generating an imine or its iminium ion. The reduced enzyme binds molecular oxygen and undergoes re-oxidation with release of hydrogen peroxide.

A typical, endogenous substrate is represented by histamine (Figure 2.55). Histamine (in the small amounts normally ingested or formed by bacteria in the GI tract) is rapidly metabolised and eliminated in the urine.

There are two major pathways of histamine metabolism in humans [127]. The more important involves ring methylation, with subsequent formation of N-methylhistamine (under the catalytic action of the well distributed N-methyltransferases). Most of this intermediate is then converted by MAO to the corresponding N-methylimidazoleacetic acid (reaction may be blocked by MAO inhibitors).

An alternative pathway involves oxidative deamination catalysed mainly by the non-specific enzyme diamine oxidase (DAO), yielding imidazoleacetic acid, subsequently converted to imidazoleacetic riboside, (metabolites that display little or no activity and are readily excreted in the urine). An account of the biological role of histamine and its relation to development of antihistamines recently appeared [128].

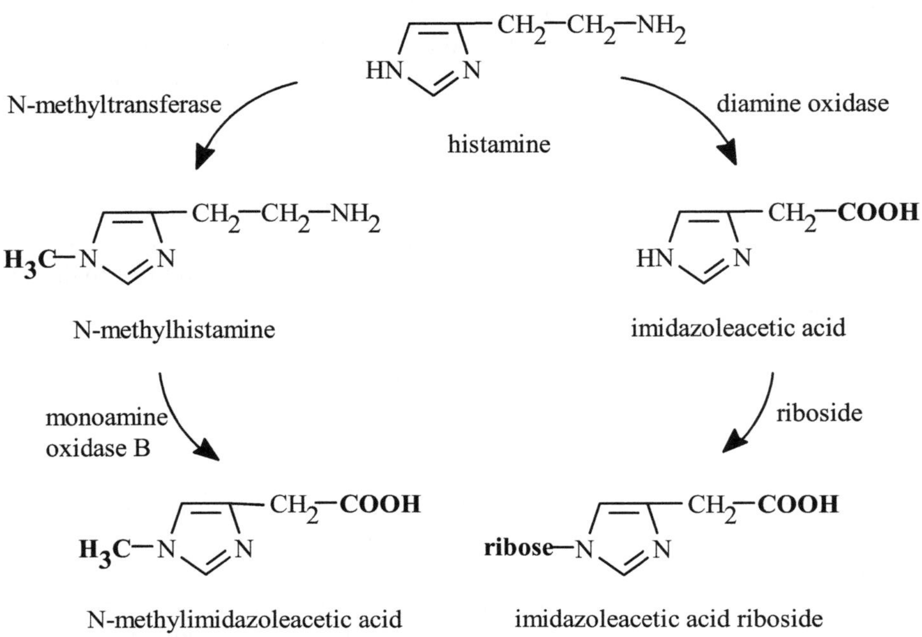

Fig.2.55 Alternative pathways in the biotransformation of histamine; note the participation of the B-form of MAO, yielding an acid that subsequently may be conjugated

It is important to mention the existence of two classes of MAO inactivators, depending on whether covalent binding occurs to FAD or to an amino side-chain in the active site.

The flavin-containing monooxygenase

The so-called FMOs are NADPH-dependent and oxygen-dependent microsomal FAD-containing enzyme systems, functioning as sulphur, nitrogen and phosphorus oxygenases. The proposed mechanism of action involves the sequential binding of NADPH and oxygen to the enzyme to generate an FAD C-4α-hydroperoxide. The general mechanism of action will be detailed in Chapter 4.

Nucleophilic substrates (organic nitrogen and sulphur compounds, including drugs such as phenothiazines, ephedrine, N-methylamphetamine, norcocaine) attack the distal oxygen of this hydroperoxide generating a hydroxyflavin species (the resultant oxygen being transferred to the substrate). The wide tissue distribution of the enzyme suggests that this enzymatic system plays a major role in the oxidative biotransformation of drugs while the broad substrate specificity is associated with the presence of multiple forms (sustained by the cloning and sequencing of distinct genes from several species and tissues) [129-131].

The prototypical FMO xenobiotic reaction pathway is considered to be the conversion of tertiary amines to highly polar N-oxides. Thus, the implication of FMOs is obvious in the metabolism of a variety of tertiary amine central nervous system-active agents (e.g. nicotine, olanzapine, clozapine). Consequently, it is readily understandable that there is considerable interest and advantage in identifying brain FMO isoforms capable of attenuating the pharmacological activity of such tertiary amines directly at their sites of action.

Prostaglandin synthetase

Present in all mammalian cells, this enzyme catalyses the oxidation of arachidonic acid to prostaglandin H_2, an important precursor in the arachidonate cascade. However, what is most important is that this fatty acid cyclooxygenase activity is coupled with a hydroperoxidase activity, resulting in some drugs being co-oxidised during arachidonic acid metabolism [132- 134].

Drugs capable of undergoing such co-oxidation biotransformations include aminopyrine, benzphetamine, oxyphenbutazone and paracetamol. It is emphasised here that the same biotransformation mechanism occurs with certain carcinogens such as benzidine or the well-known benzo[α]pyrene, a component of tobacco smoke. Further details appear in Chapter 4. However one can conclude that the prostaglandin synthetase-dependent co-oxidation of certain drugs may represent a significant metabolic pathway, playing a major role in drug biotransformation, particularly in those tissues that are low in M.F.O. activity, but rich in prostaglandin synthetase (such as the kidney, renal medulla, skin and lung) [130].

Xanthine dehydrogenase – Xanthine oxidase

These enzymes, denoted XDH and XO respectively, represent two forms of a homodimeric enzyme, the two component subunits being of equal size [135-139].

These enzymes are sometimes designated as the molybdenum hydroxylases, XO being a xanthine-oxygen oxidoreductase (or hypoxanthine oxidase), and XDH, a xanthine NAD+ dependent oxidoreductase. They are cytosolic enzymes with complementary roles to those of monooxygenases in the metabolism of both endogenous and exogenous (xenobiotic) compounds.

Each subunit of XD/XO contains as cofactors:

- one atom of molybdenum in the form of a molybdopterin cofactor, whose oxidised form can be written as $[Mo^{VI}(=S)(=O)]^{2+}$,
- one FAD molecule, and

- four non-heme iron atoms in the form of two Fe_2/S_2 centres.

The general reaction catalysed by this unique combination of prosthetic groups obeys the general equation;

$$\mathbf{SH} + H_2O \rightarrow \mathbf{SOH} + 2e^- + 2H^+ \qquad 2.7$$

where SH is a reduced substrate, and
SOH, the resulting hydroxylated metabolite.

From the equation, two conclusions can be drawn: 1) the oxygen atom transferred to the substrate is derived from water, and 2), as the reaction liberates two electrons, an electron acceptor must also be present.

In the case of XO, this electron acceptor is represented by molecular oxygen; the reaction generates uric acid (in the form of urate), plus hydrogen peroxide and superoxide (reactive oxygen species that can cause lipid peroxidation and general oxidative damage in cells).

In contrast to XO, XDH uses as electron acceptor, oxidised NAD+; urate is again generated, plus the reduced form of NAD (NADH + H^+).

Specific substrates of the molybdenum hydroxylases are characterized by having electron-deficient sp^2-hybridised carbon atoms, and belong to the following chemical classes:

- aromatic azaheterocycles (mono-, bi-, or polycyclic), containing the –CH=N- moiety
- aromatic or non-aromatic charged azaheterocycles, that contain the moiety $-CH=N^+<$, and
- aldehydes, containing the –CH=O moiety.

It is interesting to note that conversion of XDH to XO may occur *in vivo* under the influence of different metabolic states such as hypoxia and ischemia. This conversion is associated with a variety of toxicities, a consequence of increased production of reactive oxygen species and amplification of oxidative cellular damage; this explains the continued interest in the regulation of the two forms of the enzyme. Details of this aspect appear in Chapter 4.

Aldehyde oxidase

This enzyme is also known as aldehyde oxygen oxidoreductase and is designated as AO [140-142]. Human AO has a limited tissue distribution, with significant levels detected only in the liver. It is noteworthy that human AO activity appears to be rather unstable; this may be due to substrate-dependence, being at the same time an indication of the presence of multiple forms that exhibit differences in substrate specificities and stability.

Aldehyde oxidase (existing solely in its oxidase form) is a cytosolic enzyme, which although completely unrelated to the molybdenum hydroxylases, shares much similarity with the XO/XDH enzymes, yet does not participate in a dehydrogenase-oxidase transition.

The electrons received from a reducing substrate are used by the flavin to reduce dioxygen. As an electron acceptor, AO uses molecular oxygen, to yield the corresponding acid and superoxide.

Despite the fact that the dehydrogenase-oxidase transition does not occur with AO, the same pathophysiological implications that were mentioned for XO exist for AO as well (we refer especially to the reactive oxygen species generated during metabolism by AO). Details are given in Chapter 4.

Copper-containing amine oxidases

In this group are included amine oxygen oxidoreductases, diamine oxidases and histaminase [143, 144].

These enzymes are found in many tissues as well as in the plasma, the distribution being species-dependent. They are associated as an inorganic cofactor with copper, but also contain a covalently bound organic cofactor at the catalytic site (details in Chapter 4).

A common reaction that they catalyse is the oxidative deamination of primary amines:

$$RCH_2\mathbf{NH_2} + H_2O + O_2 \rightarrow R\mathbf{CHO} + NH_3 + H_2O_2 \quad 2.8$$

Unfortunately, not many drugs have been investigated for their biotransformation by copper-containing amine oxidases, but at least one interesting finding suggests that the phenomenon deserves more attention. This is the deamination of the calcium channel blocker amlodipine [145] (Figure 2.56).

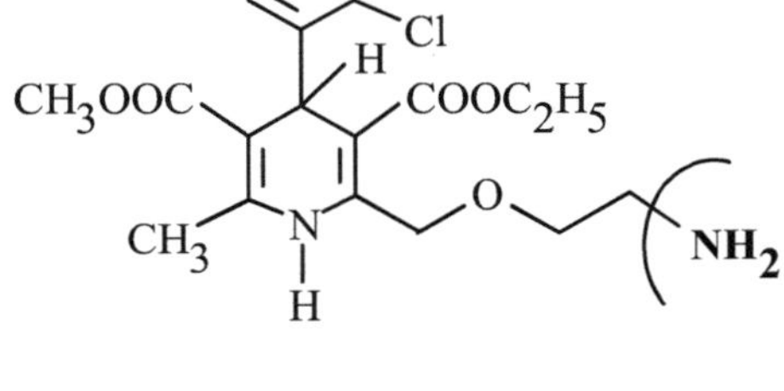

Fig.2.56 Oxidative deamination of amlodipine

The reaction is species-dependent, occurring in humans and dogs but not in rats, for instance. The reaction was shown to occur on incubation of the drug in dog plasma, and the involvement of plasma amine oxidases was suggested. Deamination leads to an aldehyde assumed to be the precursor of valproic acid, an important drug whose metabolism is of particular biochemical and toxicological interest [146].

2.4.2 Other representative examples

Phenelzine is a representative hydrazine that undergoes MAO-catalysed oxidation, yielding in the first step a diazene; this metabolite may rearrange to a hydrazone, which hydrolyses to hydrazine and an aldehyde (phenylacetaldehyde) that subsequently will yield the corresponding acid (phenylacetic acid), as the major urinary metabolite.

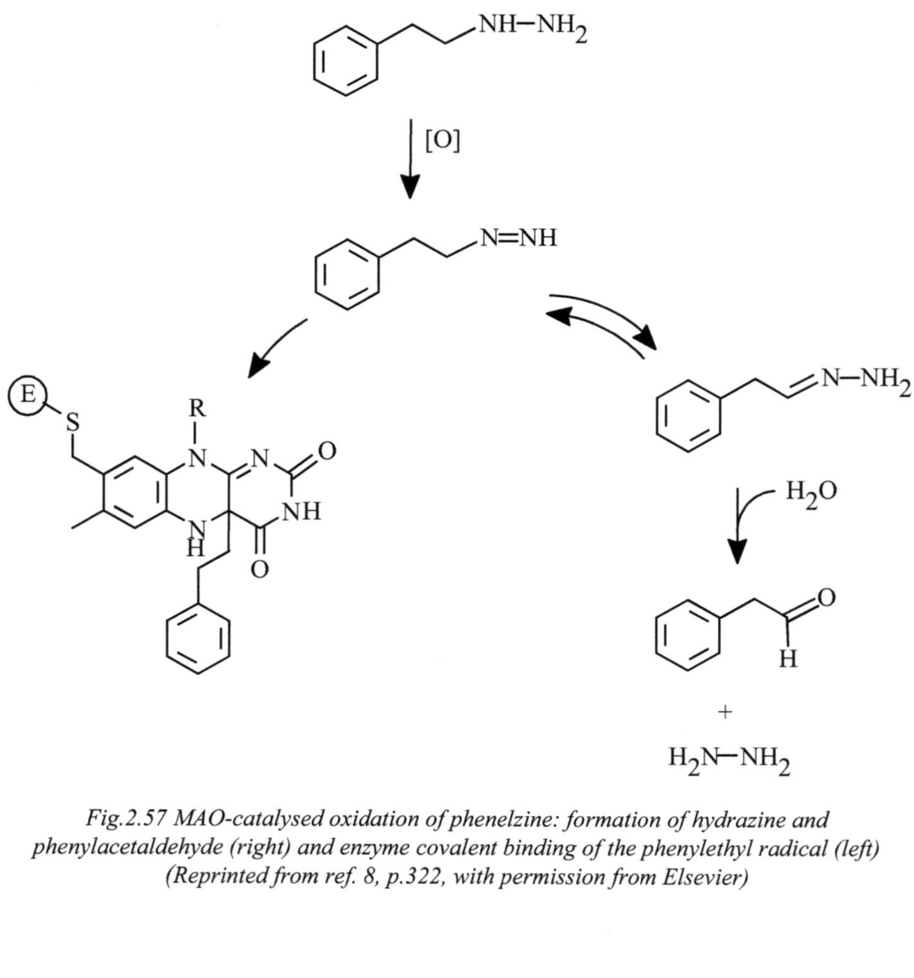

Fig.2.57 MAO-catalysed oxidation of phenelzine: formation of hydrazine and phenylacetaldehyde (right) and enzyme covalent binding of the phenylethyl radical (left) (Reprinted from ref. 8, p.322, with permission from Elsevier)

An aspect worth stressing is that phenelzine is not only an MAO substrate, but also an inactivator of the enzyme; it facilitates the breakdown of diazene to N_2 and the phenylethyl radical, which is capable of forming covalent adducts with the enzyme at the specific position C_4 of the flavin. Under these circumstances, the enzyme will be inactivated by alkylation ("suicide substrate" behaviour) [147] (Figure 2.57).

An interesting example is provided by the endogenous compound purine [148]. In the form of the N(9)-H tautomer, it undergoes xanthine-oxidase-catalysed oxidation with high affinity for different regions, as follows: at C(6), yielding hypoxanthine, then at C(2), generating xanthine, and finally at C(8), with corresponding formation of uric acid that is excreted in the urine (Figure 2.58):

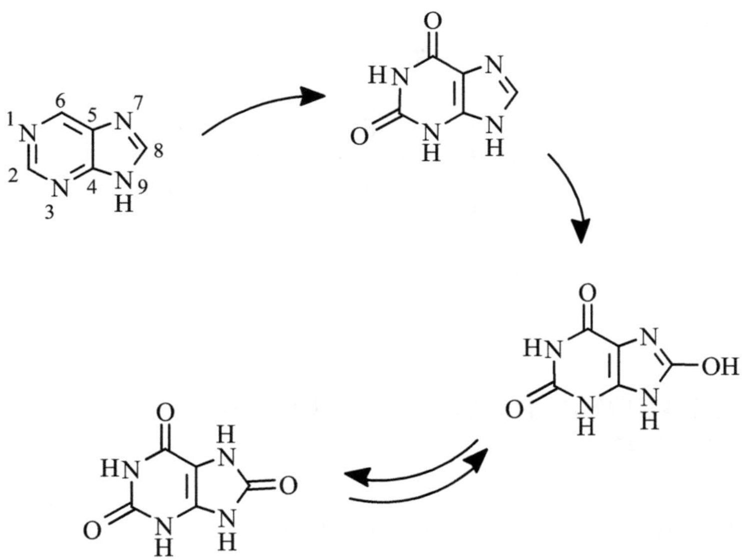

Fig.2.58 The regiospecific XO metabolization of purine

The hypoglycaemic drug, tolbutamide, affords a good example of the complexity of biotransformations in the *in vivo* metabolic context: it undergoes an initial oxidation yielding the corresponding aldehyde, which is subsequently oxidised by both XO and an NAD-linked aldehyde dehydrogenase [149] (Figure 2.59). The aldehyde metabolite is generated by the sequential action of CYTP450 and an alcohol dehydrogenase).

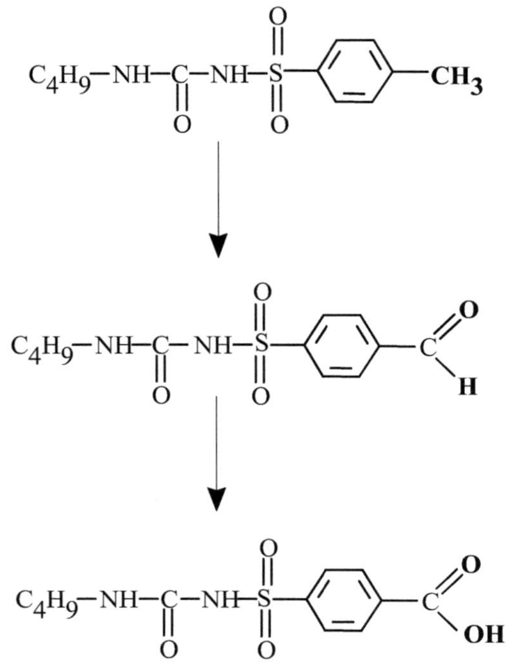

Fig.2.59 Tolbutamide metabolites

2.5 METABOLIC REACTIONS INVOLVING REDUCTION

2.5.1 Components of the enzyme system

Generally, reductive processes involve two separate enzyme systems: one is represented by the already well-known cytochrome P450, while the other involves an NADH H^+ dependent system. The latter is assumed to be a flavoprotein containing a molecule of FAD as the prosthetic group, and NADH H^+, as preferred source of the necessary reducing equivalents. This system is known as the NADH-cytochrome b_5 reductase system and its participation in physiological processes involves two main steps: acceptance of two electrons, to reach a reduced state, followed by the reduction of two equivalents of cytochrome b_5 in successive one-electron steps [150-152].

It is assumed as well that this system is also responsible for the reductive denitrosation of nitrosourea anti-tumor drugs, consequently representing an important deactivation pathway [152].

NADPH-cytochrome P450 reductase, a ferrihemoprotein oxidoreductase, also known as NADPH- cytochrome *c* reductase, is considered to be the major oxidoreductase transferring electrons to microsomal cytochrome P450. The system contains one molecule each of FAD and FMN per polypeptide chain, and the NADPH (resulting from the pentose phosphate pathway), represents the preferred source of reducing equivalents. The electron acceptors are the cytochrome P450 and a small metalloprotein, the soluble cytochrome *c*, that acts as an electron carrier in the respiratory chain of all aerobic organisms [151,153].

2.5.2 Compounds undergoing reduction

Although relatively uncommon, metabolic reduction is also an important pathway in the biotransformation of drugs. Actually, it represents the major route of metabolism for aromatic nitro- and nitroso- groups (as in chloramphenicol, nitroglycerine and organic nitrites), for the azo- group (as in prontosil) as well as for a wide variety of aliphatic and aromatic N-oxides.

Rreduction of azo- and nitro-compounds usually leads to primary amines. However, a number of azo-compounds (such as sulfasalazine) are converted to aromatic primary amines by azoreductase, an NADPH-dependent enzyme system present in the liver microsomes. The colonic metabolites of sulfasalazine are 5-aminosalicylic acid and sulfapyridine. Inflammatory bowel disease results in increase in the production of prostaglandins and leukotrienes. Consequently, the effects of sulfasalazine on the metabolism of the precursor arachidonic have attracted wide interest [154].

Nitro- compounds (chloramphenicol, for example) are reduced to aromatic primary amines by a nitroreductase, presumably through nitrosoamine and hydroxylamine intermediates.

It is important to stress that these enzymes are not solely responsible for the reduction of azo- and nitro- compounds, probably because of reduction by the bacterial flora in the anaerobic environment of the intestine.

Steps in the mechanism of reduction of an aromatic nitro- group are represented in Figure 2.60 [155].

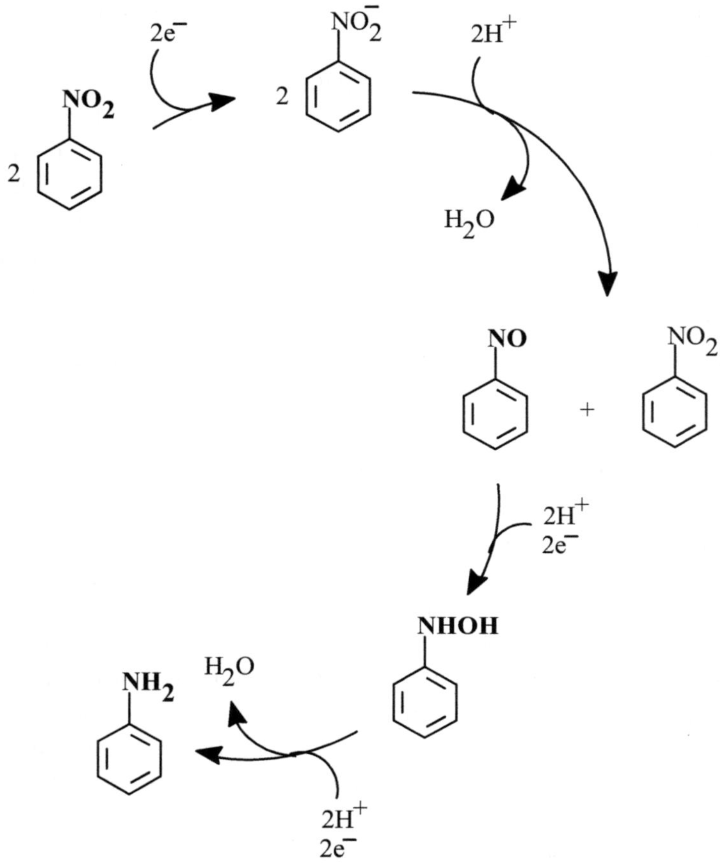

Fig.2.60 Reduction steps for an aromatic nitro-group
(Reproduced with the permission of Nelson Thornes from 'Introduction to Drug Metabolism', 2001, 3rd Ed., isbn 0 7487 6011 3 - Gibson & Skett - first published in 1986)

Examples of this type of biotransformation also include certain aldehydes which are reduced to the corresponding alcohols, as well as sulphoxides and sulphones. However, in these cases reduction is not considered to be the major metabolic pathway.

Metabolic reduction has been shown to occur mainly in liver microsomes, but occasionally takes place in other tissues as well.

Some general reactions are presented in Figure 2.61:

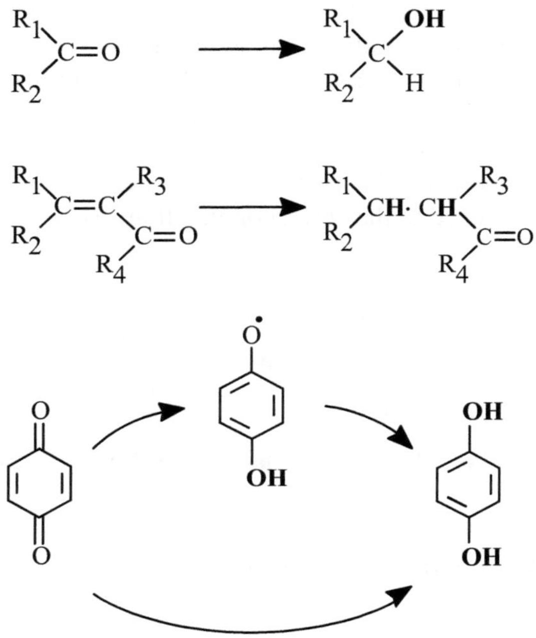

Fig. 2.61 Some representative types of reductive reactions at carbon atoms

Aldehydes and ketones are reduced to the corresponding respective primary and secondary, alcohols, while quinones may be reduced to the corresponding hydroquinones. Some of the radical species formed as intermediates may have significant toxicological potencies.

Aldehydes and ketones are widely distributed and have several biological functions. In addition to alcohol dehydrogenase (ADH), there are several enzymes in the aldo-keto reductase family that may participate in the metabolism of aldehydes and ketones in the kidney [156].

Dehalogenations may also proceed in a reductive manner, as in the case of halothane, with the intermediate formation of a radical (1-chloro-2,2,2-trifluoroethyl) [157] (Figure 2.62):

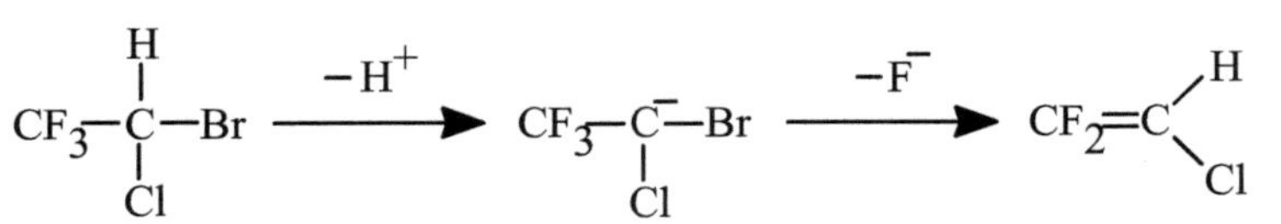

Fig.2.62 Reductive metabolism of halothane
(Reproduced with the permission of Nelson Thornes from 'Introduction to Drug Metabolism', 2001, 3rd Ed., isbn 0 7487 6011 3 - Gibson & Skett - first published in 1986)

Fluorocarbons of the halothane type can be defluorinated by liver microsomes in anaerobic conditions as shown above.

Some aromatic compounds such as nitro-, nitroso- and hydroxylamines, as well as imines and oximes, are reduced to the corresponding primary amines. Some of the azo-aromatic compounds yield by reductive metabolism primary aromatic amines that are potentially toxic. Disulphides are reduced to the corresponding thiols (Figure 2.63):

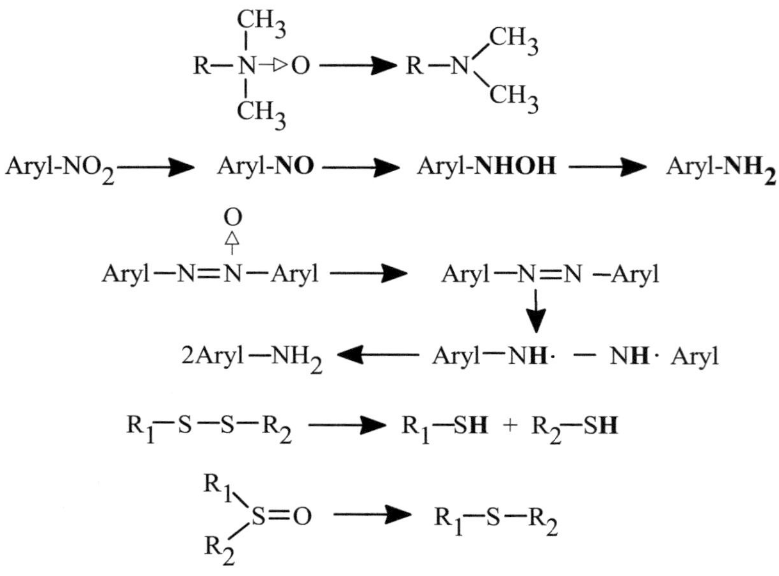

Fig.2.63 Some representative reductive reactions involving heteroatoms nitrogen and sulphur

Some non-microsomal metabolic reductions have also been found to occur, but relatively little is known concerning either the nature of the enzymatic systems involved or their location. Usually, such reductions refer to the double bond, especially in unsaturated monocyclic terpenes.

2.6 HYDROLYSIS

Hydrolysis occurs especially with esters and amides in reactions catalysed by various enzymes located in hepatic microsomes, kidneys and other tissues. Other compounds susceptible to such a biotransformation pathway are carbamates and hydrazides.

Usually, esters and amides are rapidly hydrolysed under the catalytic action of specialised carboxylesterases. Some of the resulting metabolites may be subsequently conjugated, as glucuronides for example, and so, rapidly eliminated.

Carboxylesterases include cholinesterases, pseudocholinesterases, arylcarboxylesterases, hepatic microsomal carboxylesterases and other unclassified hepatic analogues.

Besides the important group of carboxylesterases, in the category of hydrolyses involved in xenobiotic metabolism, we should also mention arylsulphatases, epoxide hydroxylases, cysteine endopeptidases and serine endopeptidases as examples.

Carboxylesterases/amidases catalyse hydrolysis of carboxylesters, carboxyamides and carboxythioesters, as seen in the equations below. The specificity of their action depends on the nature of the groups R, R', R'':

carboxylester:

$$R(CO)\mathbf{O}R' + H_2O \rightarrow R(CO)OH + \mathbf{HOR'} \qquad 2.9$$

carboxylamide:

$$R(CO)\mathbf{N}R'R'' + H_2O \rightarrow R(CO)OH + \mathbf{HNR'R}'' \qquad 2.10$$

carboxythioether:

$$R(CO)\mathbf{S}R' + H_2O \rightarrow R(CO)OH + \mathbf{HSR'} \qquad 2.11$$

The main reactions of hydrolytic cleavage are summarised in Figure 2.64:

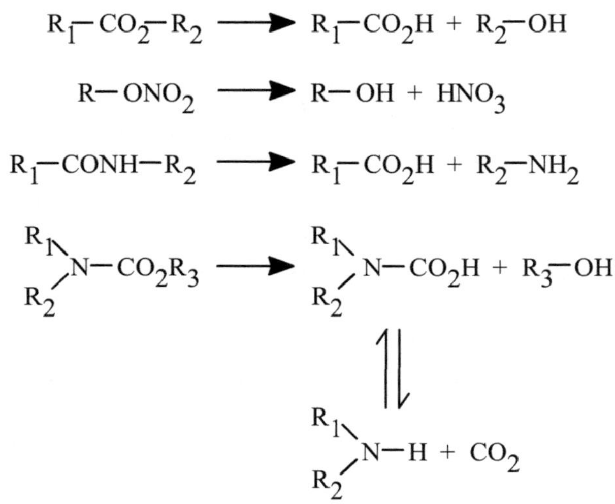

Fig.2.64 Representative types of hydrolytic reactions

2.6.1 Hydrolysis of esters

This type of reaction can take place either in the plasma (non-specific acetylcholinesterases, pseudocholinesterases and other esterases) or in the liver (specific esterases for particular groups of compounds).

A well-known example is the hydrolysis of procaine, catalysed by a plasma esterase (Figure 2.65):

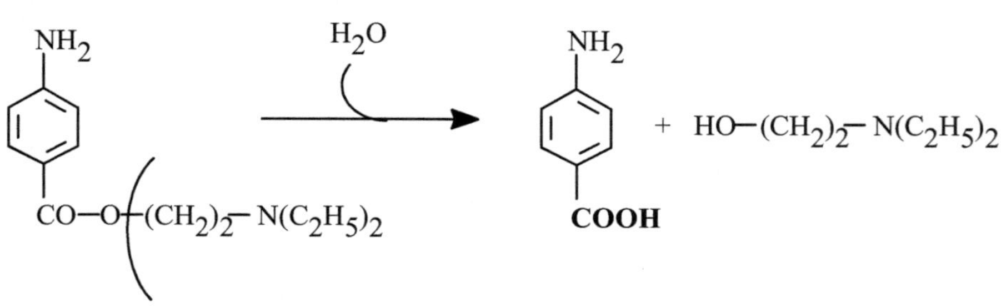

Fig.2.65 Hydrolysis of procaine
(Reproduced with the permission of Nelson Thornes from 'Introduction to Drug Metabolism', 2001, 3rd Ed., isbn 0 7487 6011 3 - Gibson & Skett - first published in 1986)

Procaine is rapidly hydrolysed, while in the case of its amide, about 60% of the administered dose has been recovered unchanged in human urine, the rest being primarily N-acetylated [158].

Of particular pharmacological interest, we mention here the hydrolysis of several prodrugs. It is emphasised that the rate of hydrolysis of such compounds is structure-dependent, sterically masked esters being more slowly hydrolysed and sometimes occurring totally unchanged in urine.

Pivampicillin is a prodrug that is enzymatically cleaved in the organism with subsequent formation of ampicillin (Figure 2.66). It is synthesised by esterification of the carboxyl group of ampicillin with a pivaloyl-oxymethyl function. Upon oral administration, pivampicillin displays a better absorption than ampicillin, ensuring superior plasma concentrations, at equivalent doses [159, 160] (See also Chapter 9).

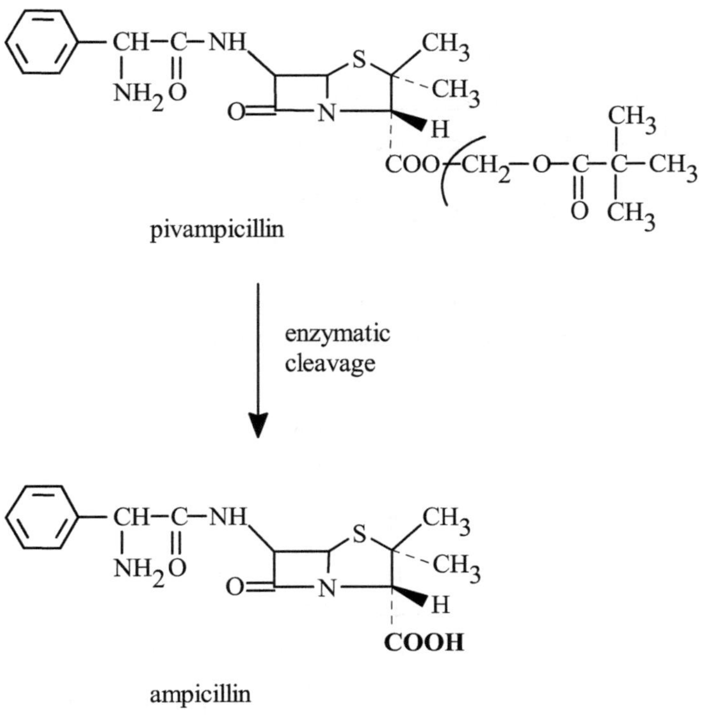

Fig.2.66 Enzymatic cleavage of pivampicillin, yielding the active drug ampicillin

Analogously, the ester of piroxicam with pivalic acid is an effective prodrug with activity comparable to that of the parent drug but with fewer of the ulcerogenic effects that are associated with the carboxyl group of the parent compound [161]. Another, but older example of a prodrug in this drug class is ampiroxicam derived by conversion of the parent drug to an ethyl carbonate ester [162]. This compound generally displayed potencies that were similar to or greater than those of the parent drug, but (as with piroxicam pivalate) with reduced tendency to form gastrointestinal lesions.

Cefuroxime axetil is a cefalosporine prodrug obtained by esterification of the carboxyl group of the parent compound with a complex group, namely 1-acetoxyethyl (designated as "axetil"); this esterification improves the lipohilicity of the parent molecule with consequent improvement in intestinal absorption [160]. The reaction of enzymatic conversion is presented in Figure 2.67:

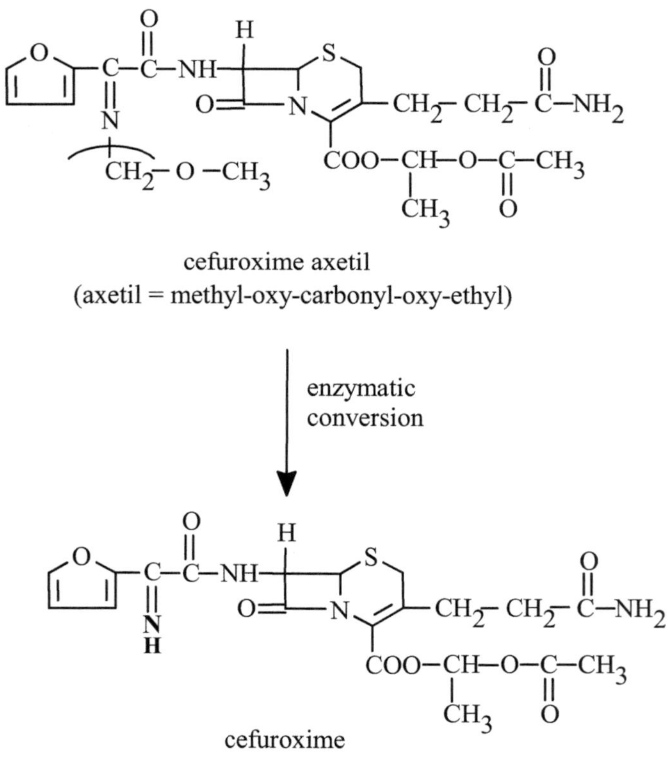

Fig.2.67 Activation of cefuroxime axetil, one of the most common prodrugs of cefuroxime

When tested in animals, low toxicity was registered for cefuroxime axetil [163].

Ritipenem acoxyl is a representative prodrug for the class of penems (synthetic β-lactamines) displaying a wide spectrum of activity and, in contrast to the un-esterified drug, can be administered orally [160]. In the organism, it is enzymatically deacetylated (Figure 2.68), yielding the parent, active drug, ritipenem.

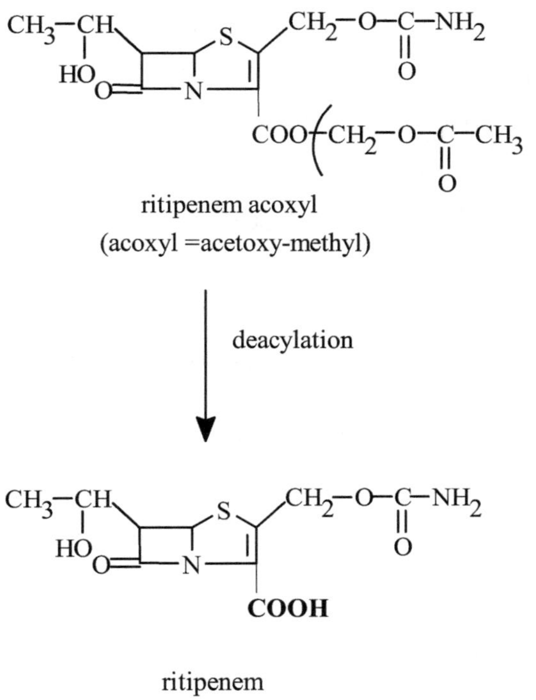

Fig.2.68 Biotransformation of ritipenem acoxyl, yielding the corresponding active form

Quantitation of ritipinem in human plasma and urine can be performed by HPLC analysis [164].

Erythromycin salts, such as the laurylsulphate or the stearate (Figure 2.69) are more lipophilic and consequently more easily absorbable than the parent drug. In these salts the dimethyl amino N atom of the desosamine residue is protonated. When other salts are used (lactobionate or glucoheptonate), water-soluble erythromycin prodrugs are obtained, allowing parenteral administration. By esterification of the hydroxyl group in position 2- of the desosamine moiety, the propionate and ethyl-succinate esters of erythromycin have been obtained. Through enzymatic cleavage, the active erythromycin is liberated. As well as displaying improved absorption relative to the parent, these prodrugs are more robust in the acidic gastric juice [160].

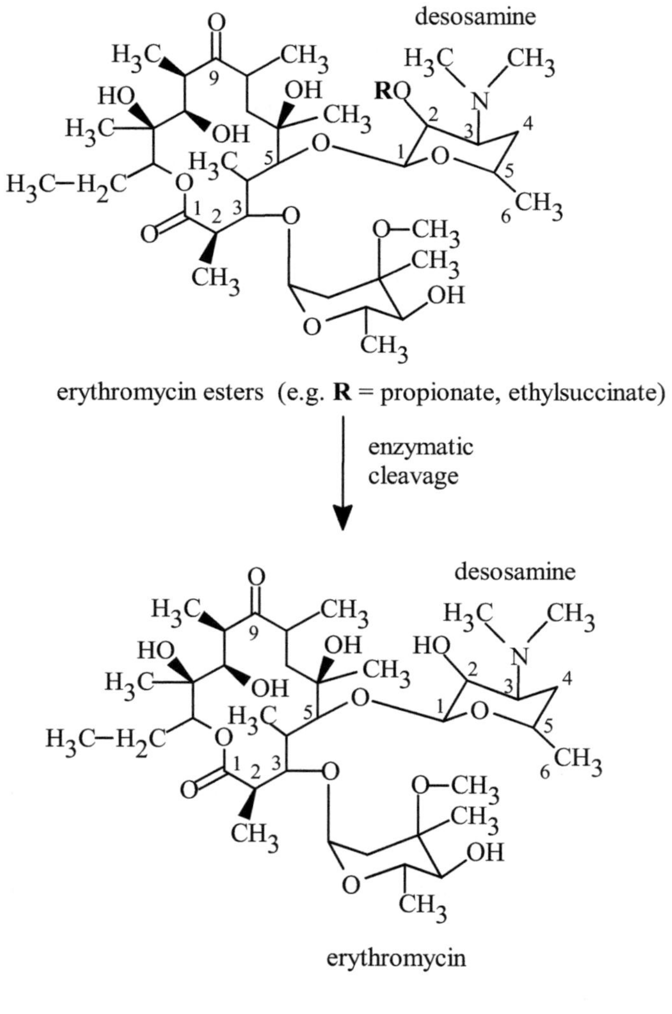

Fig.2.69 Esters of erythromycin displaying prodrug properties

Of a series of erythromycin esters assessed for bioavailability, the 3,4,5-trimethoxybenzoate ester was reported to perform similarly to the stearate salt and the estolate of the parent antibiotic [165].

Esterification of chloramphenicol at the primary alcohol group yielded a more lipophilic product, devoid of the bitter taste of the parent compound [160]. These prodrugs, through enzymatic cleavage, are converted *in vivo*, into the active chloramphenicol, as presented in Figure 2.70.

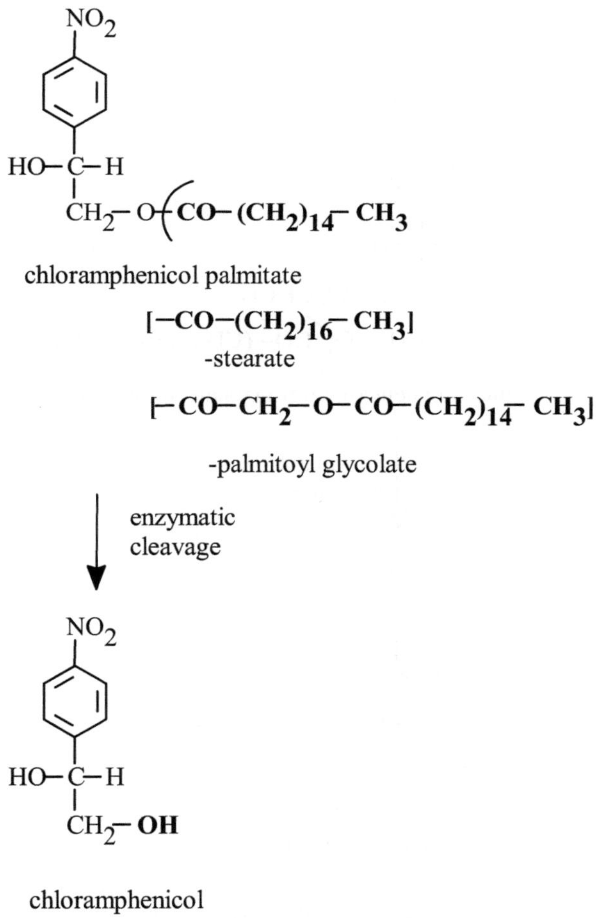

Fig.2.70 Enzymatic conversion of some esters of chloramphenicol to the parent, active drug

Moreover, the absorption of these products is slower, consequently ensuring a therapeutic plasma level of the drug for a longer period. In the organism, the esters are cleaved by specific lipases, liberating the active, parent drug.

Water-solubility improvement for chloramphenicol can be realized by esterification with dicarboxylic acids and conversion of the acid function to the salt [160]. An example of such a prodrug, that can be administered intravenously, is the sodium salt of chloramphenicol hemisuccinate, hydrolysed *in vivo* under the action of specific esterases (Figure 2.71):

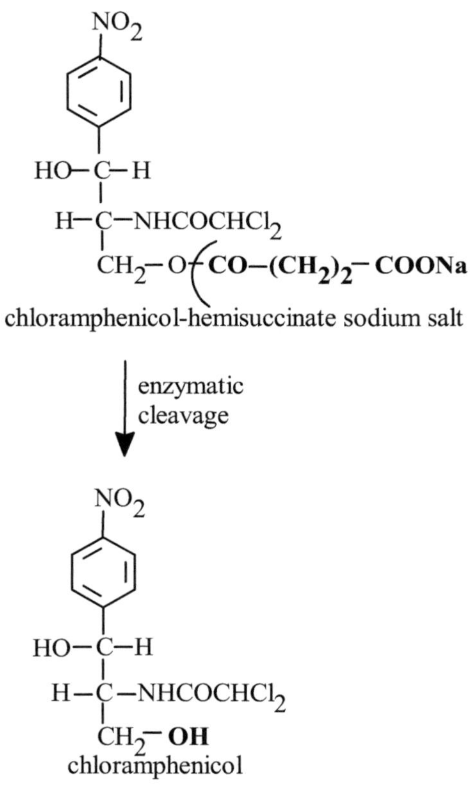

Fig.2.71 Biotransformation of chloramphenicol hemisuccinate sodium salt to the parent, active chloramphenicol

2.6.2 Hydrolysis of amides

Most amides are hydrolysed by the liver amidases. Theoretically, amides may be hydrolysed by plasma esterases too, but such reactions proceed more slowly. The deacylated metabolite of indomethacin (a tertiary amide) has been detected in human urine as one of the major metabolites of this compound [166] (Figure 2.72).

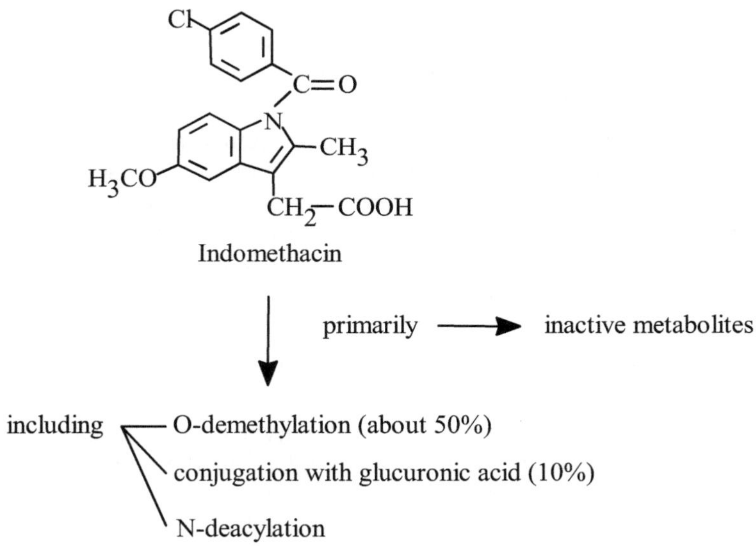

Fig.2.72 Biotransformation routes for indomethacin

Some of these metabolites are detectable in plasma, and the free and conjugated metabolites are eliminated in the urine, bile and faeces. The occurrence of enterohepatic cycling of the conjugates and probably of indomethacin itself is an important feature of the metabolism of this drug. Between 10 and 20% of the drug is excreted unchanged in the urine (in part by tubular secretion).

On the other hand, it is important to emphasise that in certain cases, the reaction of hydrolysis may liberate the active metabolite of a parent drug; a case in point is that of phthalylsulphathiazole, which, under the action of bacterial enzymes in the colon, liberates the antibacterial agent, sulphathiazole.

2.6.3 Hydrolysis of compounds in other classes

Less common functional groups in drugs, such as hydrazide and carbamate, can also be hydrolysed. A well-known example is the hydrolysis of the hydrazide group of isoniazid (Figure 2.73):

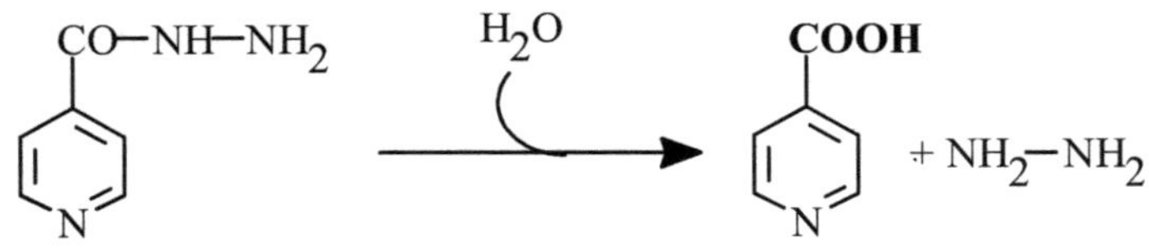

Fig.2.73 Hydrolysis of isoniazid
(Reproduced with the permission of Nelson Thornes from 'Introduction to Drug Metabolism', 2001, 3rd Ed., isbn 0 7487 6011 3 - Gibson & Skett - first published in 1986)

In this context, hydrolysis of proteins and peptides by enzymes can also be mentioned, with the qualification that these enzymes are mainly found in gut secretions and are usually involved only to a small extent in drug metabolism. Exceptions occur in the further metabolism of glutathione conjugates as well as in the metabolism of orally administered peptide/protein drugs.

2.7 MISCELLANOUS PHASE I REACTIONS

Regarded as a specialised form of hydrolysis, we can mention here the hydration reaction, where water is added to the compound without causing its dissociation. Particular substrates for this type of reaction are epoxides, yielding the corresponding dihydrodiols. The reaction is catalysed by epoxide hydratases, which are substrate-specific.

In particular, this type of biotransformation occurs with the pre-carcinogenic polycyclic hydrocarbon epoxides and forms a *trans*-diol [167], (Figure 2.74):

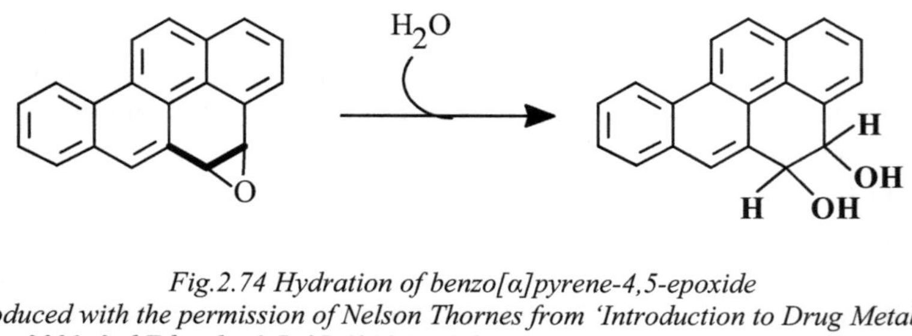

Fig.2.74 Hydration of benzo[α]pyrene-4,5-epoxide
(Reproduced with the permission of Nelson Thornes from 'Introduction to Drug Metabolism', 2001, 3rd Ed., isbn 0 7487 6011 3 - Gibson & Skett -first published in 1986)

Many other reactions that do not fall within the above-mentioned groups have indeed been proposed as possible routes of biotransformation for specific drugs. These include e.g. isomerisations, dimerisations, transamidations, N-carboxylations and ring cyclisations.

2.8 THE FATE OF PHASE I REACTION PRODUCTS

The main function of Phase I metabolism is to prepare the compound for phase II metabolism and not to prepare the drug for excretion.

All the various types of reactions are classified under the three major groups of oxidation, reduction and hydrolysis. Virtually every possible chemical reaction that a compound can undergo is catalysed by the drug-metabolising enzyme systems, yielding final products containing chemically reactive functional groups that will represent targets for the enzymes of phase II metabolism. Phase II generally represents the true "detoxication" of drugs, giving more water-soluble, and thus more easily excreted, metabolites.

Another important concluding remark of the present chapter is that many drugs can undergo a number of the reactions listed, being able to pass along several of the routes of biotransformation described above.

The following points are also noteworthy in this context:

- the significance of a particular pathway varies with many factors (details in Chapters 6 and 7).
- the general difficulty of predicting the biotransformation pathways that a given drug will undergo in the human organism (see also Chapter 9).
- Finally, we have to stress that during the Phase I metabolism, potential pharmacologically toxic intermediates may occur (e.g. free radicals, superoxides, epoxides). Taking this into account, a major concern in current drug design and the development of new therapeutic agents is the metabolism-mediated toxicity of xenobiotics. The importance of gaining more knowledge and understanding of biotransformation pathways and the factors that influence them is obvious. The increasing understanding of the metabolic fates of biologically active compounds will continue to aid identification of latent functionalities that may mediate toxic effects following bioactivation; at the same time, anticipation of the enzyme systems involved in the metabolic reactions leading to structural modifications (resulting either in bioactivation or in detoxication), as well as the factors that might influence these processes, represent other important challenges whose solution could be highly beneficial.

The culmination of this ever-expanding knowledge base should be the improvement of strategies for designing needed drugs with appropriate therapeutic effects and devoid of toxicities mediated by reactive metabolites, so that the ratio of therapeutic effect to toxic risk is maximised in the interest of providing real benefit to the patient.

References

1. Taylor JB, Kennewell PD. 1993. Biotransformation. Metabolic pathways. In: Modern Medicinal Chemistry, New York: Ellis Horwood Ltd, pp 102-108; 109-116.

2. Wilkinson GR. 2001. Pharmacokinetics: The Dynamics of Drug Absorption, Distribution, and Elimination. In: Hardman JG, Limbird LE, Gilman GA, editors. Goodman & Gilman's The pharmacological Basis of Therapeutics, 10th ed. New York: McGraw-Hill International Ltd. (Medical Publishing Division), pp 12-13.

3. Ritter JM, Lewis LD, Mant T.GK. 1999. Drug metabolism. In: Radojicic R, Goodgame F, editors. A Textbook of Clinical Pharmacology, 4th Ed. Oxford University Press Inc., pp 36-40.

4. Rang HP, Dale MM, Ritter JM. 1999. Absorption, distribution and fate of drugs. In: Pharmacology, 4th ed. Edinburgh: Churchill Livingstone, pp 74-76.

5. Gibson GG, Skett P. 1994. Pathways of drug metabolism. In: Introduction to Drug Metabolism. London: Blackie Academic & Professional, An imprint of Chapman & Hall, pp 1-13.

6. Mabic S, Castagnoli K, Castagnoli Jr, N. 1999. Oxidative Metabolic Bioactivation of Xenobiotics. In: Woolf TF editor. Handbook of Drug Metabolism. New York: Marcel Dekker Inc, pp 49-79.

7. Wingard LB Jr, Brody TM, Larner J, Schwartz A. 1991. General Principles. In: Kist K, Steinborn E, Salway J, editors. Human Pharmacology, Molecular-to-Clinical. St.Louis (Missouri): Mosby Year Book Inc, pp 55-57.

8. Testa B. 1994. Xenobiotic Metabolism: The Biochemical View. In: Testa B, Caldwell J, editors. The Metabolism of Drugs and Other Xenobiotics: Biochemistry of Redox Reactions. London: Academic Press Ltd.(Harcourt Brace and Company, Publishers), pp 15-40.

9. Smith PC. 1999. Pharmacokinetics of Drug Metabolites. In: Woolf TF editor. Handbook of Drug Metabolism. New York: Marcel Dekker Inc, pp 2-47.

10. Noriyuki K, Toshiyuki S, Raku S, Shin-ichi I, Takashi I, Maya K, Toshio O, Miho O, Kuniyo I. 2004. Sequential metabolism of 2,3,7-trichlorodibenzo-p-dioxin (2,3, 7-triCDD) by cytochrome P450 and UDP-glucuronosyltransferase in human liver microsomes. Drug Metab Dispos 32:870-875.

11. Walle T, Conradi EC, Walle UK, Fagan TC, Gaffney TE. 1980. 4-Hydroxypropranolol and its glucuronide after single and long-term doses of propranolol. Clin Pharmacol Ther 27:22-31.

12. Jann MW, Lam Y, Francis W, Gray EC, Chang W-H. 1994. Reversible metabolism of drugs. Drug Metab Interact 11:1-24.

13. Linder MW, Prough RA, Valdes Jr R. 1997. Pharmacogenetics: a laboratory tool for optimizing therapeutic efficiency. Clin Chem 43:254-266.

14. Shibata S. 1999. Biological clock and chronopharmacology. Nippon Yakuzaishikai Zasshi 51:1879-1885.

15. Gibson GG, Skett P. 1994. Induction and inhibition of drug metabolism. In: Introduction to Drug Metabolism. London: Blackie Academic & Professional, An imprint of Chapman & Hall, pp 77-106.

16. Morris JS, Stockley IH. 2000. Fundamentals of drug interactions. In: Sirtori CR, Kuhlmann J, Tillement J-P, Vrhovac B, Reidenberg MM, editors. Clinical Pharmacology. London: Mc-Graw-Hill International (UK) Ltd, pp 51-65.

17. Gibson GG, Skett P. 1994. Enzymology and molecular mechanisms of drug metabolism reactions. In: Introduction to Drug Metabolism. London: Blackie Academic & Professional, An imprint of Chapman&Hall, pp 35-76.

18. Elfarra AA. 2005. Renal cytochrome P450s and flavin-containing monooxygenases: potential roles in metabolism and toxicity of 1,3-butadiene, trichloroethylene, and tetrachloroethylene. In: Lawrence LH editor. Drug Metabolism and Transport. Humana Totowa, N.J. Press Inc., pp 1-18.

19. Gibson GG, Skett P. 1994. Pharmacological and toxicological aspects of drug metabolism. In: Introduction to Drug Metabolism. London: Blackie Academic&Professional, an imprint of Chapman&Hall, pp 157-179.

20. Ortiz de Montellano PR. 1999. The Cytochrome P450 System. In: Woolf TF editor. Handbook of Drug Metabolism. New York: Marcel Dekker Inc., pp 109-130.

21. Burchell B, Ethell B, Coffey MJ, Findlay K, Jedlitschky G, Soars M, Smith D, Hume R. 2001. Interindividual variation of UDP-glucuronosyltransferases and drug glucuronidation. Interindividual Variability in Human Drug Metabolism: 358-394.

22. Cryle MJ, Stok JE, James J. 2003. Reactions catalyzed by bacterial cytochromes P450. Aust J Chem 56:749-762.

23. Parrill AL. University of Memphis, Tennessy, USA, available on: http://www.chem.memphis.edu/parrill/chem4315/Drug%20Metabolism.pdf. For details of the NIH shift mechanism, see also: (a) Guroff G, Daly JW, Jerina DM, Renson J, Witkop B, Udenfriend S. 1967. Science 157:1524 (b) Jerina D. 1973. Chem Technol 4:120.

24. Park W, Jeon CO, Cadillo H, DeRito C, Madsen EL. 2004. Survival of naphthalene-degrading Pseudomonas putida NCIB 9816-4 in naphthalene-amended soils: toxicity of naphthalene and its metabolites. Appl Microb Biotech 64:429-435.

25. Paselk R. 2003. Biochemical Toxicology Lecture Notes, available at http://www.humboldt.edu/~rap1/C451.S03/C451LecNotes/451nFeb26.html

26. Ibuki Y, Goto R. 2004. Dysregulation of apoptosis by benzene metabolites and their relationships with carcinogenesis. Acta Biochim Biophys 1690:11-21.

27. Testa B. 1994. Xenobiotic Metabolism: The Biochemical View. In: Testa B, Caldwell J, editors. The Metabolism of Drugs and Other Xenobiotics: Biochemistry of Redox Reactions. London: Academic Press Ltd.(Harcourt Brace and Company, Publishers), p 146.

28. Parales RE, Resnick SM.2004.Aromatic hydrocarbon dioxygenases. Soil Biol 2:175-195.

29. Testa B. 1994. Xenobiotic Metabolism: The Biochemical View. In: Testa B, Caldwell J, editors. The Metabolism of Drugs and Other Xenobiotics: Biochemistry of Redox Reactions. London: Academic Press Ltd.(Harcourt Brace and Company, Publishers), pp 102, 125, 236, 348.

30. Guengerich FP, Kim DH. 1991. Enzymatic oxidation of ethyl carbamate to vinyl carbamate and its role as an intermediate in the formation of 1,N^6-etheno-adenosine. Chem Res Toxicol 4:413-421.

31. Yun CH, Okerholm RA, Guengerich FP. 1993. Oxidation of the antihistaminic drug terfenadine in human liver microsomes. Role of cytochrome P4503A4 in N-dealkylation and C-hydroxylation. Drug Metab Dispos 21:403-409.

32. Jurima-Romet M, Huang HS, Beck DJ, Li AP. 1996. Evaluation of drug interactions in intact hepatocytes: inhibitors of terfenadine metabolism. Toxicol in Vitro 10:655-663.

33. Akhtar M, Njar VCO, Wright JN. 1993. Mechanistic studies on aromatase and related C-C bond cleaving P-450 enzymes. J Ster Biochem Molec Biol 44:375-387.

34. Eberhart D, Fitzgerald K, Parkinson A.1992. Evidence for the involvement of a distinct form of cytochrome P450 3A in the oxidation of digitoxin by rat liver microsomes. J Biochem Toxicol 7:53-64.

35. Yanase T, Suzuki S, Goto K, Nawata H, Takayanagi R. 2003. DHEA and bone metabolism. Clin Calcium 13:1419-1424.

36. Tsuruo Y. 2000. The localization and function of androgen metabolizing enzymes in the brains. Denshi Kenbikyo 35:230-235.

37. Garcia-Segura LM, Veiga S, Sierra A, Melcangi RC, Azcoitia I. 2003. Aromatase: a neuroprotective enzyme. Prog Neurobiol 71:31-41.

38. Gilbert JNT, Powell JW, Templeton J. 1975. A study of the human metabolism of secbutobarbitone. J Pharm Pharmacol 27:923-927.

39. Testa B, Jenner P. 1976. The concept of regioselectivity in drug metabolism. J Pharm Pharmacol 28:731-744.

40. Al Sharifi MA, Gilbert JNT, Powell JW. 1983. 3'-hydroxylated derivatives as urinary metabolites of two barbiturates. Xenobiotica 13:179-183.

41. Fischbach T, Lenk W. 1990. Additional routes in the metabolism of phenacetin. Xenobiotica 20:209-222.

42. Sesardic D, Cole KJ, Edwards RJ, Davies DS, Thomas PE, Levin W, Boobis AR. 1990. The inducibility and catalytic activity of cytochromes P450c (P450IA1) and P450d (P450IA2) in rat tissues. Biochem Pharmacol 39:499-506.

43. Starek A. 1992. Circadian variations of phenacetin metabolism in rats *in vivo* and *in vitro*. Pol J Pharmacol Phar 44:663-670.

44. Peters JM, Morishima H, Ward JM, Coakley CJ, Kimura S, Gonzales FJ. 1999. Role of CYP1A2 in the toxicity of long-term phenacetin feeding in mice. Toxicol Sci 50:82-89.

45. Thomas RC, Judy RW. 1972. Metabolic fate of chlorpropamide in man and in the rat. J Med Chem 15:964-968.

46. Waxman DJ. 1988. Interaction of hepatic cytochromes P-450 with steroid hormones. Regioselectivity and stereospecificity of steroid metabolism and hormonal regulation of rat P-450 enzyme expression. Biochem Pharmacol 37:71-84.

47. Korzekwa KR, Trager WF, Nagata K, Parkinson A, Gillette JR. 1990. Isotope effect studies on the mechanism of the cytochrome P-450IIA1-catalyzed formation of Δ6-testosterone from testosterone. Drug Metab Dispos 18:974-979.

48. Rosenshein B, Flockhart DA, Ho H. 2004. Induction of testosterone metabolism by esomeprazole in a CYP2C19*2 heterozygote. Am J Med Sci 327:289-293.

49. Lin DCK, Fentiman Jr AF, Foltz RL, Forney RD Jr, Sunshine I. 1975. Quantification of phencyclidine in body fluids by GC/CI/MS and identification of two metabolites. Biomed Mass Spectrom 2:206-214.

50. Hallstrom G, Kammerer RC, Nguyen CH, Schmitz DA, Di Stefano EW, Cho AK. 1983. Phencyclidine metabolism in vitro. The formation of a carbinolamine and its metabolites by rabbit liver preparations. Drug Metab Dispos 11:47-53.

51. Laurenzana EM, Owens SM. 1997. Metabolism of phencyclidine by human liver microsomes. Drug Metab Dispos 25:557-563.

52. Veselovskaya NV, Savchuk SA, Izotov BN. 1999. Chromatographic analysis of phencyclidine, its metabolites and analogs in biological fluids. Sudebno-Meditsinskaya Ekspertiza 42:20-25.

53. Karam JH, Matin SB, Forsham PH. 1975. Antidiabetic Drugs After the University Group Diabetes Program (UGDP). Ann Rev Pharmacol 15:351-366.

54. Thomas RC, Duchamp DJ, Judy RW, Ikeda GJ. 1978. Metabolic fate of tolazamide in man and in the rat. J Med Chem 21:725-732.

55. Ascalone V, Flaminio L, Guinebault P, Thenot JP, Morselli PL. 1992. Determination of zolpidem, a new sleep-inducing agent and its metabolites in biological fluids – Pharmacokinetics, drug metabolism and overdosing investigations in humans. J Chromatogr-Biomed Appl 581:237-250.

56. Chouinard G, Lefko-Singh K, Teboul E Louis-H. 1999. Role of cytochrome P450 isozymes in the metabolism of benzodiazepines. Cell Mol Neurobiol 19:533-552.

57. Von Moltke LL, Greenblatt DJ, Granda BW, Duan SX, Grassi JM, Venkatakrishnan K, Harmatz JS, Shader RI. 1999. Zolpidem metabolism *in vitro:* responsible cytochromes, chemical inhibitors, and in vivo correlations. Brit J Clin Pharmacol 48:89-97.

58. Kronbach T, Mathys D, Umeno M, Gonzalez FJ, Meyer UA. 1989. Oxidation of midazolam and triazolam by human liver cytochrome P450IIIA4. Mol Pharmacol 36:89-96.

59. Fabre G, Rahmani R, Placidi M, Combalbert J, Covo J, Cano JP, Coulange C, Ducros M, Rampal M. 1988. Characterization of midazolam metabolism using human hepatic microsomal fractions and hepatocytes in suspension obtained by perfusing whole human livers. Biochem Pharmacol 37:4389-4397.

60. Ito K, Ogihara K, Kanamitsu S-I, Itoh T. 2003. Prediction of in vivo interaction between midazolam and macrolides based on in vitro studies using human liver microsomes. Drug Metab Dispos 31:945-954.

61. Shetty HU, Nelson WL. 1988. Chemical aspects of metoprolol metabolism. J Med Chem 31:55-59.

62. Wan J, Xie Y-H, Xia H, Lu Y-Q. 1997. Age might influence the frequency distribution of metoprolol hydroxylation polymorphism in a Chinese population. Pharmacol Toxicol (Copenhagen) 80:167-170.

63. Van der Graaff M, Vermeulen NPE, Breimer DD. 1988. Disposition of hexobarbital: 15 years of an intriguing model substrate. Drug Metab Rev 19:109-164.

64. Kennedy KA, Ambre JJ, Fischer LG. 1978. A selected ion monitoring method for glutethimide and six metabolites: application to blood and urine from humans intoxicated with glutethimide. Biomed Mass Spectrom 5:679-685.

65. Shulgin AT. 1975. Drugs of Abuse in the Future. Clin Toxicol 8:405-456.

66. Silver EH, Kuttab SH, Hasan T, Hassan M. 1982. Structural considerations in the metabolism of nitriles to cyanide *in vivo*. Drug Metab Dispos 10:495-498.

67. Testa B. 1994. Xenobiotic Metabolism: The Biochemical View. In: Testa B, Caldwell J, editors. The Metabolism of Drugs and Other Xenobiotics: Biochemistry of Redox Reactions. London: Academic Press Ltd. (Harcourt Brace and Company, Publishers), pp 138-139.

68. Bartok M, Lang KL. 1985. Oxiranes. In: Hassner A, editor. Small Ring Heterocycles. (Part 3). Wiley: NY, pp 1-196.

69. Faigle JW, Böttcher I, Godbillon J, Kriemler HP, Schlumpf E, Schneider W, Schweizer A, Stierlin H and Winkler T. 1988. A new metabolite of diclofenac sodium in human plasma. Xenobiotica 18:1191-1197.

70. Yasumori T, Yamazoe Y and Kato R. 1991. Cytochrome P-450 human-2 (P-450IIC9) in mephenytoin hydroxylation polymorphism in human livers: Differences in substrate and stereoselectivities among microheterogeneous P-450IIC species expressed in yeast. J Biochem 109:711-717.

71. Meier UT, Kronbach T, Meyer UA. 1985. Assay of mephenytoin metabolism in human liver microsomes by high-performance liquid chromatograpy. Anal Biochem 151:286-291.

72. Talaat RE, Nelson WL. 1988. Regiosomeric aromatic dihydroxylation of propranolol. Drug Metab Dispos 16:207-216.

73. Masubuchi Y, Hosokawa S, Horie T, Suzuki T, Ohmori S, Kitada M. 1994. Cytochrome P450 isozymes involved in propranolol metabolism in human liver microsomes. The role of CYP2D6 as ring-hydroxylase and CYP1A2 as N-desisopropylase. Drug Metab Dispos 22:909-915.

74. Butler TC, Dudley KH, Johnson D, Roberts SB. 1976. Studies on the metabolism of 5, 5-diphenylhydantoin relating principally to the stereoselectivity of the hydroxylation reactions in man and the dog. J Pharmacol Exp Ther 199:82-92.

75. Moustafa MAA, Claesen M, Adline J, Vandevorst D, Poupaert JH. 1983. Evidence for an arene-3,4-oxide as a metabolic intermediate in the *meta*- and *para*-hydroxylation of phenytoin in the dog. Drug Metab Dispos 11:574-580.

76. Cuttle L, Munns AJ, Hogg NA, Scott JR, Hooper WD, Dickinson RG, Gillam EMJ. 2000. Phenytoin Metabolism by Human Cytochrome P450: Involvement of P450 3A and 2C Forms in Secondary Metabolism and Drug-Protein Adduct Formation. Drug Metab Dispos 28:945-950.

77. Howell SR, Kotsoskie LA, Dills RL, Klaassen CD. 1988. 3-Hydroxylation of salicylamide in mice. J Pharm Sci 77:309-313.

78. Cavalieri EL, Rogan EG. 1992. The approach to understanding aromatic hydrocarbon carcinogenesis – the central role of radical cations in metabolic activation. Pharmacol Therapeut 55:183-199.

79. Lowe JP, Silverman BD. 1981. Simple molecular orbital explanation for 'bay-region' carcinogenic reactivity. J Amer Chem Soc 103:2852-2855.

80. Fitzpatrick FA, Murphy RC. 1989. Cytochrome P-450 metabolism of arachidonic acid: Formation and biological actions of "epoxygenase"-derived eicosanoids. Pharmacol Rev 40:229-241.

81. Testa B. 1990. Mechanisms of inhibition of xenobiotic-metabolizing enzymes. Xenobiotica 20:1129-1137.

82. Rambeck B, May T, Juergens U. 1987. Serum concentrations of carbamazepine and its epoxide and diol metabolites in epileptic patients: the influence of dose and comedication. Therap Drug Monit 9:298-303.

83. Furst SM, Uetrecht JP. 1993. Carbamazepine metabolism to a reactive intermediate by the myeloperoxidase system of activated neutrophils. Biochem Pharmacol 45:1267-1275.

84. Gatti G, Furlanut M, Perucca E. 2001. Interindividual variability in the metabolism of anti-epileptic drugs and its clinical implications. In: Interindividual Variability in Human Drug Metabolism, pp 157-180.

85. Slack JA, Ford-Hutchinson AW, Richold M, Choi BCK. 1981. Some biochemical and pharmacological properties of an epoxide metabolite of alclofenac. Chem-Biol Interact 34:95-107.

86. Brown LM, Ford-Hutchinson AW. 1982. The destruction of cytochrome P-450 by alclofenac: possible involvement of an epoxide metabolite. Biochem Pharmacol 31:195-199.

87. Ortiz de Montellano PR, Kunze KL.1981. Cytochrome P-450 inactivation: structure of the prosthetic heme adduct with propyne. Biochemistry-US 20:7266-7271.

88. Schmid SE, Au WYW, Hill DE, Kadlubar FF, Slikker Jr W. 1983. Cytochrome P-450-dependent oxidation of the 17 α-ethynyl group of synthetic steroids. Drug Metab Dispos 11:531-536.

89. Beckett AH, Belanger PM. 1978. The disposition of phentermine and its N-oxidized metabolic products in the rabbit. Xenobiotica 8:555-560.

90. Mori MA, Uemura H, Kobayashi M, Miyahara T, Kozuka H. 1993. Metabolism of phentermine and its derivatives in the male Wistar rat. Xenobiotica 23:709-716.

91. Budinski RA, Roberts SM, Coats EA, Adams L, Hess EV. 1987. The formation of procainamide hydroxylamine by rat and human liver microsomes. Drug Metab Dispos 15:37-43.

92. Uetrecht JP. 1985. Reactivity and possible significance of hydroxylamine and nitroso metabolites of procainamide. J Pharmacol Exp Ther 232:420-425.

93. Testa B, Jenner P. 1976. Chemical and Biochemical Aspects. In: Drug Metabolism. Decker, New York, pp 61-73.

94. Patterson LH, Hall G, Nijar BS, Khatra PK, Cowan DA. 1986. *In vitro* metabolism of lignocaine to its N-oxide. J Pharm Pharmacol 38:326-331.

95. Bargetzi MJ, Aoyama T, Gonzales FJ, Meyer UA. 1989. Lidocaine metabolism in human liver microsomes by cytochrome P450IIIA4. Clin Pharmacol Ther 46:521-527.

96. Tanaka E, Inomata S, Yasuhara H. 2000. The clinical importance of conventional and quantitative liver function tests in Liver Transplant. J Clin Pharm Therapeut 25:411-419.

97. Duquette PH, Erickson RR, Holtzman JL. 1983. Role of substrate lipophilicity on the N-demethylation and type I binding of 3-O-alkylmorphine analogues. J Med Chem 26:1343-1348.

98. McQuay HJ, Moore RA. Opioid problems, and morphine metabolism and excretion. Available at: http://www.jr2.ox.ac.uk/bandolier/booth/painpag/wisdom/c14.html#RTFT.

99. Oguri K. 2000. An active metabolite of morphine and the responsible glucuronosyltransferase for its formation. Yakubutsu Dotai 15:136-142.

100. Sharke C, Loetsch J. 2002. Morphine metabolites: clinical implications. Semin Anesth Periop Med Pain 21:258-264.

101. Bäärnhielm C, Weterlund C. 1986. Quantitative relationships between structure and microsomal oxidation rate of 1,4-dehydropyridines. Chem Biol Interact 58:277-288.

102. Bäärnhielm C, Skanberg I, Borg KO. 1984. Cytochrome P-450-dependent oxidation of felodipine-a 1,4-dihydropyridine-to the corresponding pyridine. Xenobiotica 14: 719-726.

103. Gorrod JW. 1985. Amine-imine tautomerism as a determinant of the site of biological N-oxidation. In: Gorrod J W, Damani LA, editors. Biological Oxidation of Nitrogen in Organic Molecules. Chemistry, Toxicology and Pharmacology. Horwood, Chichester, pp 219-230.

104. Berndt C, Thomas K. 1990. Hepatic microsomal N-hydroxylation of adenine to 6-N-hydroxylamineopurine. Biochem Pharmacol 39:925-933.

105. Clement B, Kunze T. 1990. Hepatic microsomal N-hydroxylation of adenine to 6-hydroxyaminopurine. Biochem Pharmacol 39:925-933.

106. Ford GP, Griffin GR, Galen R. 1992. Relative stabilities of nitrenium ions derived form heterocyclic amine food carcinogens. Relationships to mutagenicity. Chem Biol Interact 81:19-33.

107. Dias RMB, Vieira AJSC. 1997. Effect of oxygen on the hydroxylation of adenine by photolytically generated hydroxyl radical. J Photoch Photobio A 109:133-136.

108. LaCagnin LB, Colby HD, O'Donnell JP. 1986. The oxidative metabolism of hydralazine by rat liver microsomes. Drug Metab Dispos 14:549-554.

109. Weinkaum RJ. Shiba DA. 1978. Metabolic activation of procarbazine. Life Sci 22: 937-946.

110. Inaba T, Tait A, Nakano M, Mahon WA, Kalow W. 1988. Metabolism of diazepam *in vitro* by human liver. Independent variability of N-demethylation and C3-hydroxylation. Drug Metab Dispos 16:605-608.

111. Marcucci F, Airoldi L, Zavattini G, Mussini E. 1981. Metabolism of pinazepam by rat liver microsomes. Eur J Drug Metab Ph 6:109-114.

112. Bertilsson L, Baillie TA, Reviriego J. 1990. Factors influencing the metabolism of diazepam. Pharmacol Ther 45:85-91.

113. Janbroers JM. 1984. Pinazepam: review of pharmacological properties and therapeutic efficacy. Clin Ther 6:434-450.

114. Berthou F, Flinois JP, Ratanasavanh D, Beaune P, Riche C, Guillouzo A. 1991. Evidence for the involvement of several cytochromes P-450 in the first steps of caffeine metabolism by human liver microsomes. Drug Metab Dispos 19:561-567.

115. Ashihara H, Crozier A. 1999. Biosynthesis and metabolism of caffeine and related purine alkaloids in plants. Adv Bot Res 30:117-205.

116. Nelson WL, Bartels MJ. 1984. N-Dealkylation of propranolol in rat, dog and man. Chemical and stereochemical aspects. Drug Metab Dispos 12:345-352.

117. 2000. In: Gennaro AR editor. Remington: The Science and Practice of Pharmacy, 20th ed. Philadelphia: Lippincott Williams&Wilkins, pp 1326, 1757.

118. Sipes IG, Gandolfi AJ, Pohl LR, Krishna G, Brown BR Jr. 1980. Comparison of the biotransformation and hepatotoxicity of halothane and deuterated halothane. J Pharmacol Exp Therap 214:716-720.

119. Kenna JG, Van Pelt FNAM. 1994. The metabolism and toxicity of inhaled anesthetic agents. Anaesth Pharmacol Rev 2:29-42.

120. Bedford CT, Crayford JV, Hutson DH, Wiggins DE. 1978. An example of the oxidative de-esterification of an isopropyl ester. Its role in the metabolism of the herbicide flampropisopropyl. Xenobiotica 8:383-395.

121. Testa B. 1995. Reaction Catalyzed by monoamine oxidase. In: Testa B, Caldwell J, editors. The Metabolism of Drugs and Other Xenobiotics: Biochemistry of Redox Reactions. London: Academic Press Ltd. (Harcourt Brace and Company, Publishers), pp 313-323.

122. Gerlach M, Riederer P. 1993. Human brain MAO. In: Yasuhara H, Parvez SH, Oguchi K, Sandler M, Nagatsu T, editors. Monoamine oxidase: Basic and Clinical Aspects. Utrecht (Netherlands), pp 147-158.

123. Shih JC. 1991. Molecular basis of human MAO A and B. Neuropsychopharmacol 4:1-7.

124. Silverman RB, Zelechonok Y. 1992. Evidence for a hydrogen atom transfer mechanism or a proton/fast electron transfer mechanism for monoamine oxidase. J Org Chem 57:6373-6374.

125. Weiler W, Hsu YPP, Breakfield XO. 1990. Biochemistry and genetics of monoamine oxidase. Pharmacol Therapeut 47:391-417.

126. Edmondson DE, Mattevi A, Binda C, Li M, Hubalek, F. 2004. Structure and mechanism of monoamine oxidase. Curr Med Chem 11:1983-1993.

127. 2001. In: Hardman JG, Limbird LE, Gilman GA, editors. Goodman&Gilman's The Pharmacological Basis of Therapeutics, 10th ed. New York: McGraw-Hill (Medical Publishing Division), pp 313, 643, 645-651.

128. Repka-Ramirez MS, Baraniuk JN. 2002. Histamine in health and disease. Clin Allergy Immunol 17:1-25.

129. Testa B. 1995. Cytochromes P450 and flavin containing monooxygenases. In: Testa B, Caldwell J, editors. The Metabolism of Drugs and Other Xenobiotics: Biochemistry of Redox Reactions. London: Academic Press Ltd. (Harcourt Brace and Company, Publishers), pp 109-111.

130. Rettie AE, Fisher MB. 1999. Transformation Enzymes: Oxidative; Non-P450. In: Woolf TF editor. Handbook of Drug Metabolism. New York: Marcel Dekker Inc., pp 132-137.

131. Ziegler, Daniel M. 2002. An overview of the mechanism, substrate specificities, and structure of FMOs. Drug Metab Rev 34:503-511.

132. Guengerich FP. 1999. Inhibition on Drug Metabolizing Enzymes: Molecular and Biochemical Aspects. In: Woolf TF editor. Handbook of Drug Metabolism. New York: Marcel Dekker Inc., pp 215-216.

133. Gibson GG, Skett P. 1994. Enzymology and molecular mechanisms of drug metabolism reactions. In: Introduction to Drug Metabolism. London: Blackie Academic & Professional, An Imprint of Chapman & Hall, pp 35-76.

134. Testa B. 1994. Reactions catalyzed by peroxidases. In: Testa B, Caldwell J, editors. The Metabolism of Drugs and Other Xenobiotics: Biochemistry of Redox Reactions. London: Academic Press Ltd. (Harcourt Brace and Company, Publishers), pp 353-363.

135. Rettie AE, Fisher MB. 1999. Transformation Enzymes: Oxidative; Non-P450. In: Woolf TF editor. Handbook of Drug Metabolism. New York: Marcel Dekker Inc., pp 137-141.

136. Testa B. 1994. Oxidations catalyzed by various oxidases and monooxygenases. In: Testa B, Caldwell J, editors. The Metabolism of Drugs and Other Xenobiotics: Biochemistry of Redox Reactions. London: Academic Press Ltd. (Harcourt Brace and Company, Publishers), pp 323-334.

137. Lowe DJ, Richards RL, Bray RC. 1997. Role of Mo-C bonds in xanthine oxidase action. Biochem Soc T 25:774-778.

138. Hille R, Massey V. 1985. Molybdenum-containing hydroxylases: xanthine oxidase, aldehyde oxidase, and sulfite oxidase. Met Ions Biol 7 443-518.

139. Beedham C. 1987. Molybdenum hydroxylases: biological distribution and substrate-inhibitor specificity. Progr Med Chem 12:35-48.

140. Testa B. 1994. Oxidations catalyzed by various oxidases and monooxygenases. In: Testa B, Caldwell J, editors. The Metabolism of Drugs and Other Xenobiotics: Biochemistry of Redox Reactions. London: Academic Press Ltd. (Harcourt Brace and Company, Publishers), pp 323-334.

141. Rettie AE, Fisher MB. 1999. Transformation Enzymes: Oxidative; Non-P450. In: Woolf TF editor. Handbook of Drug Metabolism. New York: Marcel Dekker Inc., pp 142-151.

142. Wermuth B, Omar A, Forster A, di Francesco Ch, Wolf M, von Wartburg JP, Bullock B, Gabbay KH. 1987. Primary structure of aldehyde reductase from human liver. In: Weiner H, Flynn TG, editors. Enzymology and Molecular Biology of Carbonyl Metabolism. New York: Liss, pp 297-307.

143. Testa B. 1994. Oxidations catalyzed by various oxidases and monooxygenases. In: Testa B, Caldwell J, editors. The Metabolism of Drugs and Other Xenobiotics: Biochemistry of Redox Reactions. London: Academic Press Ltd. (Harcourt Brace and Company, Publishers), pp 334-337.

144. Buffoni F, Ignesti G. 2003. Biochemical aspects and functional role of the copper-containing amine oxidases. Inflammopharmacol 11:203-209.

145. Beresford AP, Macrae PV, Stopher DA. 1988. Metabolism of amlodipine in the rat and the dog: a species difference. Xenobiotica 18:169-182.

146. Baillie TA. 1992. Metabolism of valproate to hepatotoxic intermediates. Pharm Weekbl Sci Ed 14:122-125.

147. Kyburz E. 1990. New developments in the field of MAO inhibitors. Drug News Perspect 3:592-599.

148. Krenitski TA, Neil SM, Elion GB, Hitchings GC. 1972. A comparison of the specificities of xanthine oxidase and aldehyde oxidase. Arch Biochem Biophys 150: 585-599.

149. McDaniel HG, Podgainy H and Bressler R. 1969. The metabolism of tolbutamide in the rat liver. J Pharmacol Exp Therap 167:91-97.

150. Yubisui T, Shirabe K, Takeshita M, Kobayashi Y, Fukumaki Y, Sakaki Y, Takano T. 1991. Structural role of serine 127 in the NADH-binding site of human NADH-cytochrome b_5 reductase. J Biol Chem 266:66-77.

151. Testa B. 1994. Oxidations catalyzed by various oxidases and monooxygenases. In: Testa B, Caldwell J, editors. The Metabolism of Drugs and Other Xenobiotics: Biochemistry of Redox Reactions. London: Academic Press Ltd. (Harcourt Brace and Company, Publishers), pp 105-106.

152. Gibson GG, Skett P. 1994. Enzymology and molecular mechanisms of drug metabolism reactions. In: Introduction to Drug Metabolism. London: Blackie Academic & Professional, An Imprint of Chapman & Hall, pp 54-57.

153. Zubay GL, Parson WW, Vance DE. 1995. In: Sievers EM, editor. Principles of Biochemistry. Dubuque, Iowa: Wm C Brown Publishers, pp 308-309.

154. Hoult JRS. 1986. Pharmacological and biochemical actions of sulfasalazine. Drugs 32:18-26.

155. Gibson GG, Skett P. 1994. Pathways of drug metabolism. In: Introduction to Drug Metabolism. London: Blackie Academic & Professional, An Imprint of Chapman & Hall, pp 9-10.

156. Lohr JW, Willsky GR, Acara MA. 1998. Renal Drug Metabolism. Pharmacol Rev 50:118-119.

157. Baker MT, Nelson RM, van Dyke RA. 1983. The release of inorganic fluoride from halothane and halothane metabolites by cytochrome P-450, hemin, and haemoglobin. Drug Metab Dispos 11:308-311.

158. 2000. In: Gennaro AR editor. Remington: The Science and Practice of Pharmacy, 20th ed. Philadelphia: Lippincott Williams&Wilkins, p 1404.

159. Von Daehne W, Godtfredsen WO, Roholt K, Tybring L. 1971. Pivampicillin, a new orally active ampicillin ester. Antimicrob Agents Ch Vol 1971:431-437.

160. Ovidiu O, Tiperciuc B. 2003. Antibiotice antibacteriene, In: "I. Haţieganu" Universitary Medical Printing House, Cluj-Napoca, Romania, pp 58, 83-84, 97, 121-125, 208-212.

161. Caira MR, Zanol M, Peveri T, Gazzaniga A, Giordano F. 1998. Structural Characterization of Two Polymorphic Forms of Piroxicam Pivalate. J Pharm Sci 87:1608-1614.

162. Yamanaka K, Munehasu S, Suzuki M, Ishiko J. 1991. Pharmacological actions of ampiroxicam, a new prodrug of nonsteroidal anti-inflammatory agent. Oyo Yakuri 41:597-612.

163. Spurling NW, Harcourt RA, Hyde JJ. 1986. An evaluation of the safety of cefuroxime axetil during six months oral administration to beagle dogs. J Toxicol Sci 11:237-77.

164. Matsuoka M, Hosomi R, Maki T, Banno K, Sato T. 1995. Determination of ritipenem in human plasma and urine by high performance liquid chromatography. Nippon Kagaku Ryoho Gakkai 43:91-96.

165. Dall'Asta L, Comini A, Garegnani E, Alberti D, Coppi G, Quadro G. 1988. Studies on the bioavailability of some new erythromycin esters. J Antibiot 41:139-141.

166. 2000. In: Gennaro AR editor. Remington: The Science and Practice of Pharmacy, 20th ed. Philadelphia: Lippincott Williams & Wilkins, p 1457.

167. Gibson GG, Skett P. 1994. Pathways of drug metabolism. In: Introduction to Drug Metabolism. London: Blackie Academic & Professional, An Imprint of Chapman & Hall, pp 9-10.

Chapter 3

PATHWAYS OF BIOTRANSFORMATION – PHASE II REACTIONS

3.1 INTRODUCTION

This chapter addresses the Phase II biotransformation reactions that a drug or its metabolite typically undergo. In these so-called 'conjugation reactions', mediated by the appropriate enzymes, the drug becomes linked to an endogenous moiety through one or more functional groups, that may either be present on the parent drug, or which may have resulted from a phase I reaction of oxidation, reduction or hydrolysis.

A characteristic of most conjugation reactions is the replacement of a hydrogen atom present in a hydroxyl, amino or carboxyl group, by the conjugating agent. In general, the resulting conjugated metabolites have no pharmacological activity, are highly water-soluble and therefore subsequently readily excreted in the urine.

These reactions are usually considered as detoxication reactions, but in certain cases, toxication has been recorded, and examples of both are treated below.

Major phase II reactions include glucuronidation, sulphation, acetylation, and conjugation with glutathione or amino acids. Detailed examples of all of these are provided below, with an account of the relevant enzymes involved.

3.2 GLUCURONIDATION

Glucuronidation represents the major route of sugar conjugation, although conjugation with xylulose and ribose are also possible [1-12].

Quantitatively, glucuronide formation is the most important form of conjugation both for drugs and endogenous compounds and can occur with very different substrates. The synthesis of ether, ester, carboxyl, carbamoyl, carbonyl, sulphuryl and nitrogenyl glucuronides generally leads to an increase in their polarity, and consequently their aqueous solubility and thus suitability for excretion.

Mechanistically, glucuronidation is an S_N2 reaction in which an acceptor nucleophilic group on the substrate attacks an electrophilic C-1 atom of the pyranose acid ring of UDPGA (uridine 5′-diphosphate-glucuronic acid) which results in the formation of a glucuronide, a β-D-glucopyranosiduronic acid conjugate [5]. Thus, many electrophilic groups such as hydroxyl, carboxyl, sulphhydryl (thiol), or phenol can serve as acceptors. N-glucuronides may be formed by certain nitrogen containing groups such those in tertiary or aromatic amines.

Esterification of the hemiacetyl hydroxyl group of glucuronic acid to organic acids forms acyl or ester glucuronides. The acyl glucuronides, unlike glucuronides formed with alcohols and phenols, have a great susceptibility to nucleophilic substitution and intramolecular rearrangement. It has even been proposed that the formed acyl glucuronides, acting as electrophiles and reacting with thiol and hydroxyl groups of cell macromolecules, might be responsible for toxicity of some compounds [5]. Renewed interest in this process from pharmaceutical companies has focused on development of drugs that avoid glucuronidation as a biotransformation pathway, thereby improving bioavailability.

Glucuronidation is conjugation with α-D-glucuronic acid and is indeed the most widespread of the conjugation reactions, probably due to the relative abundance of the cofactor for the reaction, UDP-glucuronic acid.

3.2.1 Enzymes involved and general mechanism

The transfer of glucuronic acid from UDP- glucuronic acid (UDPGA) to an aglycone is catalysed by a family of enzymes generally designated as UDP-glucuronosyltransferases (UGTs) [5,7,9]. These ubiquitous microsomal enzymes are present principally in the liver, but also occur in a variety of extrahepatic tissues. Their location in the endoplasmic reticulum has important physiological effects in the neutralisation of reactive intermediates generated by the CYTP450 enzyme system and in controlling the levels of reactive metabolites present in these tissues.

There are more then 50 known microsomal membrane-bound isoenzymes in humans, found in liver, lung, skin, intestine, brain and olfactory epithelium; however, the major site of glucuronidation is the liver. Thus the liver, being the central organ for a variety of anabolic and catabolic functions, plays a significant role in drug metabolism, toxicity and especially in detoxication processes [12].

Structural and functional aspects of human UDP-glucuronosyltransferases (UGTs) have been reviewed with details of the mechanisms of glucuronidation of both drugs and endogenous compounds [13]. Characterization of the active site in terms of amino acids and peptide domains that bind substrates and effectors in such reactions is also discussed. Genetic differences in the expression of UDP-glucuronosyltransferases in humans result in interindividual variations. This topic has also been reviewed recently [14]. Characterization of genetic multiplicity and regulatory patterns of UGTs is being aided by new developments in the field of genetics. An account of recent findings relating to this topic has appeared [15].

The different isoenzymes of the UGT family have high organ specificity locations: for example, bilirubin UGT is highly expressed in human liver, but is absent in human kidney, whereas phenol UGT is highly expressed in both organs.

Individual UGTs are subject to differential induction by hormones, leading to tissue-specific regulated expression. In addition, the spectrum of UGTs in different tissues can be differentially altered by exposure to drugs and other xenobiotics.

Glucuronidation requires an adequate supply of UDPGA and its concentration in cytosol may determine the transferase activity. This may be more critical in extrahepatic tissues than in the liver. The concentration of UDPGA in the kidney has been estimated to be one-fifteenth that in the liver in humans [8].

As mentioned above, the glucuronidation mechanism involves a nucleophilic substitution [5], illustrated in Figure 3.1 for a phenol as substrate.

The resultant glucuronide has the β-configuration at the C-1 atom of the glucuronic acid. With the attachment of the hydrophilic carbohydrate moiety, containing an easily ionisable carboxyl group, a lipid-soluble compound is thus converted into a conjugate that is poorly reabsorbed by the renal tubules from the urine, and therefore more rapidly excreted, predominantly *via* the kidneys.

Nonetheless, it should be noted that certain high molecular weight glucuronides are excreted *via* the bile into the gastrointestinal tract where subsequent hydrolysis may result in reabsorption of drug or metabolites (biliary recirculation) or excretion in the faeces.

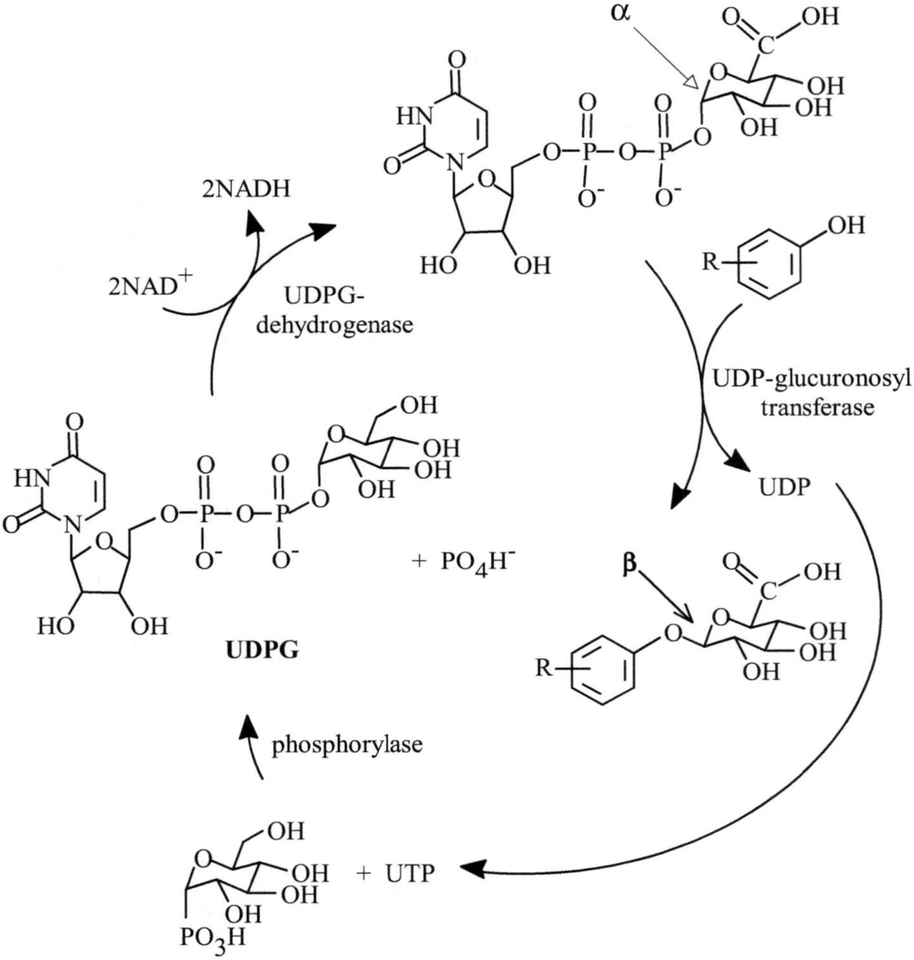

Fig.3.1 The general mechanism of glucuronidation

Functional groups susceptible to glucuronidation are presented in Figure 3.2 with GLU representing glucuronic acid.

As seen from the latter figure, alcohols and phenols form ether glucuronides; aromatic and some aliphatic carboxylic acids form ester (acyl) glucuronides; aromatic amines form N-glucuronides, and thiol compounds form *S*-glucuronides, both of these being more labile to acid than are the O-glucuronides. Some tertiary amines have been found to form quaternary ammonium N-glucuronides.

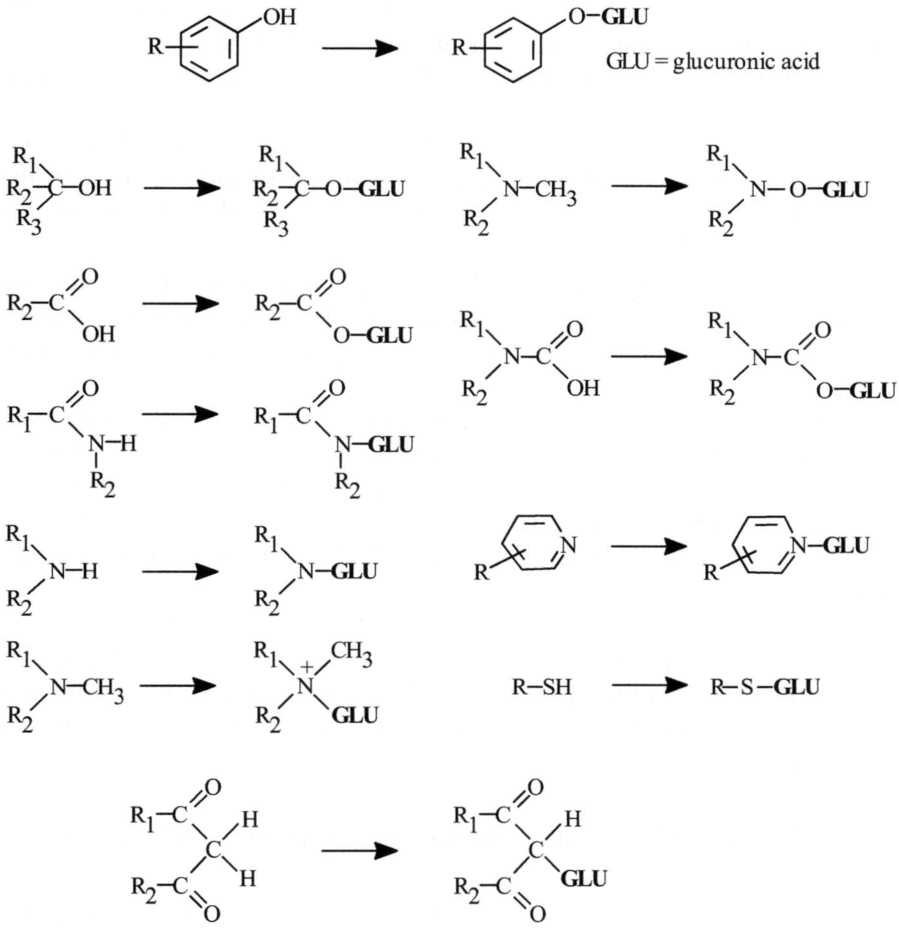

Fig.3.2 The most common functional groups undergoing glucuronidation

Compounds containing a 1,3-dicarbonyl system (e.g. phenylbutazone) can form C-glucuronides by direct conjugation, bypassing prior metabolism. The degree of C-glucuronide formation is determined by the acidity of the functional group separating the carbonyl groups.

Drug-acyl glucuronides are reactive conjugates at physiological pH. The acyl group of the C^1-acyl glucuronide can migrate via transesterification from the original C-1 position of the glucuronic acid to the C-2, C-3, or C-4 positions. The resulting positional isomers are not hydrolysable by β-glucuronidase. Under physiological or weakly alkaline conditions, however, the C^1-acyl glucuronide can hydrolyse in the urine to the parent compound (aglycone) or effect acyl migration to an acceptor macromolecule.

The pH-catalysed migration of the acyl group from the drug C^1-O-acyl glucuronide to a protein or other cellular constituent occurs with the formation of a covalent bond to the protein [5]. Further details of this process are given below.

Endogenous compounds undergoing glucuronoconjugation include steroids, bilirubin and thyroxine. In the case of bilirubin, this pathway of detoxification is a major one, mediated by UGTs located in numerous tissues [16].

It should be noted that not all glucuronide conjugates are excreted by the kidneys; some may be excreted in the intestinal tract together with bile (they undergo enterohepatic cycling). Under the action of β-glucuronidase present in the intestinal flora, the C^1-O-acyl glucuronide will be hydrolysed back to the aglycone (drug or its metabolite) for re-absorption into the portal circulation.

A very important aspect that merits emphasis is that, besides leading to "detoxication" for many drugs, glucuronidation is also capable of promoting cellular injury (hepatotoxicity, carcinogenesis) by facilitating the formation of reactive electrophilic intermediates and their transport into target tissues [17-20] (details in Chapter 8).

3.2.2 Glucuronidation at various atomic centres (O, S, N)

Drugs from almost all therapeutic classes are glucuronidated. For those having narrow therapeutic indices (e.g. morphine, chloramphenicol), glucuronidation is therefore likely to have important consequences in their clinical use.

O-glucuronidation of phenolic drugs (or other xenobiotics) is often in competition with O-sulphation, which has been demonstrated to be predominant at low doses of the administered drug, while glucuronidation prevails only with high doses. It is well established that sulphation and glucuronidation occur in parallel, often competing for the same substrate (most commonly phenols) the balance between sulphation and glucuronidation being influenced by different factors such as species, doses, availability of co-substrates, and inhibition or induction of the respective transferases.

Another major group of substrates for glucuronidation is represented by alcohols (primary, secondary and tertiary). An interesting example is given by codeine (Figure 3.3) which, following demethylation to morphine, can undergo glucuronidation either at the phenolic, or at the secondary alcohol group, with concomitant formation of two distinct metabolites with different pharmacological activities. The pharmacokinetics of morphine, morphine-3-glucuronide (M3G) and morphine-6-glucuronide (M6G) in newborn infants receiving diamorphine infusions was reported earlier [21]. A very recent study concludes that both compounds display opiate agonistic behaviour [22]. These metabolites and further references to them were introduced in Chapter 2.

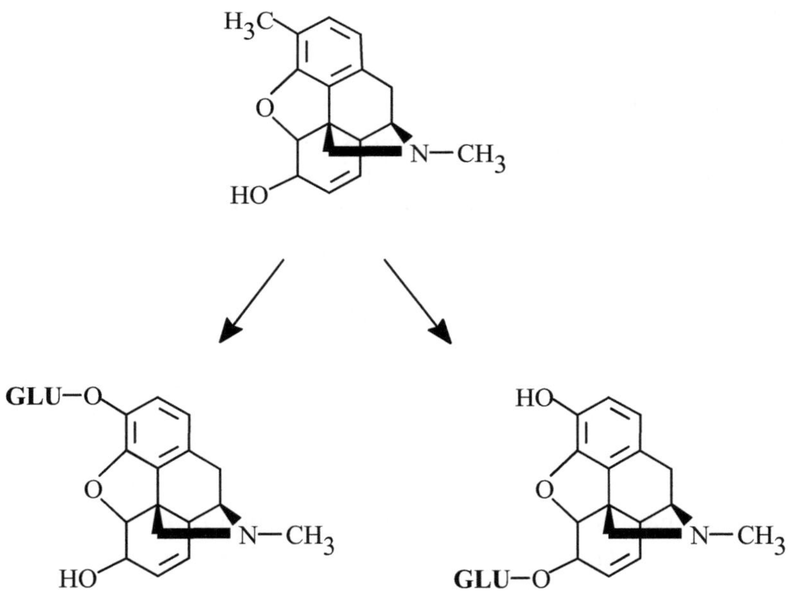

Fig.3.3 Specific positions in glucuronidation of codeine: the phenolic hydroxyl (after demethylation to yield morphine) and the secondary alcohol group

Another major pathway of O-glucuronidation is represented by the formation of acylglucuronides [23], the ideal substrates for this alternative pathway being aliphatic and aromatic acids (acidic drugs, such as NSAIDs). However, as already stressed, drug-acyl glucuronides are reactive conjugates at physiological pH, able to undergo different intramolecular rearrangements. This rearrangement of the glucuronide is actually an acyl migration, a process whereby the aglycone moves from the 1-hydroxyl group of the glucuronic acid sugar to the 2, 3-, or 4-hydroxyl groups. The rate of acyl migration differs from one compound to the next, and their stabilities are also highly variable [23]. Since the resulting positional isomers are not hydrolysable by β-glucuronidase, acyl migration to an acceptor macromolecule (e.g. protein in plasma or in tissue) may occur, resulting in covalent bond formation. Acylated proteins (usually designated 'haptens') thus formed, can stimulate an immune response against the drug, resulting in hypersensitivity reaction or other forms of immunotoxicity. For this reason, conjugation of this type is not, in the strict sense, a process of inactivation and so cannot necessarily be considered a safe detoxification pathway. In recent years, the potential toxicity of certain glucuronides, such as the morphine metabolite M6G mentioned above, has been well recognized [24]. Interestingly, although several drug glucuronides bind irreversibly to proteins both *in vitro* and *in vivo* (Figure 3.4 [5]), the parent drug alone has been ineffective.

Other relatively common substrates undergoing glucuronoconjugation are hydroxylamines and hydroxylamides. A recent example in the former category refers to the identification of an O-glucuronide metabolite of an aryl piperazine oral hypoglycaemic agent [25]. A relatively small number of aromatic amines are first N-hydroxylated, and then undergo O-glucuronidation. The reactivity of the resulting N-O-glucuronides and their potential for hydrolytic cleavage with subsequent formation of nitrenium ions are, however, still subjects of on-going investigation.

N-glucuronidations are considered to be of secondary importance; substrates undergoing this type of reaction are carboxyamides, sulphonamides, as well as different types of amines. The special relevance of this reaction in the case of antibacterial sulphonamides (particularly the older ones) is that because of the consequent production of highly water-soluble metabolites, crystallization of the parent compound at renal level (crystalluria) is avoided.

N-glucuronidation of aliphatic and aromatic amines as well as of some compounds with pyrimidynic structure has also been mentioned. Special significance was attributed to N-glucuronidation of lipophilic, basic tertiary amines, containing one or two methyl groups in their structure.

Aliphatic and aromatic thiols, as well as dithiocarboxylic acids undergo S-glucuronidation.

In the case of some 1,3-dicarbonyl compounds, such as sulphinpyrazone, C-glucuronidation has been reported [26].

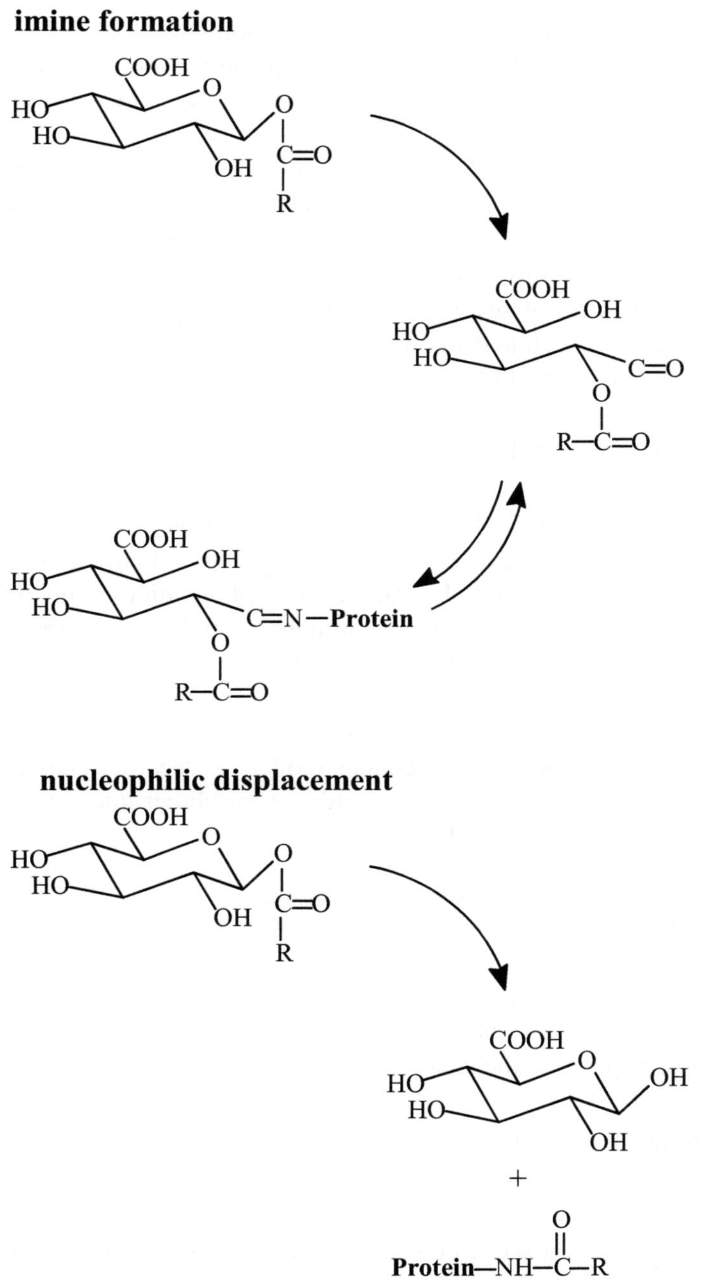

Fig.3.4 Irreversible binding of glucuronides to proteins by two main mechanisms: imine formation and nucleophilic displacement

Polymorphism of drug glucuronidation in humans
Various mutations within the UGT-1 gene and consequently, within the corresponding encoded isoforms, give rise to the hereditary hyperbilirubinemias.

The *in vitro* analyses of hepatic samples from patients with severe hyperbilirubinemia revealed that UGT activities toward certain drugs (e.g. propofol, ethinylestradiol, phenols) are severely reduced. Gilbert's disease, a mild familial hyperbilirubinemia, is a well-known syndrome associated with decreased clearance of several drugs such as rifamycin, acetaminophen and tolbutamide; it is assumed that a decreased rate of glucuronidation for this condition occurs as well [27, 28].

3.3 ACETYLATION

Acetylation is a Phase II reaction of amino groups and it involves the transfer of acetyl-coenzyme A (acetyl CoA) to an aromatic primary or aliphatic amine, amino acid, hydrazine, or sulphonamide group. The primary site of acetylation is the liver, although extrahepatic sites have been identified as well (e.g. spleen, lung and gut).

Acetylation reactions require a specific co-factor, acetyl-CoA, which is obtained mainly from the glycolysis pathway (breakdown of glucose yielding pyruvate and its subsequent oxidative decarboxylation), or from catabolism of fatty acids or amino acids, or *via* direct interaction of acetate and coenzyme A [29] (Eq.3.1):

$$CH_3\text{-}COO^- + CoASH \xrightarrow{\text{CoA-S-acetyltransferase}} CH_3\text{-}CO\text{-}\mathbf{S\text{-}CoA} \qquad 3.1$$

Genetic polymorphism affecting the rate of acetylation has important consequences in drug therapy and tumorigenicity of certain xenobiotics (details in Chapter 7, subchapter 7.1.1).

3.3.1 Role of acetyl-coenzyme A

Coenzyme A (A standing for 'acyl') participates in activation of acyl groups in general, including the acetyl group derived from pyruvate (by oxidative decarboxylation)[29,30].

The coenzyme is derived metabolically from the vitamin pantothenic acid, β-mercaptoethylamine and ATP (Figure 3.5).

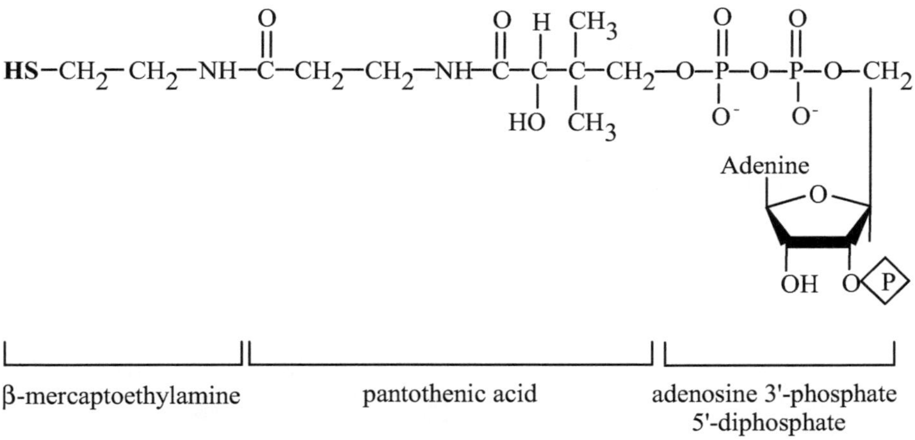

Fig.3.5 Structure of CoA

Initially, the 4-phosphopantethine is formed under the catalytic action of a specific kinase and consumption of an ATP molecule. Then follows a sequence of reactions, with the consumption of two more ATP molecules, yielding finally CoA. This molecule is an important coenzymatic factor, participating in both biosynthetic and biodegradative reactions.

The functionally significant part of the coenzyme molecule is the free thiol on the β-mercaptoethylamine moiety, the rest of the molecule providing enzyme binding sites.

In acylated derivatives, such as acetyl-coenzyme A, the acyl group is linked to the thiol group, with consequent formation of an 'energy-rich' thioester (Figure 3.6):

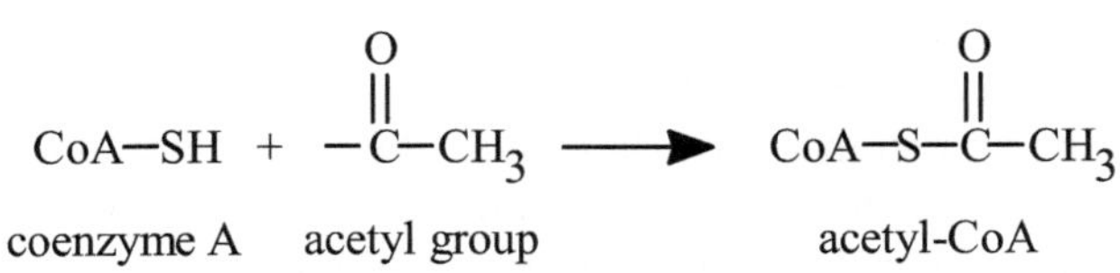

Fig.3.6 Formation of the energy-rich thioester

Usually, the unbranched form is designated as CoA-SH, and the acylated forms as acyl-CoA or,

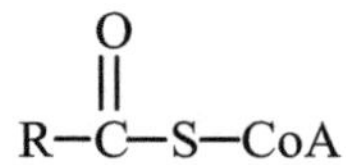

The energy-rich nature of thioesters, as compared with ordinary esters, is related primarily to resonance stabilisation [30], shown in Figure 3.7:

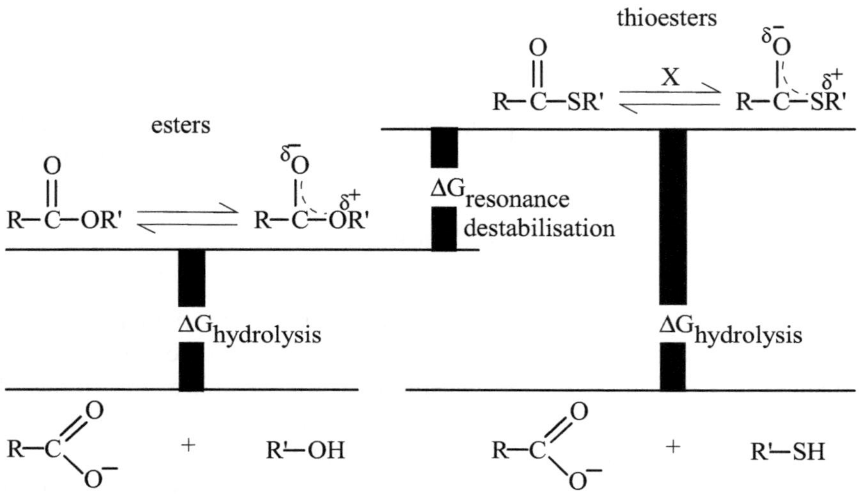

Fig.3.7 Resonance stabilisation explaining the energy-rich nature of thioesters (Fig.14.9, p.494 from BIOCHEMISTRY, 3rd ed. by Christopher K. Mathews, K.E. van Holde and Kevin G. Ahern. Copyright © 2000 by Addison Wesley Longman, Inc. Reprinted by permission of Pearson Education, Inc.)

Ordinary esters have two resonance forms, their stabilisation involving π-electron overlap, giving partial double-bond character to the C-O link. However, in thioesters, because of the larger atomic size of S *vs* O, there is reduced π-electron overlap and the C-S structure does not significantly contribute to resonance stabilisation. The thioester is thus destabilised relative to an ester and consequently the free energy change for its hydrolysis is enhanced.

The chemical consequences are important: the lack of double-bond character in the C-S bond of acyl-CoAs makes this bond weaker than the corresponding C-O bond in ordinary esters. This renders the thioalkoxide ion ($R\text{-}S^-$) a good leaving group in nucleophilic displacement reactions, allowing the acyl group to be consequently readily transferred to other metabolites in so-called 'transacylation reactions'.

Because of the important biological roles played by acetyl-CoA and related species, studies probing their structure and function are ongoing. Mechanistic aspects of the action of acetyl-CoA in modulating protein structure have been discussed in a recent paper [31]. The crystal structure of the β-subunit of the enzyme acyl-CoA carboxylase has been reported [32] with the aim of understanding its substrate specificity; this would assist in the development of therapeutics against diseases such as obesity and diabetes. Another recent crystallographic study investigated the carboxyltransferase domain of acetyl-coenzyme A carboxylase in complexed form with an inhibitor [33]. Regions for drug binding in the active site were established in this way. The catalytic action of acetyl-CoA synthase, a bifunctional Ni-Fe-S containing enzyme that catalyses the synthesis of acetyl-CoA, has been reviewed [34]. The possibility of involvement of zero-valent Ni (unusual in biology) in the catalytic action of this enzyme was raised.

3.3.2 Acetylation of amines, sulphonamides, carboxylic acids, alcohols and thiols

The general mechanism of acetyl transfer catalysed by N-acetyltransferases involves a double displacement – a so-called 'ping-pong' mechanism:

$$\text{Ac-CoA} + \text{isoniazid} \longrightarrow \text{Ac-isoniazid} + \text{CoA} \qquad 3.2$$

The reaction actually proceeds in two steps, namely transfer of the acetyl group from Ac-CoA with formation of an acetyl-enzyme intermediate and the subsequent acetylation of the arylamine with regeneration of the enzyme [6].

Because of their structural similarities to the substrates, some compounds act as reversible inhibitors towards N-acetyltransferases whereas others, such as iodoacetate and p-chloromercurybenzoate, are irreversible inhibitors.

The principal types of acetylation are summarised in Figure 3.8:

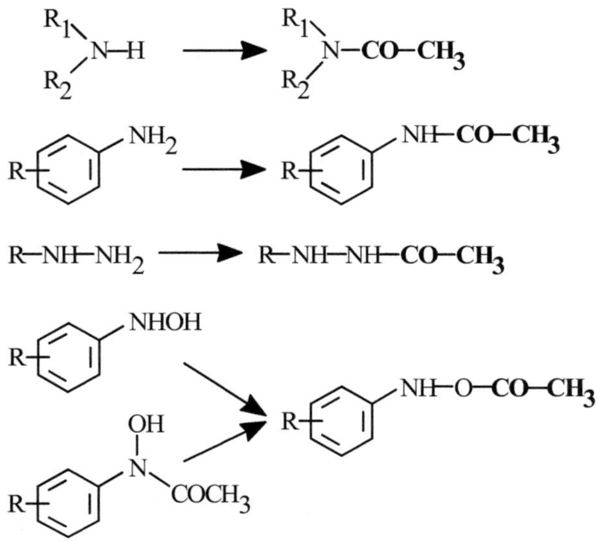

Fig.3.8 Major types of acetylations

As representative examples we illustrate the N-acetylation of sulphanilamide and isoniazid (Figure 3.9):

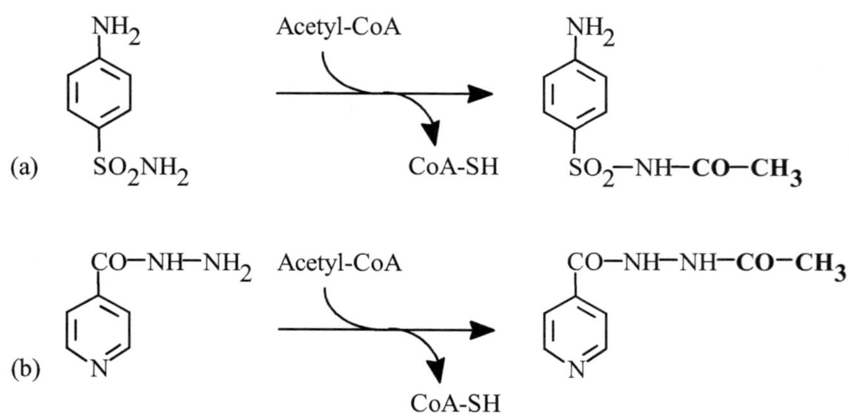

Fig.3.9 N-acetylation of (a) sulphanilamide and (b) isoniazid
(Reproduced with the permission of Nelson Thornes from 'Introduction to Drug Metabolism', 2001, 3rd Ed., isbn 0 7487 6011 3 - Gibson & Skett - first published in 1986)

Sulphanilamide may also undergo acetylation on the amine nitrogen atom, with consequent formation of a diacetylated metabolite.

The formation of acetyl-sulphonamides is of particular toxicological interest as these metabolites are less soluble in water than the parent drug and the renal toxicity of sulphonamides has been directly attributed to precipitation of these conjugates in the kidney. Secondary amines are not acetylated.

Acetylation may produce conjugates that retain the pharmacological activity of the parent drug (e.g. N-acetylprocainamide).

Variability in human drug acetylation was noted many years ago [14] with individuals designated as 'rapid' or 'slow' acetylators, based on their blood levels after administration of isoniazid. Only recently, however, has it been demonstrated that such differences are caused by genetic variability. The relevant human arylamine acetyltransferases are termed NAT1 and NAT2. Details concerning isoforms appear in Chapters 4 and 7.

The role of genetic polymorphism in the rate of acetylation has important consequences in drug therapy and tumorigenity of certain xenobiotics, including drugs. The two acetylator phenotypes may determine significant differences in human drug toxicity, as follows:

- Slow acetylators accumulate higher blood concentrations of the unacetylated drug than rapid acetylators. Such individuals are thus more prone to drug-induced toxicities such as sulphasalazine-induced hematologic disorders, procainamide-induced lupus erythematosus and isoniazid-induced peripheral nerve damage.
- Fast acetylators eliminate certain drugs more rapidly, which presents a greater risk of liver toxicity. As a current example we mention the hepatotoxic monoacetylhydrazine metabolite formed by acetylation of isoniazid (Figure 3.9 above).

Another noteworthy aspect is the difference in susceptibility to chemical carcinogenicity from arylamines directly related to differences in acetylating capacity resulting from genetic polymorphism. Apparently, the tumorigenity of arylamines may be the result of a complex series of sequential metabolic reactions beginning with N-acetylation. By the end of the sequence, an arylnitrenium ion is formed; this is a reactive species capable of covalent binding to proteins and even nucleic acids [5].

For the rapid acetylator phenotype, the rate of forming the acetoxyarylamine metabolite and consequent loss of the acetoxy group to form the reactive species, is greater than for slow acetylators, thereby presenting a greater risk for the development of bladder and liver tumors [5,10,14].

3.4 GLUTATHIONE CONJUGATION

Glutathione [N-(N-*L*-γ-glutamyl-L-cysteinyl)glycine], an atypical tripeptide (Figure 3.10), is an endogenous compound, recognized as playing a protective role within the body in removal of potentially toxic electrophilic compounds.

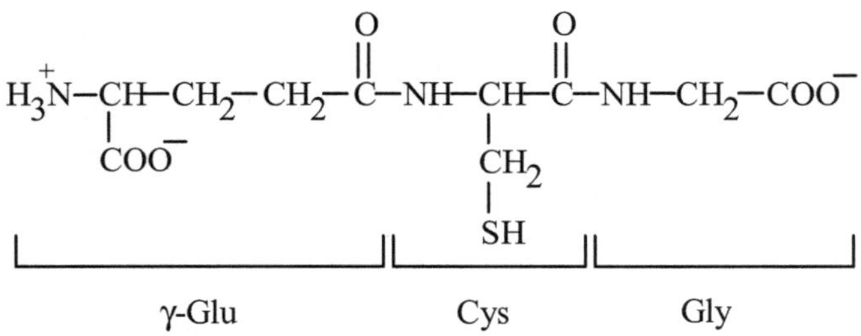

Fig.3.10 Structure of glutathione

Glutathione (GSH) is present at highest concentration in the liver, with higher values in the cortex than in the medulla, but is also present in cytosol, mitochondria and nucleus [29,30]. In the blood, it is present at a relative concentration of about 20 μM.

GSH conjugation involves the formation of a thioether link between the GSH and electrophilic compounds. The reaction can be considered as the result of nucleophilic attack by GSH on electrophilic carbon atoms, with leaving functional groups such as halogen, sulphate and nitro, ring opening (in the case of small ring ethers – epoxides, β-lactones), and the addition to the activated β-carbon of an α,β-unsaturated carbonyl compound.

Thus, conjugation with glutathione usually results in detoxication of the electrophilic compounds by preventing their reaction with nucleophilic centres in macromolecules such as proteins and nucleic acids. The electrophilic substrates for glutathione are commonly generated by prior metabolism of the xenobiotics, or by displacement of suitable electron withdrawing groups in nitro or halo-alkanes, benzenes and sulphonic acid esters by the sulphur atoms of glutathione, and it is usually eliminated as mercapturic acid after further metabolism of the S-substituted glutathione.

Major types of reaction are summarised in Figure 3.11:

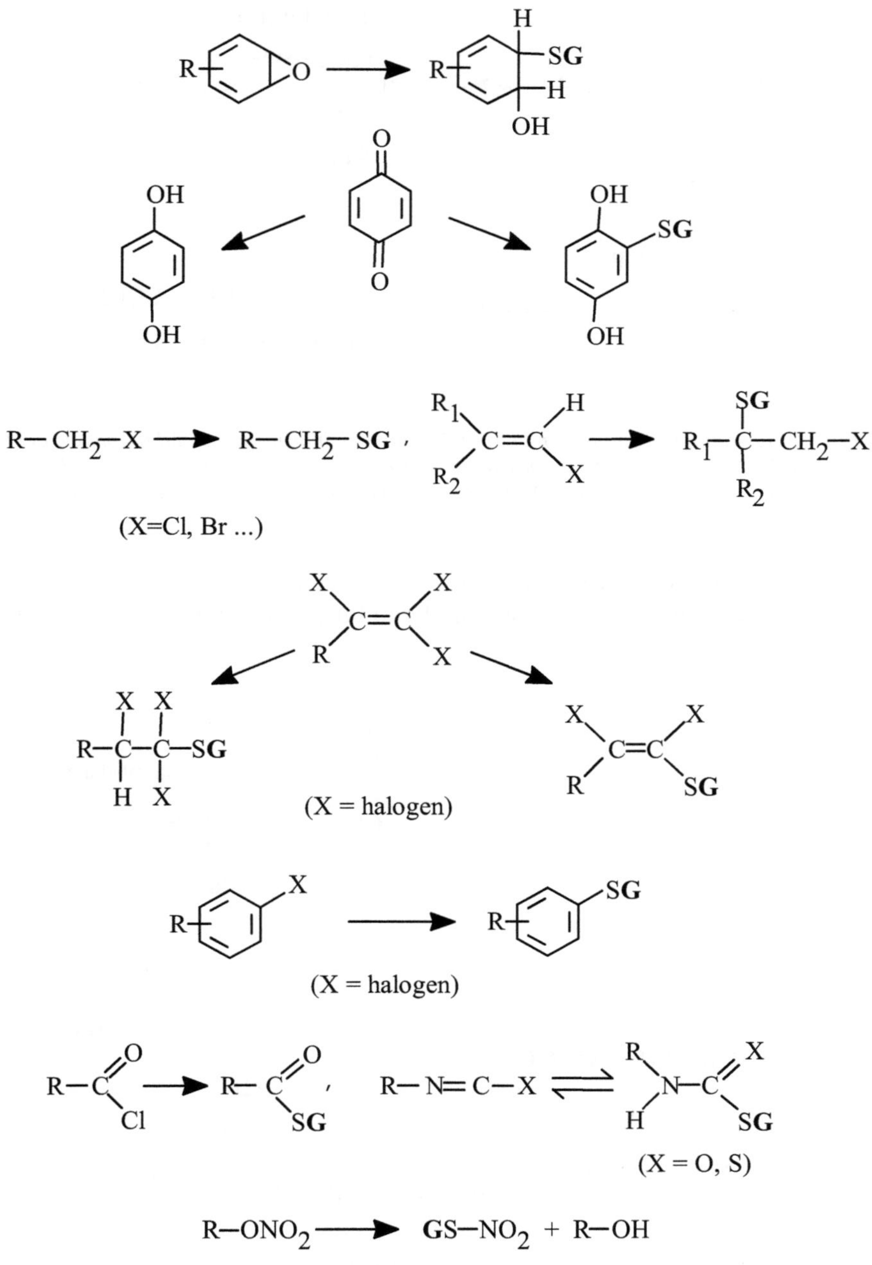

Fig.3.11 Representative types of glutathione conjugation

Two specific examples [6] are shown in Figure 3.12:

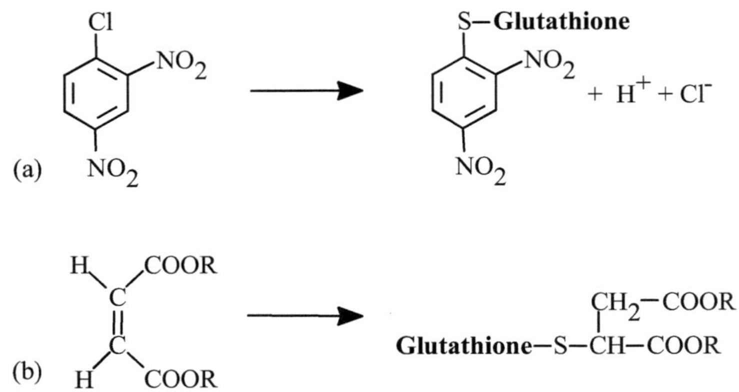

Fig.3.12 Glutathione conjugation of (a) 2,4-dinitro-1-chlorobenzene and (b) maleic acid esters (Reproduced with the permission of Nelson Thornes from 'Introduction to Drug Metabolism', 2001, 3rd Ed., isbn 0 7487 6011 3 - Gibson & Skett - first published in 1986)

The glutathione conjugates may be excreted directly in urine (or more usually in bile) but more commonly undergo further metabolism (Figure 3.13):

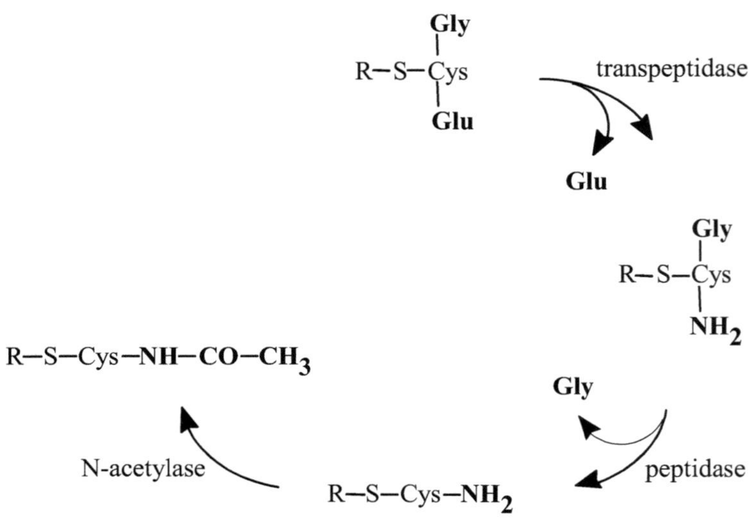

Fig.3.13 Further possible biotransformation pathways of glutathione conjugates (Reproduced with the permission of Nelson Thornes from 'Introduction to Drug Metabolism', 2001, 3rd Ed., isbn 0 7487 6011 3 - Gibson & Skett - first published in 1986)

Many GSH conjugates undergo further enzymatic modification by hydrolysis of the glutathione-S-conjugate at the γ-glutamyl bond. This specific reaction is catalysed by the enzyme γ-glutamyl transferase, well known in the clinical laboratory as γ-GT. As can be seen from the above figure, the tripeptide glutathione (Gly-Cys-Glu), once attached to the acceptor molecule, can be attacked by this specific enzyme, which removes the glutamate yielding a dipeptide; the latter may be further attacked by a peptidase which removes the glycine, thus forming the cysteine conjugate of the xenobiotic. In the final step, mediated by specific N-acetylases, the cysteine conjugate previously formed may undergo N-acetylation (*via* the normal acetylation pathway already described), yielding the N-acetylcysteine conjugate of mercapturic acid. The first two enzymes involved are most commonly found in the liver and kidney cytosol, while the highest N-acetyltransferase (NAT) activity is found in the proximal tubules. Depending on the nature of the substrate and the species investigated, each of the three conjugated metabolites (glycylcysteine, cysteine and mercapturic acid conjugates) may appear as excretion products.

Conjugation reactions of GSH were reviewed earlier [35] as was the role of GSH S-transferases in the detoxification of reactive metabolites of benzo[a]pyrene-7,8-dihydrodiol [36].

3.5 OTHER CONJUGATIVE REACTIONS

Methylation

O-and N-methylation are common biochemical reactions but appear to be of greater significance in the metabolism of endogenous compounds than for drugs or other xenobiotics. However, some drugs may also undergo methylation by non-specific methyltransferases found in the lung, or by the physiological methyltransferases.

For example, histamine N-methyltransferase, (HMT), is a primary enzyme effecting degradation of histamine in the body. Its role in the regulation of airway functions has been discussed [37]. Another example is phenylethanolamine N-methyltransferase (PNMT), found in the adrenal medulla and many other tissues. This enzyme methylates noradrenaline yielding the product adrenaline. An account of the location and activity of extra-adrenal PNMT has appeared recently [38]. High levels of PNMT in the adrenal depend critically on glucocorticoids [39]. The three-dimensional structures of rabbit and human indolethylamine N-methyltransferases (INMTs) have been predicted from their amino acid sequences to bridge the gap between structure and pharmacogenetic aspects of their function [40].

The co-factor required to form methyl conjugates is S-adenosylmethionine (SAM), produced from L-methionine and ATP under the influence of the enzyme L-methionine adenosyltransferase, as presented in Figure 3.14:

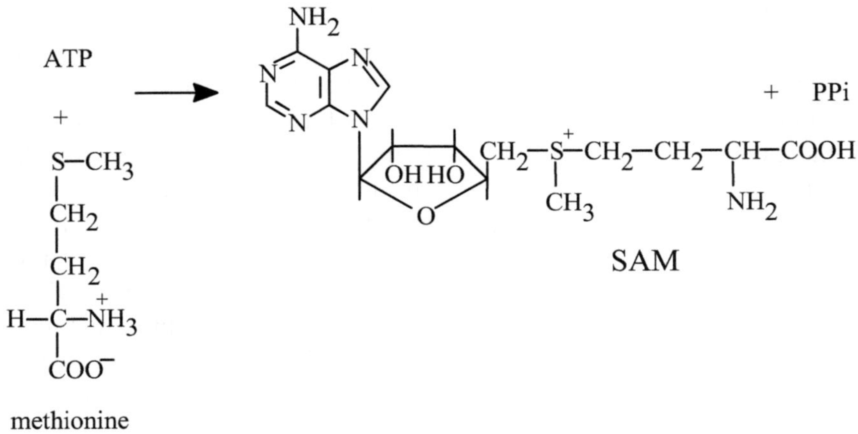

Fig.3.14 The formation of S-adenosylmethionine (SAM)

As can be seen from the above figure, methionine is involved in the methylation of endogenous and exogenous substrates, by transferral of its methyl group *via* the activated, high-energy intermediate SAM, to different substrates under the influence of specific methyl transferases.

The general mechanism can be presented as follows (Figure 3.15):

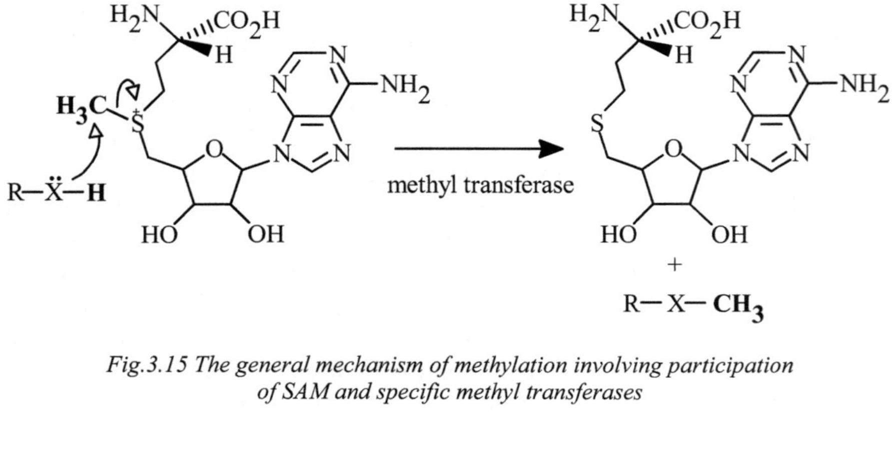

Fig.3.15 The general mechanism of methylation involving participation of SAM and specific methyl transferases

Reaction results mainly in the formation of O-methylated, N- methylated, and S-methylated products. This differs from other conjugation processes in that the O-methyl metabolites that form may, in certain instances, possess equal, or even enhanced pharmacological activity and lipophilicity, than the parent molecule.

The process of O-methylation is catalysed by a magnesium-dependent enzyme, generically designated as catechol-O-methyltransferase (COMT). The reaction involves the transfer of a methyl group to either the meta- or less frequently, to the para-phenolic hydroxyl group of catecholamines, and their deaminated metabolites. O-methylation of phenolic groups is important in the metabolism of neurotransmitters such as the catecholamines and structurally related drugs. The most representative example is afforded by norepinephrine (Figure 3.16).

It must be stressed that the meta/para product ratio is greatly dependent on the type of substituent attached to the catechol ring. Specific substrates for COMT include:

- the catecholamines: norepinephrine, epinephrine and dopamine;
- some specific aminoacids: L-DOPA, and α-methyl-DOPA, as well as,
- the 2- and 4-hydroxy- metabolites of estradiol.

Monohydric or other dihydric phenols are not methylated.

COMT is present both in kidney and liver, with the kidney activity present at about a quarter of the level found in the liver. Pharmacogenetic studies have revealed differences in inherited phenotype activities (details in Chapter 7, subchapter 7.1); at the same time, ageing has been associated with a decrease in COMT affinity for a particular substrate (details in Chapter 6).

The N-methylation of various amines is among several conjugate pathways for metabolising amines. The transfer of active methyl groups from SAM to the acceptor substrate is catalysed by specific N-methyl-transferases. There are three important N-methyltransferases, namely:

- histamine N-methyltransferase (HMT), a cytoplasmic enzyme that methylates histamine and similar amine compounds in which positions 3- and 12- are unsubstituted and there is a positive charge on the side chain. Methylation of histamine leads to the inactive metabolite N^1-methyl-histamine. In this context, it is important to mention the existence of a great number of HMT inhibitors, including H_1 and H_2 receptor antagonists, diuretics and some local anaesthetics. Details regarding this enzyme appear in Chapter 4.
- phenylethanolamine N-methyltransferase (PNMT), requiring the presence of phenylethanolamine compounds as substrate acceptors for the methyl group; endogenous substrates include norepinephrine and epinephrine (see again Figure 3.16); further details regarding the enzyme are given in Chapter 4;

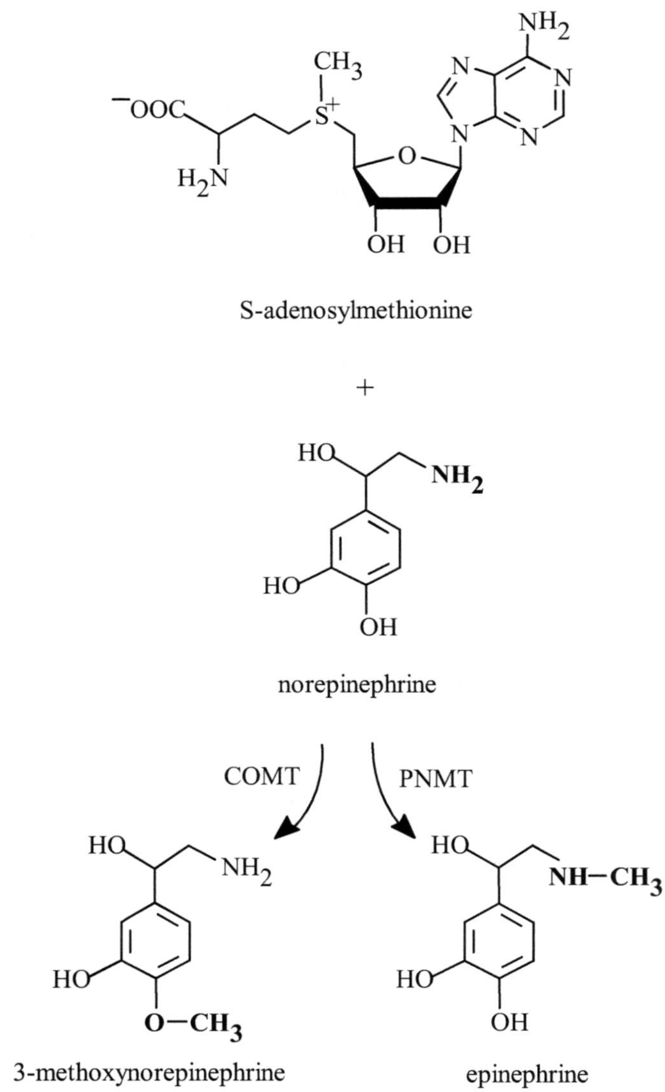

Fig.3.16 Methylation pathways for norepinephrine (PNMT = phenylethanolamine N-methyltransferase)

- amine-N-methyltransferases (also known as indolethylamine N-methyltransferases), which catalyse the transfer of a methyl group from SAM to the amino group of indoleamines; these enzymes will N-methylate a variety of primary and secondary amines, including endogenous biogenic amines (e.g. serotonin, tyramine, dopamine) and drugs such as amphetamine, normorphine and desmethylimipramine.

Amine-N-methyltransferases evidently have a role in recycling N-demethylated drugs (Figure 3.17):

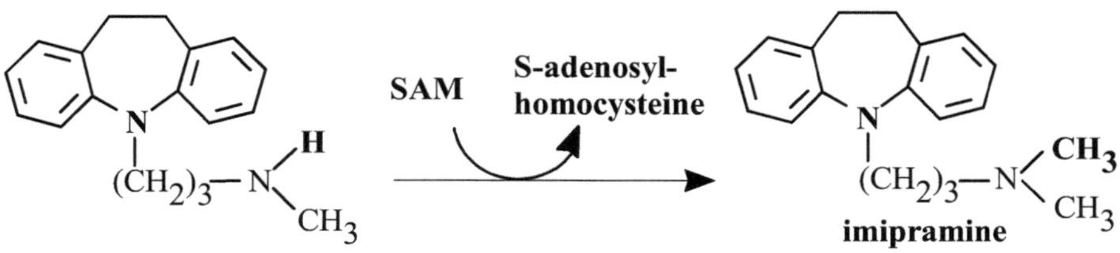

Fig.3.17 The recycling of demethylated imipramine

Other substrates for methylation reactions include thiols, which are generally considered as toxic. Thiol S-methyl transferases thus play a role among other detoxication pathways for these compounds. In this specific reaction of S-methylation on sulphhydryl groups, the microsomal enzymes involved also require participation of SAM. Thiol methylation is important in the metabolism of captopril, D-penicillamine, azathioprine and 6-mercap topurine (6MP), which, following this process will be excreted as sulphoxides or sulphones.

In the context, we should also mention the so-called thiomethyl shunt, acting on compounds in which sulphur has been added from glutathione. It begins with the addition of glutathione, followed by conversion to the cysteine conjugate. Sequential steps include the cleavage of the conjugate by a cysteine conjugate β-lyase to pyruvate, ammonia and thiol, and subsequent methylation of the thiol formed.

There are three enzymes with different characteristics involved in this process:

- the microsomal thiol-methyl transferase (TMT), a membrane-bound enzyme that catalyses demethylation of aliphatic sulphhydryl compounds (such as captopril and D-penicillamine);
- the cytosolic thioether-S-methyltransferase (TEMT), and
- the soluble thiopurine-methyl transferase (TPMT). This is a cytoplasmic enzyme catalysing the S-methylation of aromatic and heterocyclic sulphhydryl compounds, such as 6MP and azathioprine preferentially. Further details are provided in Chapter 4.

S-methylation of sulphhydryl compounds also requires the presence and participation of SAM. It is important to mention that none of the endogenous sulphhydryl compounds (e.g. cysteine, GSH) can function as substrates, although a wide variety of exogenous sulphhydryl compounds may be S-methylated by the microsomal S-methyl transferases.

Amino acid conjugation reactions generally involve one or two amino acids, *viz.* glycine or glutamic acid, the former being the more common. These reactions can occur with substrates containing either an alcohol, or a carboxyl moiety, especially substrates that contain aromatic groups.

The general reaction is given in Figure 3.18:

$$\mathrm{R{-}COOH + ATP \longrightarrow R{-}CO{-}AMP + PPi}$$

$$\mathrm{R{-}CO{-}AMP + CoASH \longrightarrow R{-}CO{-}\mathbf{S{-}CoA} + AMP}$$

$$\mathrm{R{-}CO{-}S{-}CoA + R'{-}\mathbf{NH_2} \longrightarrow R{-}\mathbf{CO{-}NH}{-}R' + CoASH}$$

Fig.3.18 The amino acid conjugation of carboxylic acids

It is evident from the above figure that amino acid conjugation is a special form of N-acylation, where the drug, and not the endogenous co-factor is activated. Exogenous carboxylic acids can form CoA derivatives in the body and can then react with endogenous amines, forming an amide (or peptide) bond. The general reaction can be written more succinctly as follows:

$$\mathrm{R(CO)\mathbf{OH} + \mathbf{NH_2}CH_2(CO)OH \rightarrow R\mathbf{(CO)NH}CH_2(CO)OH + H_2O} \qquad 3.3$$

Such conjugation with amino acids represents an important metabolic pathway in the eventual elimination of drug (and other xenobiotic) carboxylic acids. The substrates may be aromatic, arylaliphatic, and heterocyclic carboxylic acids and the resulting metabolites are water-soluble ionic conjugates. Usually, these amino acid conjugates are less toxic then their precursor acids and are readily excreted into the urine or bile. Conjugation reactions with amino acids may be limited by the amount of endogenous glycine, or by the amount of enzyme available to catalyse the reaction.

The metabolic fate of the aforementioned carboxylic acids depends strongly on the size and type of substituents adjacent to the carboxylic group, according to the following guidelines:

- most unbranched aliphatic acids are completely oxidized and do not usually form conjugates;

• branched aliphatic and aryl aliphatic acids resist β-oxidation, forming glycine or glucuronide conjugates, with glycine conjugation preferred for xenobiotic carboxylic acids at low doses. It is of interest to note that substituents on the α-carbon atom favour glucuronidation over glycine conjugation.

Glycine conjugation is not confined to xenobiotics but also occurs with endogenous compounds; conjugates of bile acids are formed by enzymatic action in the microsomal fraction. Some examples will be given at the end of the chapter.

Besides glycine conjugation, other amino acids yielding conjugated metabolites are glutamine and cysteine. Glutamine conjugation reactions are, however, limited in the body to specific arylacetic acids, a representative case being phenylacetic acid, which can be converted to indolacetyl-glutamine in various species (including monkeys and humans).

In the case of cysteine, the substrates are aromatic drugs, and the subsequent metabolites are the corresponding mercapturic acid.

In contrast to the enhanced reactivity and toxicity of the various glucuronide, acetyl and glutathione conjugates, amino acid conjugates have not proven to be toxic. Moreover, it has been proposed that amino acid conjugation is an important detoxication pathway for reactive acyl CoA thioesters [6].

Sulphation

This is a major conjugation pathway for phenols, but also contributes to the biotransformation of alcohols, amines, and to a lesser extent, thiols. It is also relevant in the metabolism of endogenous compounds such as catecholamine neurotransmitters, steroid hormones, thyroxine and bile acids. Moreover, the tyrosinyl group of peptides and proteins may represent sites of sulphation, resulting in possible changes in their properties. The resulting compounds are generally less active, and more polar, thus more readily excreted in the urine.

Sulphate conjugation is a multistep process, comprising activation of inorganic sulphate, first, by converting it *via* ATP to adenosine-5'-phosphosulphate (APS), and further to the activated form, known as PAPS, 3'-phosphoadenosine-5'-phosphosulphate, as shown in equations 3.4 and 3.5. Each step involves a specific enzyme, present in cytosol.

$$ATP + SO_4^{2-} \xrightarrow{\text{ATP-sulphurase}} APS + PPi \qquad 3.4$$

$$APS + ATP \xrightarrow{\text{APS-phosphokinase}} PAPS + ADP \qquad 3.5$$

The reaction by which sulphotransferases catalyse the transfer of a sulphuryl group from PAPS to an acceptor molecule is shown in the following reaction:

$$\text{R-}\mathbf{OH} + \text{PAPS} \xrightarrow{\text{sulphotransferases}} \text{R-}\mathbf{OSO_3H} + \text{PAP} \qquad 3.6$$

The availability of PAPS and its precursor inorganic sulphate strongly determine the rate of reaction. In humans, sulphotransferases are found in the liver, small intestine, brain, kidneys, and platelets. Two forms of sulphotransferases are known to exist, namely a "thermolabile" (TL) form, responsible for the sulphation of dopamine (and other monoamines), and a "thermostable" (TS) form, which catalyses the sulphation of a variety of phenolic compounds. Further details concerning sulphotransferases appear in Chapter 4.

It is important to note that the total pool of sulphate in the body is normally limited and can be easily exhausted. Thus, with increasing doses of a drug, sulphate conjugation will become a less significant pathway. For a competing substrate, at high doses glucuronidation usually predominates over sulphation, which instead prevails at low substrate doses.

Sulphate conjugation is most common for phenols, and to a lesser extent for alcohols, yielding highly ionic and polar sulphates, metabolites that are readily excreted in the urine.

In contrast, N-sulphates, analogous to the N-glucuronides, are able to promote cytotoxicity by facilitating the formation of reactive electrophilic intermediates. Sulphation of N-oxygenated aromatic amines is an activation process for some arylamines that can eliminate the sulphate to an electrophilic species capable of reacting with proteins or DNA.

Different drug sulphate conjugates are excreted mostly in the urine, but in the case of steroids, biliary elimination is more prevalent. However, in the small intestine, through mediation of certain sulphatases, the parent drug or its metabolites may be reabsorbed into the portal circulation. The rate of sulphation varies inversely with an individual's age.

It should be noted that, especially after oral administration of a drug, the intestine represents an important site of sulphation. For drugs whose primary metabolic pathway is sulphation, the result is a pre-systemic first pass effect, which decreases the bioavailability. Some drugs in this category are acetaminophen, steroid hormones, α-methyldopa, isoproterenol and albuterol. This feature is also important when one considers co-administration of certain drugs, where competition for intestinal sulphation might influence their bioavailability, either enhancing or reducing their therapeutic effects. (Examples and details are provided in Chapter 8).

Fatty acid conjugation

Fatty acid conjugation with stearic and palmitic acids has been shown to occur for 11-hydroxy-Δ^9-tetrahydrocannabinol (THC) (Figure 3.19):

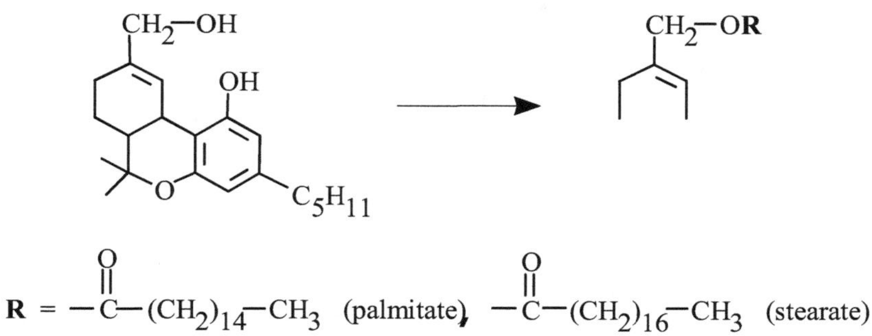

Fig.3.19 The conjugation of 11-hydroxy-Δ^9-tetrahydrocannabinol to the corresponding stearic and palmitic acids (Reproduced with the permission of Nelson Thornes from 'Introduction to Drug Metabolism', 2001, 3rd Ed., isbn 0 7487 6011 3 - Gibson & Skett - first published in 1986)

These reactions are catalysed by the microsomal fraction from liver. However, the mechanistic aspects are still unknown and the range of compounds that could be involved in such conjugations is uncertain.

Amino acid conjugation

Results from the reaction of the carboxylic group of a xenobiotic with an amino acid (most frequently, glycine, glutamine, alanine and histidine). The reaction is given only by a relatively small group of substrate structures such as aromatic, heteroaromatic and arylacetic acids, and results in enhanced elimination and decreased toxicity of the parent drug, although not as effective as the main conjugation reactions (glucuronidation, GSH-conjugation).

Condensation reactions

Condensation reactions may not proceed under enzymatic control and have been found for amine and aldehyde substrates. A representative case is the condensation of dopamine and its own metabolite, 3,4-dihydroxy phenylethanal, to form an alkaloid that is a potent dopamine antagonist (Figure 3.20):

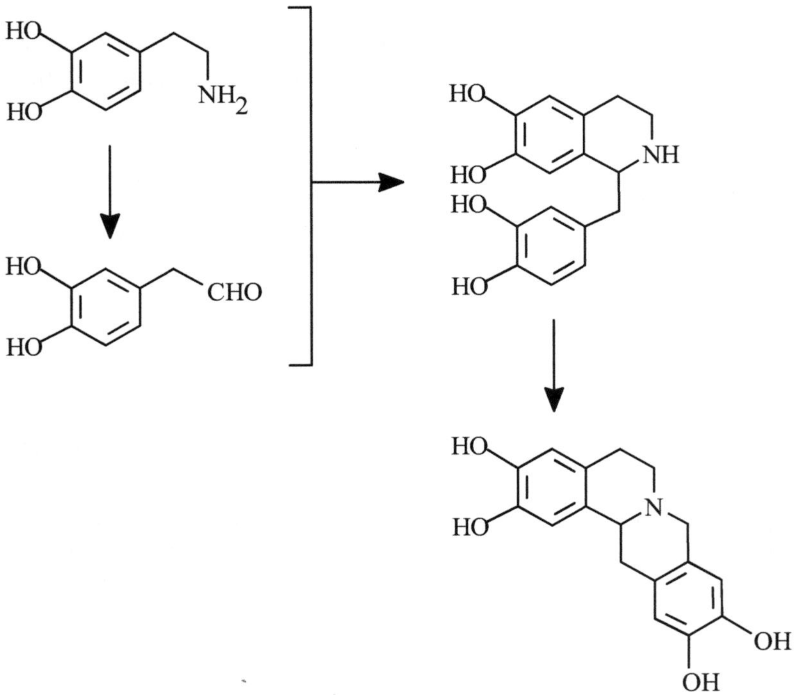

Fig.3.20 Example of condensation reaction
(Reproduced with the permission of Nelson Thornes from 'Introduction to Drug Metabolism', 2001, 3rd Ed., isbn 0 7487 6011 3 - Gibson & Skett -first published in 1986)

Representative examples of combined phase I and phase II reactions

Phenylbutazone

The first biotransformation reactions are CYTP450-mediated hydroxylations: the aromatic hydroxylation yields an active metabolite, namely oxyphenbutazone, which is even more active than the parent drug, possessing potent anti-inflammatory effects (Fig. 3.21). In contrast, under conditions of aliphatic hydroxylation, the resulting metabolite, γ-hydroxyphenylbutazone, is also active, but presents a different type of activity, namely uricosuric effects. Both hydroxylated metabolites subsequently undergo 4-glucuronoconjugation, the resulting metabolites being excreted in urine. Alternative, minor pathways, include either direct glucuronidation (bypassing phase I reactions) or a second hydroxylation of the metabolite that is hydroxylated on the aliphatic chain, yielding the inactive, dihydroxylated species [41].

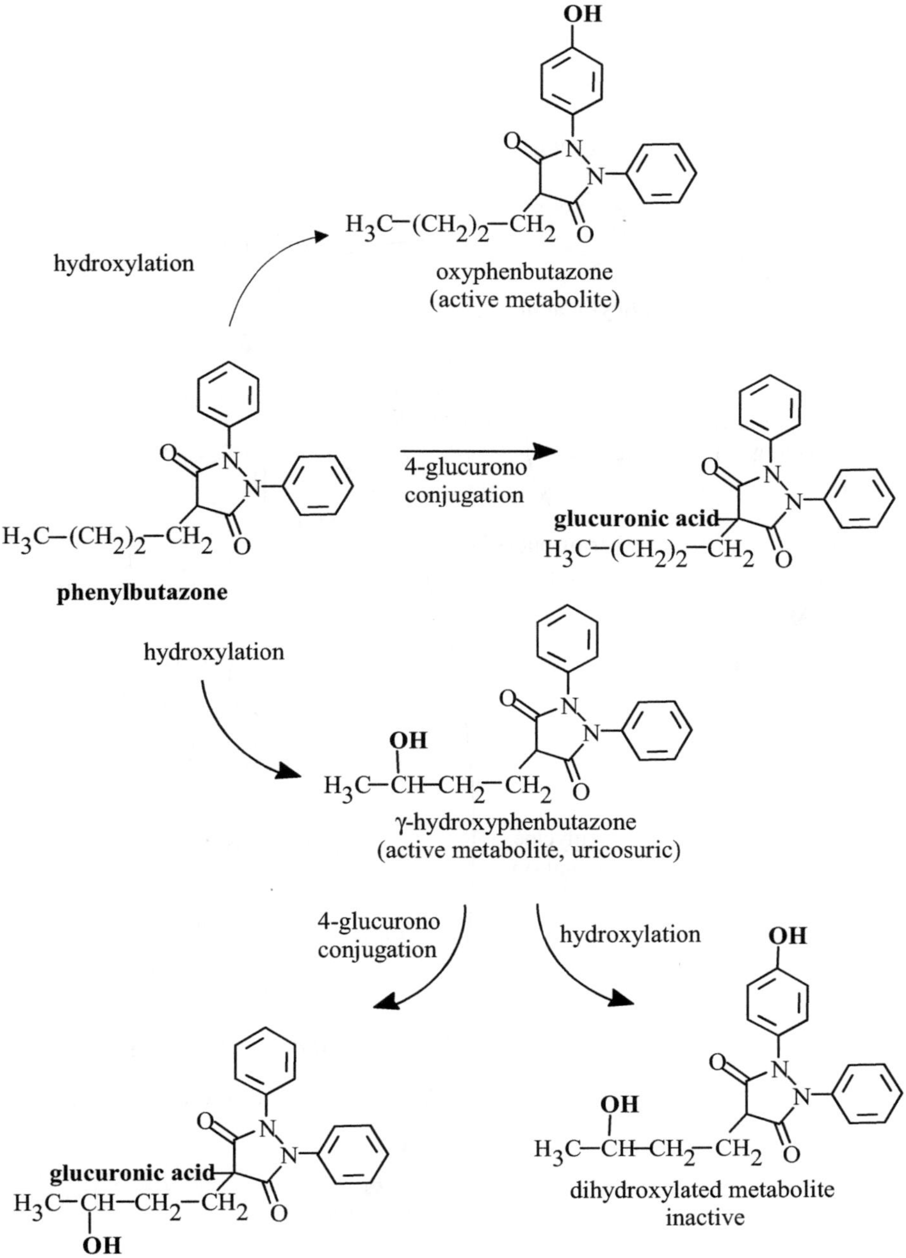

Fig.3.21 Metabolization pathways of phenylbutazone

Salicylate

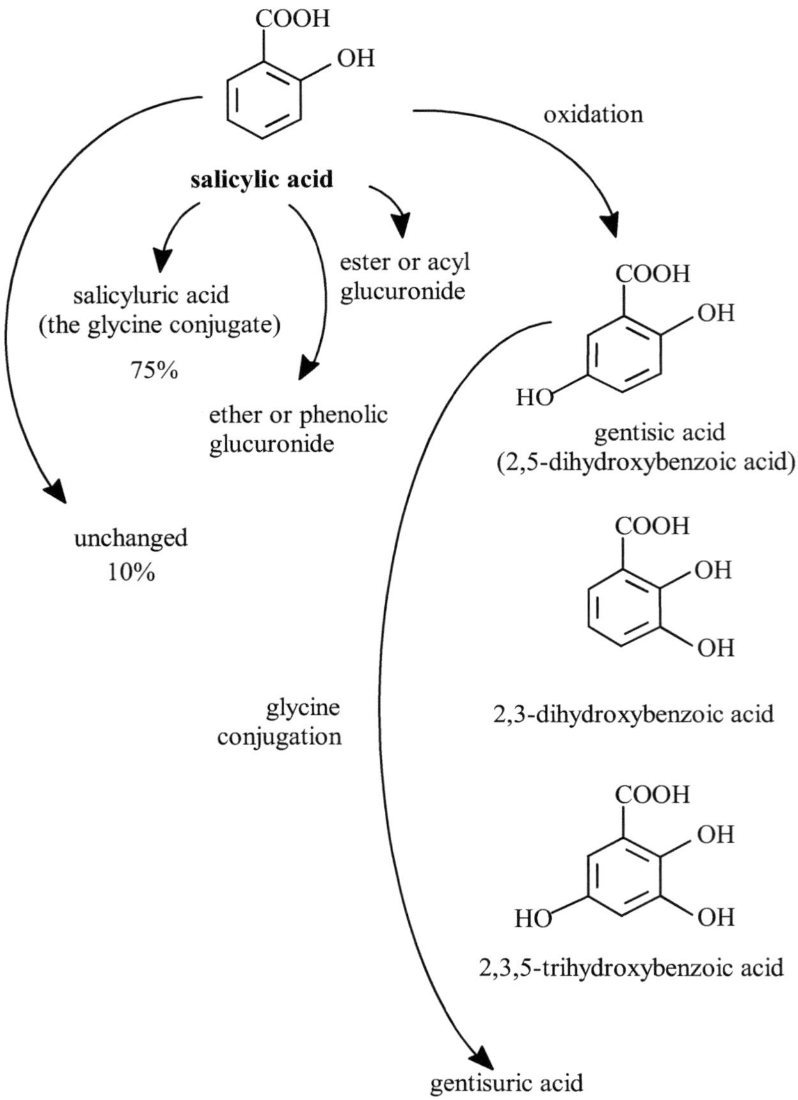

Fig.3.22 Biotransformation of salicylate

Salicylate biotransformation (Fig. 3.22 above) takes place in many tissues, but particularly in the hepatic endoplasmic reticulum and mitochondria [42]. The three major metabolites are salicyluric acid (representing in fact the glycine conjugate), the ether or phenolic glucuronide and the ester or acyl glucuronide. A small fraction is oxidized to gentisic acid (2,5-dihydroxybenzoic acid), 2,3-dihydroxybenzoic acid and the 2,3,5-trihydroxylated acid. The gentisic acid may subsequently undergo glycine conjugation, as indicated above. It is remarkable that in this system, conjugation (Phase II) reactions take place without prior Phase I reactions (see the left-hand part of Fig. 3.22).

Salicylates are excreted in the urine mostly as salicyluric acid (about 75%) and as free salicylic acid (about 10%). However, excretion of free salicylates is extremely variable, depending on both the dose and the urinary pH. An alkaline pH is favourable, leading to about 30% of the ingested drug being eliminated, whereas in acidic urine, elimination may be less than or equal to 2% [42].

Indomethacin

Indomethacin is converted primarily to inactive metabolites, including those formed by O-demethylation (about 50%), conjugation with glucuronic acid (about 10%) and N-deacylation (Fig. 3.23). Some of these metabolites are detectable in plasma, and the free and conjugated metabolites are eliminated in the urine, bile and faeces. A noteworthy feature is that enterohepatic cycling of the conjugates, and probably of indomethacin itself, occurs. Between 10 and 20% of the drug is excreted unchanged in the urine (in part by tubular secretion) [43].

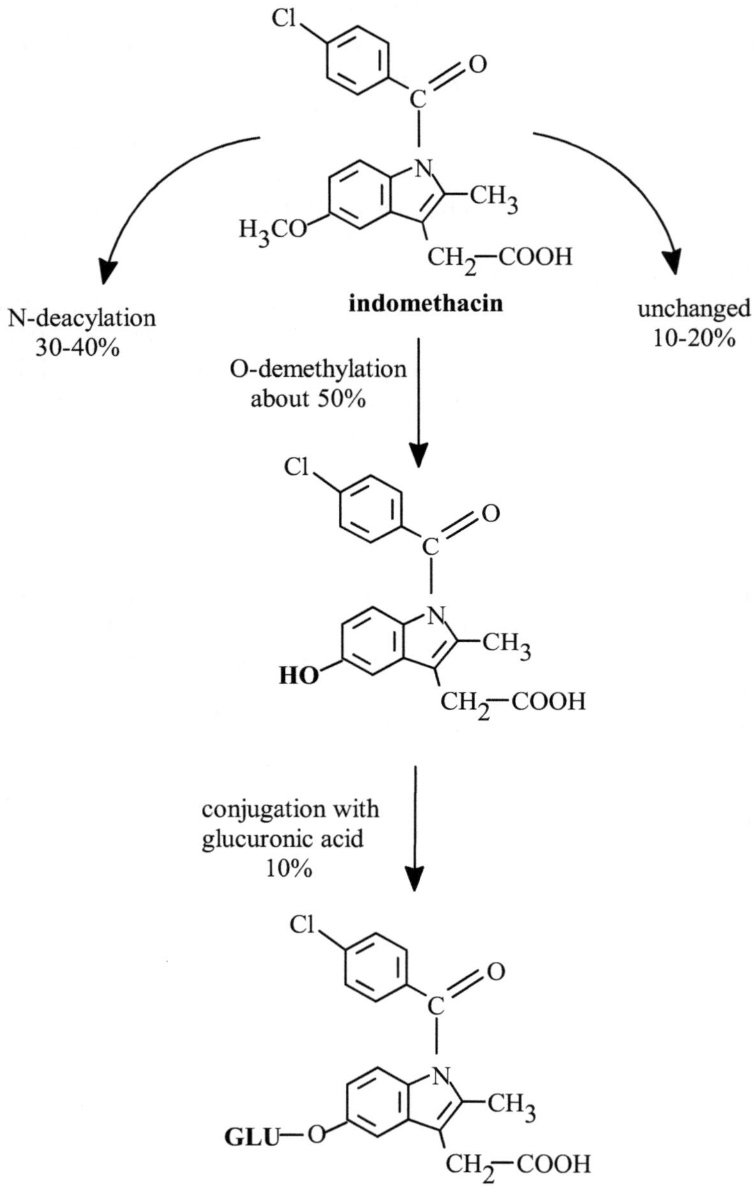

Fig.3.23 Pathways in the metabolism of indomethacin

Sulindac

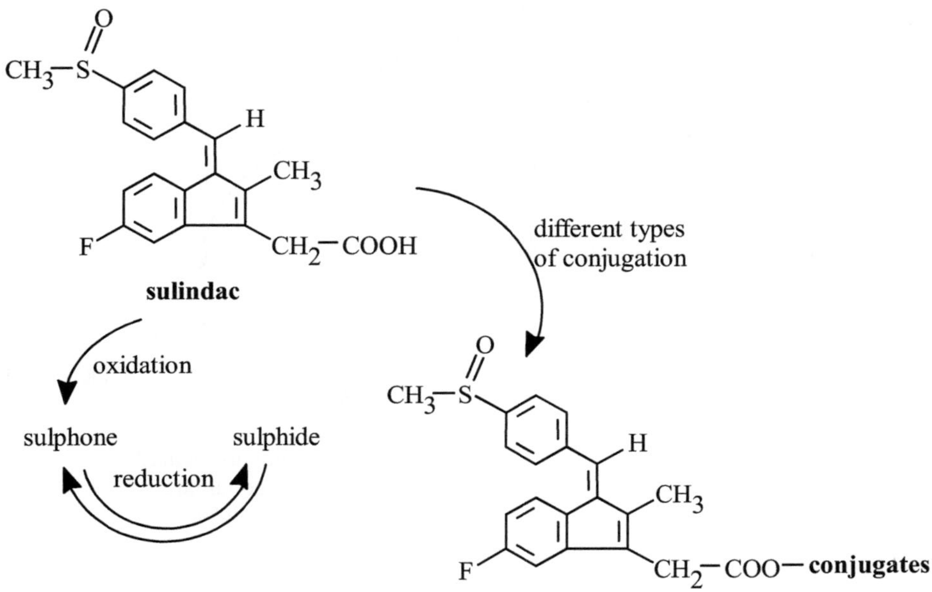

Fig.3.24 Biotransformation of sulindac

The metabolism of sulindac (Fig. 3.24) is complex and highly species dependent variable. It undergoes two major biotransformations in addition to conjugation reactions. It is oxidised to the sulphone and then reversibly reduced to the sulphide. It is this latter metabolite that is the active moiety, although all three compounds are found in comparable concentrations in human plasma [44].

Ketorolac

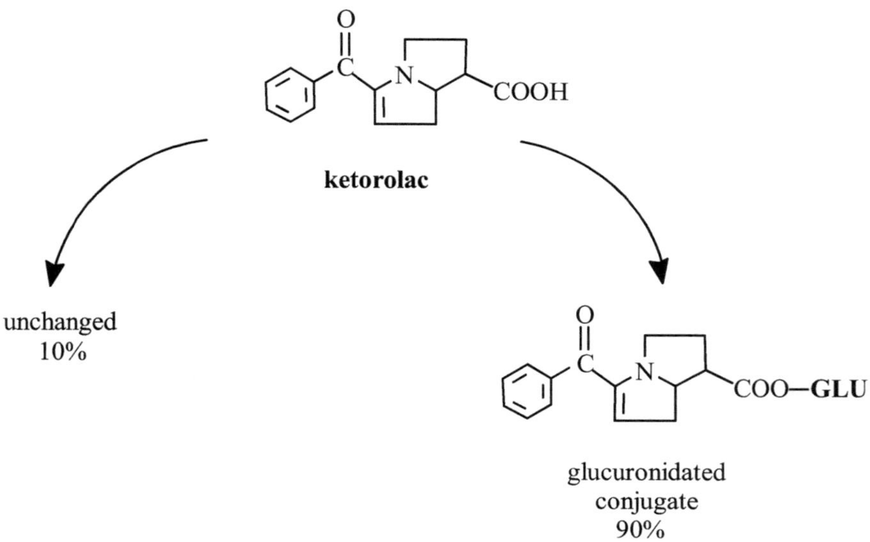

Fig.3.25 Metabolism of ketorolac

This drug is rapidly absorbed with an oral bioavailability of about 80%. Urinary excretion accounts for ~90% of eliminated drug, with the rest excreted unchanged and/or as a glucuronidated conjugate [45] (Fig. 3.25). The rate of elimination is reduced in the elderly and in patients with renal failure.

Diclofenac

Diclofenac is rapidly and completely absorbed after oral administration. There is a substantial first-pass effect, such that only about 50% of diclofenac is systemically available. Diclofenac is metabolised in the liver by a CYTP450 isozyme of the CYP2C subfamily, to the 4-hydroxy- metabolite, and other hydroxylated forms (Fig. 3.26). After glucuronidation and sulphation, the respective metabolites are excreted in the urine (65%) and bile (35%) [46].

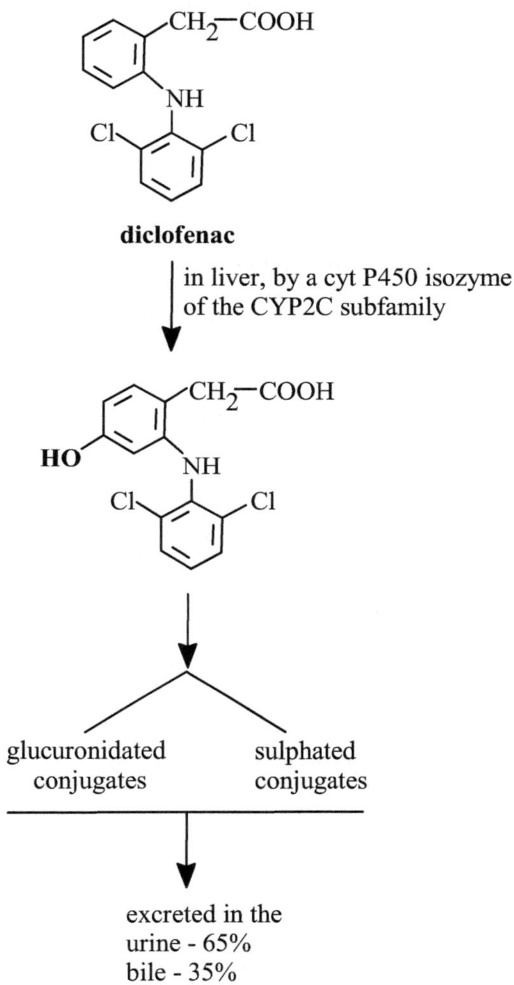

Fig.3.26 Main pathways in the biotransformation of diclofenac

Piroxicam

Piroxicam is also completely absorbed after oral administration and then extensively bound to plasma proteins (about 95%). The major metabolic transformation in humans is CYTP450-mediated hydroxylation of the pyridyl ring (predominantly by an isozyme of the CYP2C subfamily). The inactive metabolite and its glucuronide conjugate account for about 60% of the drug excreted in the urine or faeces [47] (Fig. 3.27).

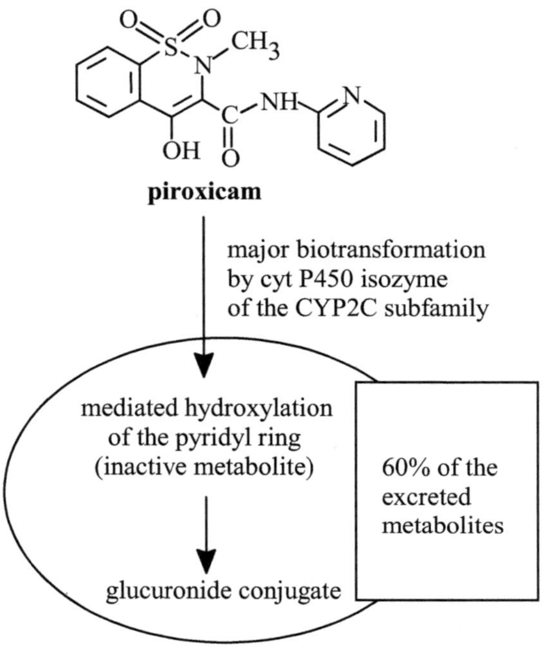

Fig.3.27 Metabolism of piroxicam

Tolmetin

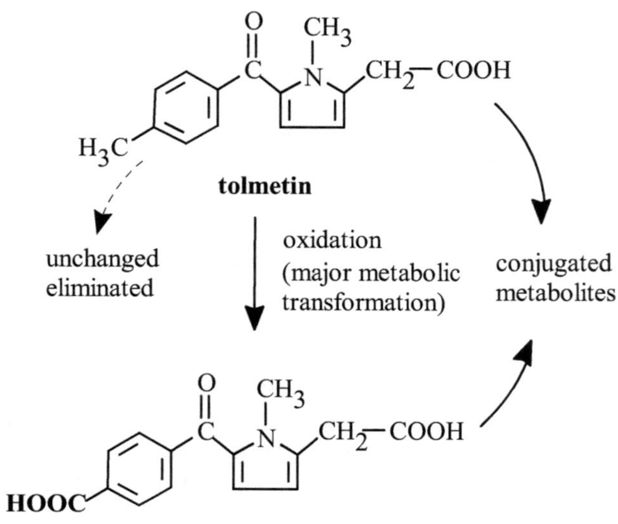

Fig.3.28 Pathways in the biotransformation of tolmetin

After absorption, tolmetin is extensively (~99%) bound to plasma proteins. Virtually all of the drug can be recovered in the urine after 24 hours; some is unchanged, but most is conjugated or otherwise metabolised (Fig. 3.28). The major metabolic transformation involves oxidation of the *p*-methyl group to a carboxylic acid [48].

3.6 CONCLUDING REMARKS

In summary, drug metabolism is generally an extremely complicated process. Often, a drug is metabolised into many products, some major, others minor; furthermore, as indicated in some of the examples above, the drug may be excreted unchanged.

Drug biotransformation may not necessarily produce a metabolite that is devoid of pharmacological activity. In the case of e.g. the antiarrhythmic encainide, hepatic oxidation produces two active metabolites, so both the parent compound and its products of metabolism contribute to the therapeutic effects produced [49].

Metabolism may convert an inactive agent (a prodrug) into the active agent responsible for producing the therapeutic effect. A representative example is given by enalapril [50]. As such it is inactive, but after serum hydrolysis, it is converted into the active, antihypertensive agent, enalaprilate, an inhibitor of angiotensin converting enzyme (See also Chapter 9).

Most drugs, however, require structural modification to facilitate excretion, and the sum of these modification processes is called drug metabolism. The latter can be considered a detoxification function that the human body possesses to defend itself from environmental hostility.

Drugs are usually lipophilic substances, so they can pass plasma membranes and reach the site of action. Drug metabolism is basically a process that introduces hydrophilic functionalities onto the drug molecule to facilitate excretion. These 'functionalized' intermediates are substrates for the phase II enzymes, generating conjugates that are more hydrophilic and thus excreted more rapidly.

Drugs often undergo both Phase I and II reactions before excretion. Nevertheless, there are certain instances where the drug is directly conjugated, or even eliminated in an unchanged form.

Although the liver is the primary site of metabolism, virtually all tissue cells have some metabolic activities. Other organs having significant metabolic activities include the gastrointestinal tract, kidneys and lung. When a drug is administered orally, it usually undergoes first-pass metabolism (it is metabolised in the GI tract or liver, before reaching the systemic circulation). First-pass metabolism limits the oral bioavailability of drugs, sometimes significantly.

The number of enzymes and enzyme systems is vast and their manifold functions lead to a wide range of products when they act on both xenobiotics as well as endogenous compounds. It follows that a drug and an endogenous substance might compete for the same enzyme. Likewise, different enzymes might compete for the same substrate. The complexity of these interactions must be considered in accounting for both toxic and therapeutic actions of drugs [51].

Ultimately, drugs are excreted from the body through various routes, the major organ for drug excretion being the kidney, which excretes hydrophilic drugs and drug metabolites through glomerular filtration. Lipophilic drug molecules can be excreted through the kidneys into urine only after they are metabolised into more hydrophilic molecules. Drugs and their metabolites may also be excreted into the bile, this process usually being mediated by protein transporters. Some drugs may be reabsorbed into the body from the intestine. Also, some drug metabolites such as glucuronide conjugates, may be converted back to the "parent" drug in the intestine (through glucuronidase enzyme), and then reabsorbed into the systemic circulation. This drug recycling process is called enterohepatic recycling. This process, if extensive, may prolong the half-life of the drug. Also, a variety of orally administered drugs are excreted through faeces because they are not absorbed through the intestine.

As a final conclusion, we underscore the highly complex nature of drug metabolism; in many cases, a complete profile of the metabolism of a drug is not attainable.

The study of drug metabolism serves primarily two purposes, namely to elucidate the function and fate of the drug, and, in connection with drug design, to manipulate the metabolic process of a potential drug. The latter theme is explored in Chapter 9.

The Phase I and Phase II reactions whose overall chemistry was described in this and the previous chapter take place under enzymatic control. In the next chapter, the focus turns to details of the interaction between a substrate and its metabolising enzyme, both from structural and kinetic viewpoints.

References

1. Taylor JB, Kennewell PD. 1993. Drug metabolism. In: Modern Medicinal Chemistry. London: Ellis Harwood Ltd, p 110.

2. Ritter JM, Lewis LD, Mant T.GK. 1999. Drug metabolism. In: Radojicic R, Goodgame F, editors. A Textbook of Clinical Pharmacology, 4th Ed. Oxford University Press Inc., pp 40-43.

3. Rang HP, Dale MM, Ritter JM. 1999. Absorption, distribution and fate of drugs. In: Pharmacology, 4th ed. Edinburgh: Churchill Livingstone, pp 85-86.

4. Wingard LB Jr, Brody TM, Larner J, Schwartz A. 1991. Biotransformation of Drugs. In: Kist K, Steinborn E, Salway J, editors. Human Pharmacology, Molecular-to-Clinical. St. Louis: Mosby Year Book, Inc., pp 66-75.

5. Burchell B. 1999. Transformation Reactions: Glucuronidation. In: Woolf TF, editor. Handbook of Drug Metabolism. New York: Marcel Dekker Inc., pp 153-173.

6. Gibson GG, Skett P. 1994. Pathways of drug metabolism. Phase II metabolism. In: Introduction to Drug Metabolism. London: Blackie Academic & Professional, An Imprint of Chapman & Hall, pp 13-34.

7. Bechtel P, Testa B. 1990. Biotransformation des Médicaments: Voies Biomoléculaires et Pharmacologie Clinique. In: Giroud JP, Maphé P, Meyniel G, editors. Pharmacologie Clinique Bases de la Therapeutique, pp 25-53.

8. Lohr GW, Willsky GR, Acara MA. 1998. Renal Drug Metabolism. Pharm Rev 50:121-132.

9. 1994. Drug Metabolism. In: Williams DA, Lemke TL, Foye W, editors. Foye's Principles of Medicinal Chemistry. London: Hardcover, pp 118-129.

10. Mulder GJ, Coughtrie MWH, Burchell B. 1990. Glucuronidation, Conjugation Reactions. In: Mulder GJ, editor. Drug Metabolism: An Integrated Approach. Taylor Francis: London, pp 51-105.

11. Kroemer HK, Klotz U. 1992. Glucuronidation of drugs: a re-evaluation of the pharmacological significance of the conjugates and modulating factors. Clin Pharmacokinet 23:292-310.

12. Radominska-Pandya A, Czernik PJ, Little JM, Battaglia E, Mackenzie PI. 1999. Structural and functional studies of UDP-Glucuronosyltransferases. Drug Metab Rev 31:817-899.

13. Ouzzine M, Barre L, Netter P, Magdalou J, Fournel-Gigleux S. 2003. The human UDP-glucuronosyltransferases: Structural aspects and drug glucuronidation. Drug Metab Rev 35:287-303.

14. Burchell B, Ethell B, Coffey MJ, Findlay K, Jedlitschky G, Soars M, Smith D, Hume R. 2001. Interindividual variation of UDP-glucuronosyltransferases and drug glucuronidation. Interindividual Variability in Human Drug Metabolism:358-394.

15. Tukey RH, Strassburg CP. 2001. Genetic multiplicity of the human UDP-glucuronosyltransferases and regulation in the gastrointestinal tract. Mol Pharmacol 59:405-414.

16. Jansen PLM, Bosna PJ, Chowdhury JR. 1995. Molecular biology of bilirubin metabolism. Prog Liver Dis 13:125-150.

17. Sphann-Langguth H, Benet LZ. 1992 Acyl glucuronides revisited: Is the glucuronidation process a toxification as well as detoxification mechanism? Drug Metab Rev 24:5-48.

18. Berkes EA. 2003. Anaphylactic and anaphylactoid reactions to aspirin and other NSAIDs. Clin Rev Allerg Immu 24:137-147.

19. Bock KW. 1991. Roles of UDP-glucuronosyltransferases in chemical carcinogenesis. Crit Rev Biochem Mol Biol 26:129-150.

20. Boelsterli UA. 1993. Specific targets of covalent drug-protein interactions in hepatocytes and their toxicological significance in drug-induced liver-injury. Drug Metab Rev 25: 395-451.

21. Barrett DA, Barker DP, Rutter N, Pawula M, Shaw PN. 1996. Morphine, morphine-6-glucuronide and morphine-3-glucuronide pharmacokinetics in newborn infants receiving diamorphine infusions. Brit J Clin Pharmacol 41:531-537.

22. Ulens C, Baker L, Ratka A, Waumans D, Tytgat J. 2001. Morphine-6-β-glucuronide and morphine-3-glucuronide, opioid receptor agonists with different potencies. Biochem Pharmacol 62:1273-1282.

23. Fenselau C. 1994. Acyl glucuronides as chemically reactive intermediates. In: Handbook of Experimental Pharmacology, 112:367-389.

24. Paul D, Standifer KM, Inturrisi CE, Pasternak JW. 1989. Pharmacological characterization of morphine-6-β-glucuronide, a very important morphine metabolite. J Pharmacol Exp Ther 251:477-483.

25. Miller RR, Doss GA, Stearns RA. 2004. Identification of a hydroxylamine glucuronide metabolite of an oral hypoglycemic agent. Drug Metab Dispos 32:178-185.

26. Dieterle W, Faigle JW, Moppert J. 1980. New metabolites of sulfinpyrazone in man. Arznei-Forschung 30:909-993.

27. Tukey RH, Strassburg CP. 2000. Human UDP-glucuronosyltransferases. Metabolism, expression, and disease. Annu Rev Pharmacol 40:581-616.

28. Rauchschwalbe SK, Zuehlsdorf MT, Wensing G, Kuhlmann J. 2004. Glucuronidation of acetaminophen is independent of UGT1A1 promoter genotype. Int J Clin Pharm Th 42:73-77.

29. Zubay GL, Parson WW, Vance DE. 1995. Priciples of Biochemistry. Dubuque, Iowa: Wm. C. Brown Publishers, pp 287-289, 301, 418.

30. Zubay GF, Parson WW, Vance DE. 1994. How Enzymes Work. In: Sievers EM, editor. Dubuque, Iowa: Wm C Brown Publishers, pp 154-174.

31. Sueda S, Islam MN, Kondo H. 2004. Protein engineering of pyruvate carboxylase. Investigation on the function of acetyl-CoA and the quaternary structure. Eur J Biochem 271:1391-1400.

32. Diacovich L, Mitchell DL, Pham H, Gago G, Melgar, Melrose M, Khosla C, Gramajo H, Tsai S-C. 2004. Crystal Structure of the β-Subunit of Acyl-CoA Carboxylase: Structure-Based Engineering of Substrate Specificity. Biochemistry US 43:14027-14036.

33. Zhang H, Tweel B, Li J, Tong L. 2004. Crystal Structure of the Carboxyltransferase Domain of Acetyl-Coenzyme A Carboxylase in Complex with CP-640186. Structure 12:1683-1691.

34. Lindahl PA. 2004. Acetyl-coenzyme A synthase: the case for a Ni0p-based mechanism of catalysis. J Biol Inorg Chem 9:516-524.

35. Ozawa N, Watabe T. 1985. Glutathione conjugations. Tokishikoroji Foramu 8:291-304.

36. Hesse S, Jernstroem B. 1984. Role of glutathione S-transferases: detoxification of reactive metabolites of benzo[a]pyrene-7,8-dihydrodiol by conjugation with glutathione. Biochem Basis Chem Carcinog, Workshop Conf. Hoechst, 13th ed., pp 5-12

37. Okinaga S, Ohrui T, Nakazawa H, Yamauchi K, Sakurai E, Watanabe T, Sekizawa K, Sasaki H. 1995. The role of HMT (histamine N-methyltransferase) in airways: a review. Method Find Exp Clin 17:16-20.

38. Ziegler MG, Bao X, Kennedy BP, Joiner A, Enns R. 2002. Location, development, control, and function of extraadrenal phenylethanolamine N-methyltransferase. Ann NY Acad Sci 971:76-82.

39. Hodel A. 2001. Effects of glucocorticoids on adrenal chromaffin cells. J Neuroendocrinol 13:217-221.

40. Thompson MA, Weinshilboum RM, El Yazal J, Wood TC, Pang Y-P. 2001. Rabbit indolethylamine N-methyltransferase three-dimensional structure prediction: a model approach to bridge sequence to function in pharmacogenomic studies. J Mol Model [online computer file] 7:324-333.

41. Hanson GR. Analgesic, Antipyretic, and Anti-Inflammatory Drugs. 2000. In: Gennaro AR editor. Remington: The Science and Practice of Pharmacy, 20th ed. Philadelphia: Lippincott Williams & Wilkins, pp 1459-1460.

42. Knutson K. Topical Drugs. 2000. In: Gennaro AR editor. Remington: The Science and Practice of Pharmacy, 20th ed. Philadelphia: Lippincott Williams & Wilkins, p 1212.

43. Hanson GR. Analgesic, Antipyretic, and Anti-Inflammatory Drugs. 2000. In: Gennaro AR editor. Remington: The Science and Practice of Pharmacy, 20th ed. Philadelphia: Lippincott Williams & Wilkins, p 1457.

44. Hanson GR. Analgesic, Antipyretic, and Anti-Inflammatory Drugs. 2000. In: Gennaro AR editor. Remington: The Science and Practice of Pharmacy, 20th ed. Philadelphia: Lippincott Williams & Wilkins, p 1460.

45. Hanson GR. Analgesic, Antipyretic, and Anti-Inflammatory Drugs. 2000. In: Gennaro AR editor. Remington: The Science and Practice of Pharmacy, 20th ed. Philadelphia: Lippincott Williams & Wilkins, p 1445.

46. Hanson GR. Analgesic, Antipyretic, and Anti-Inflammatory Drugs. 2000. In: Gennaro AR editor. Remington: The Science and Practice of Pharmacy, 20th ed. Philadelphia: Lippincott Williams & Wilkins, p 1456.

47. Hanson GR. Analgesic, Antipyretic, and Anti-Inflammatory Drugs. 2000. In: Gennaro AR editor. Remington: The Science and Practice of Pharmacy, 20th ed. Philadelphia: Lippincott Williams & Wilkins, p 1460.

48. Hanson GR. Analgesic, Antipyretic, and Anti-Inflammatory Drugs. 2000. In: Gennaro AR editor. Remington: The Science and Practice of Pharmacy, 20th ed. Philadelphia: Lippincott Williams & Wilkins, p 1460.

49. D'Souza MJ, DeSouza PJ, Anderson R, Pollock SH. 1989. Encainide Metabolism. Pharm Res 6:28-33.

50. 2000. In: Gennaro AR editor. Remington: The Science and Practice of Pharmacy, 20th ed. Philadelphia: Lippincott Williams & Wilkins, pp 517; 1281.

51. Gibson GG, Skett P. 1994. Pathways of drug metabolism. Phase II metabolism. In: Introduction to Drug Metabolism. London: Blackie Academic & Professional, An Imprint of Chapman & Hall, p 33.

Chapter 4

ENZYMATIC SYSTEMS INVOLVED IN DRUG BIOTRANSFORMATION

4.1 INTRODUCTION

Chapters 2 and 3 effectively described a vast array of *overall* drug biotransformation reactions mediated by enzymes. In the first part of the present chapter, the basic structural and dynamic features associated with enzyme activity are discussed. This includes the essential concept of specificity, the hallmark of enzymatic action, as well as the roles of coenzymes and effectors in the catalytic process. Dependence of the rate of enzyme-catalysis on various factors is discussed and the relevant basic kinetic expressions are presented. Mechanistic aspects at the molecular level are briefly explored. Finally, the main strategies by which cells regulate enzyme activities are described.

The second part of the chapter focuses on the nature and role of specific enzyme systems and is logically divided into those mediating Phase I and Phase II biotransformations. In the former category, we meet the cytochrome P450-dependent MFO system, the microsomal flavin-containing monooxygenase system and several other key enzyme systems. Classification of cytochrome P450 isoforms and the nomenclature used to describe them are also presented. Several examples of specific cytochrome P450 subfamilies are chosen to illustrate their particular metabolic functions as well as typical drugs that serve as substrates for them.

Enzyme systems that mediate Phase II reactions are exemplified in the final part of this chapter by the UDP-glucuronosyl transferases and glutathione-S-transferase.

4.2 INTERACTION BETWEEN A DRUG SUBSTRATE AND AN ENZYME

All of the thousands of drug biotransformation reactions (as well as all normal metabolic processes) are catalysed by enzymes [1-6].

The drug substance that is acted on by an enzyme is called the substrate of that enzyme. On the other hand, the enzyme, representing a compound that increases the rate, or velocity of a biochemical reaction is called the catalyst.

Several aspects should be emphasised from the outset:

- most (but not all) biological catalysts are proteins (these are called enzymes);
- a catalyst, though it participates in the reaction process, is unchanged by it, at the end of reaction being found again in exactly the same state as before, ready for another cycle of biotransformation;
- catalysts change rates of processes but do not affect the position of equilibrium of a reaction. This means that a thermodynamically favourable process is not made more favourable, nor is an unfavourable process made favourable, by the presence of a catalyst. Instead, the equilibrium state is simply approached more rapidly.

It being generally accepted that for a reaction to take place energy is needed, we present an explanation of how enzymes function. The barrier preventing a chemical reaction from occurring is called the activation energy and refers to a high-energy transition state that a reactant molecule has to pass through in order to form products.

Catalysts function by lowering this activation energy, binding the substrate in an intermediate conformation that resembles the transition state but which has a lower energy. In enzyme catalysis, one or more substrates are bound at the active site of an enzyme to form the enzyme-substrate complex, which is a highly reactive species that promotes the reaction and releases the product(s) (Figure 4.1).

It is important to stress the fact that the active site portion of the enzyme molecule is not one continuous sequence of the protein. Because of the coiling of the molecule, portions of the amino acid sequence that are far removed from one another if the protein were to be stretched linearly come into close proximity when the molecule folds into its proper conformation.

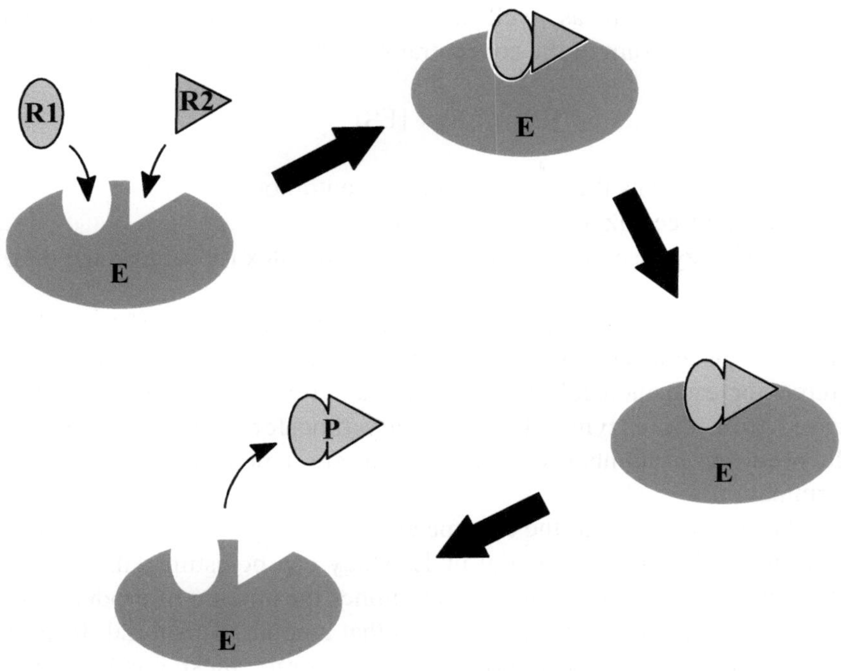

Fig.4.1 General mechanism of enzyme action; two reactants are bound to the same enzyme, which ensures their correct mutual orientation and proximity and binds them strongly

The simplest equation to describe a one-substrate, one-product reaction catalysed by an enzyme is the following:

$$\mathrm{E} + \mathbf{S} \underset{\leftarrow}{\xrightarrow{k_1}} |\mathrm{ES}| \xrightarrow{k_2} \mathrm{E} + \mathbf{P} \qquad 4.1$$

As implied by Eq.4.1, the process involves a molecule of substrate binding to an enzyme molecule, the substrate being subsequently converted to product, and the latter being released from the enzyme.

If we assume that conditions are such that the reverse reaction between E and P is negligible, then the catalytic formation of the product (with enzyme regeneration) will be a simple first-order process. Consequently, the rate will be determined only by the concentration of [ES] and the corresponding value of k_2. However, [ES] is usually not a measurable concentration; what is measurable is either the substrate or product concentration as well as the total concentration of enzyme, represented by the sum of the concentrations of free and occupied enzyme:

$$[E]_t = [E] + [ES] \qquad 4.2$$

where $[E]_t$ represents the total enzyme concentration,
[E] the free enzyme concentration, and
[ES] enzyme concentration in the ES complex (occupied enzyme).

Another aspect merits emphasis, namely that not all of the substrate molecules instantaneously change to product. There is a certain time required for each molecule to bind, be catalytically converted, and finally released from the enzyme. The necessary time for each transformation is influenced by a number of factors that are amenable to experimental determination.

The reaction rate of the enzyme and (indirectly) the amount of enzyme present in the biological material under study can be estimated. Usually, in measuring the rate of reaction, one determines the amount of product formed and divides by the time required to form that amount of material. Inspection of Eq.4.1 shows that each molecule of substrate must combine with a molecule of enzyme to form a molecule of product. However, in measuring enzyme activity, we assume that that there are many more substrate molecules than enzyme molecules. In this situation, each enzyme molecule will bind substrate and convert it, then accept another substrate molecule for further reaction, and so on. It follows that the substrate cannot be converted any faster than the number of enzyme molecules present allows.

The enzyme level is therefore said to be 'rate-limiting'. However, when an enzyme reaction is studied, several parameters are involved: the time of contact between enzyme and substrate, the concentrations of substrate and enzyme, type of buffer, pH, temperature, necessary co-factors, presence of enzyme effectors – all of which affect the rate of the reaction.

Time of contact between enzyme and substrate

The rate of an enzyme-catalysed reaction (v) evolves as a function of time. According to Figure 4.2, the reaction rate is initially high (the steep linear segment corresponding to the initial rate) and decreases as equilibrium is attained, when $v = 0$.

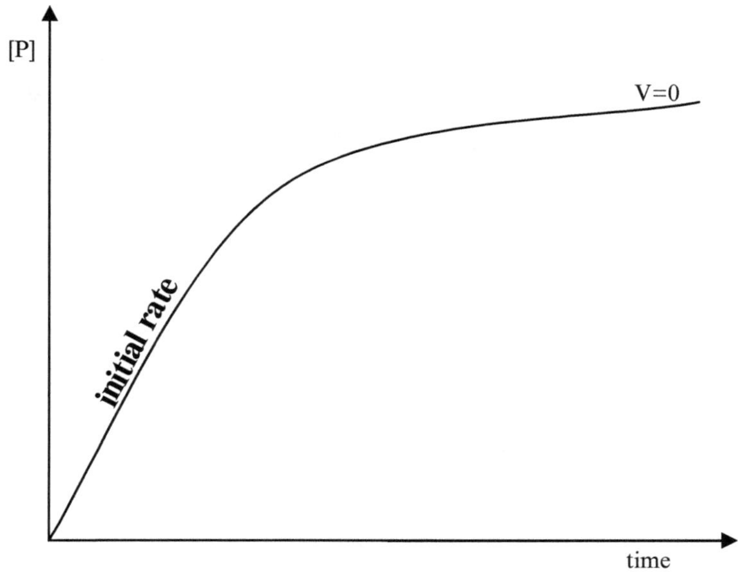

Fig.4.2 The rate evolution of an enzyme catalysed reaction

In the evolution of such a reaction rate several factors are involved; among them it is assumed that reduction in the substrate concentration concomitant with an increase in product concentration could favour the reverse reaction, P→S; the same effect is caused by enzyme denaturation.

In view of the above, it is important to stress the recommendation that the rate of such a reaction be determined before any of the phenomena mentioned above intervenes. In other words, the initial rate represents the most correct experimental datum relating to the amount of active enzyme present in the reaction environment.

Substrate concentration

If the enzyme itself is present in sufficiently high amounts, the rate of the reaction is determined by the concentration of the substrate present. As the substrate level increases, the enzyme reaction rate also increases. The reaction velocity, v_0 is a function of the substrate concentration [S] for the enzyme-catalysed reaction. At high substrate concentrations the reaction velocity reaches a limiting value, V_m; K_m is the substrate concentration at which the rate is at the half-maximum value (Figure 4.3).

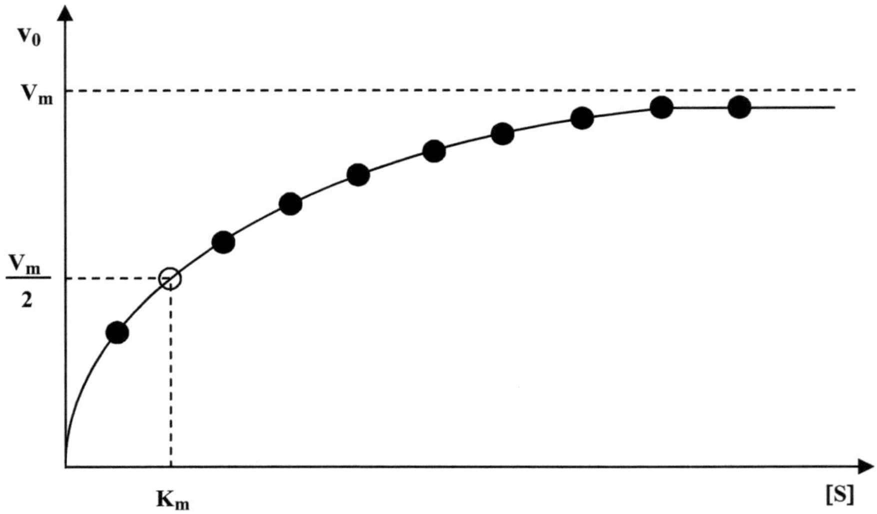

Fig.4.3 Effect of substrate concentration on reaction rate

where
v_0 = the reaction rate for a certain substrate concentration (the initial rate)
V_m = the maximum limiting rate
[S] = the substrate concentration
K_m = Michaelis constant, representing in a reversed relation the affinity of the enzyme for that substrate.

It is assumed that when the rate attains the value V_m, the enzyme is 'saturated' with substrate. The active site concept provides a simple explanation of what is taking place. A certain number of available active sites are present. When adding a low concentration of substrate, each substrate molecule can eventually bind to the active site of an enzyme molecule. If the substrate concentration is increased, the probability of substrate molecules colliding with enzyme molecules yielding the [ES] complex also increases, thus increasing reaction rate.

The parameters in the figure are related in the Michaelis-Menten equation:

$$v = V_m \cdot \frac{[S]}{[S] + K_m} \qquad 4.3$$

For the purpose of graphical representation of experimental data, it is convenient to rearrange equation 4.3. Taking the reciprocal of both sides of equation 4.3 gives:

$$\frac{1}{v} = \frac{1}{V_m} + \frac{K_m}{V_m} \cdot \frac{1}{[S]} \quad 4.4$$

The plot of the reciprocal of the rate ($1/v_0$) as a function of the reciprocal of the substrate concentration ($1/[S]$) gives a straight line, known as the Lineweaver-Burk, or double-reciprocal plot (Figure 4.4):

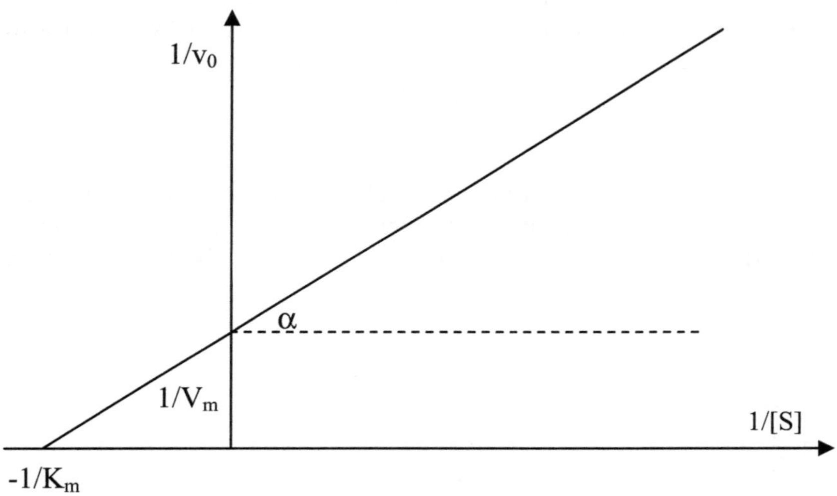

Fig.4.4 Lineweaver-Burk plot

V_m and K_m can be readily determined from the graph. The slope of the graph is given by:

$$\tan \alpha = \frac{K_m}{V_m} \quad 4.5$$

The maximum number of molecules of substrate that can be converted to product each second per active site is known as the 'turnover number' of the particular enzyme involved, designated k_{cat} (the catalytic constant). Because the maximum rate is obtained at high substrate concentrations, when all the active sites are occupied with substrate, the turnover number is a measure of how rapidly an enzyme can operate once the active site is filled ($k_{cat} = V_m / [E]_t$).

However, since enzymes usually do not operate at saturating substrate concentrations under physiological conditions, another parameter needs to be introduced, namely the specificity constant. This is the k_{cat}/K_m ratio, representing a measure of how rapidly an enzyme can function at low substrate concentrations. The specificity constant is useful for comparing the relative abilities of different compounds to serve as substrates for the same enzyme.

Enzyme concentration

If the substrate is present in sufficiently high amounts, the rate of reaction will become a function of the enzyme concentration. As the enzyme level increases, for a defined volume of body fluid, the rate will increase as well.

In some specific situations, other reaction conditions should also be taken into account. For instance, dilution could lead to a false increase in the amount of enzyme supposedly present in the system.

Effect of temperature

For most enzymes, the turnover number increases with temperature; however, beyond a certain point, further increase in temperature does not lead to further increase in rate, but to loss of enzyme activity (Figure 4.5):

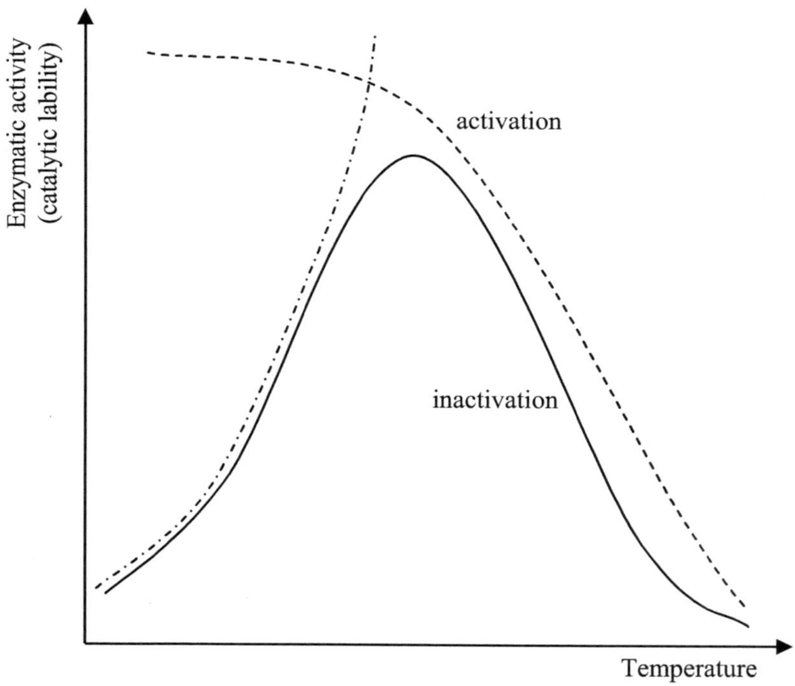

Fig.4.5 Effect of temperature upon enzymatic activity

The effect of increasing temperature is to increase the kinetic energies of molecules. This in turn results in higher frequencies of collision between enzyme and substrate molecules and thus higher reaction rate. Within a narrowly defined range of temperature, enzyme activity approximately doubles for every 10°C rise in temperature (which translates into an activation energy of ~50 kJ mol^{-1}, according to the well-known Arrhenius equation). However, a point is reached where another factor comes into play: the increase in temperature leads to an increase in the rate of unfolding of the enzyme molecule. Consequently, from its initial, rather tight globular structure, the protein begins to spread out into a more linear configuration, leading to loss of enzyme activity. This is the 'denaturation' process and its characteristics vary from one enzyme to another.

At lower temperatures, the temperature dependence of k_{cat} can be related to the activation energy of the slowest (rate-limiting) step in the catalytic pathway.

Effect of pH

Enzymes, like other proteins, are stable only over a limited range of pH. A change in the hydrogen ion concentration of the reaction medium can have profound effects on the rate of an enzyme reaction. Changes in the charges on ionisable amino acid residues result in modifications in the tertiary structure of the protein and eventually lead to denaturation.

At the pH extremes, the reaction rate is rather low and gradually increases to a pH optimum, the point at which the reaction rate is greatest for the conditions.

Several factors determine the pH optimum. If the substrate can be ionised at a certain pH, the degree of ionisation may affect binding to the active site and the resultant activity. At the same time, the active site may be able to exist in an ionised or unionised form, the presence or absence of charge on the active site affecting substrate binding and reactivity as well.

The pH optimum for different enzymes is quite variable (Figure 4.6):

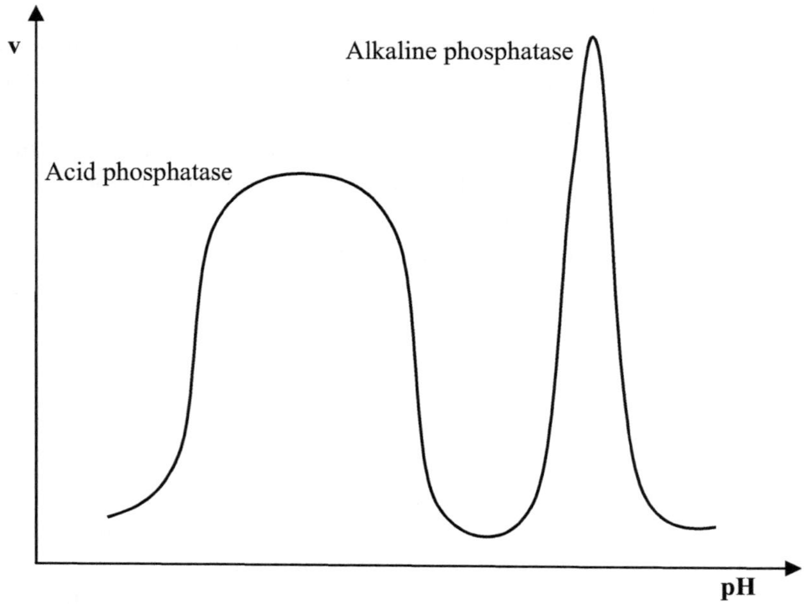

Fig.4. 6 The difference in pH optimum for the two physiological phosphatases

Coenzymes

Many enzymes require additional partners called co-factors for their activity. These act in concert with the enzymes to catalyse biochemical reactions.

Co-factors may be simple inorganic ions (such as Mg^{2+}) or complex organic molecules known as coenzymes. Many of these organic molecules are modified forms of vitamins, with the modifications taking place in the organism after their ingestion.

The co-factor usually binds tightly to a special site on the enzyme, sometimes referred to as prosthetic group. An enzyme lacking an essential co-factor is called an apoenzyme, while the intact enzyme with the bound co-factor is called the holoenzyme.

The most common enzymatic system in mammals and humans is known to be the M.F.O., the array of catalysed reactions including also oxidative reactions of drug biotransformations. For this type of reaction to proceed, the enzyme (the CYTP450) requires reducing equivalents (NADPH $+H^{+}$) and molecular oxygen. During the reaction, reducing equivalents are consumed and one atom of molecular oxygen is incorporated into the substrate, whereas the other oxygen atom is reduced to water.

Another component of the same system is represented by a flavin-containing enzyme, consisting of one mole of flavin adenine dinucleotide

(FAD) and one mole of flavin mononucleotide (FMN) per mole of apoprotein. There is strong evidence to support the role of FAD as the acceptor flavin from NADPH $+H^+$ and FMN as the donating flavin to CYTP450 in the electron transfer events.

An NADH $+H^+$ dependent system is also required in reductive drug metabolism.

Effectors

Effectors can be either inhibitors or activators.

Molecules which decrease the rate of an enzyme reaction (if they are present in the reaction system) are called inhibitors. Most enzymes are sensitive to inhibition by specific agents that interfere either with the binding of a substrate at the active site, or with the conversion of the enzyme-substrate complex into products.

If substrate is present in too high a concentration (substrate inhibition), a decrease in the rate may be seen. In other instances, the product concentration could become sufficiently high to provide product inhibition of the enzyme.

A feature that warrants emphasis is that there is increasing evidence to show that the human body has endogenous inhibitors for some enzymes. These substances, produced by the organism, exert regulatory control as a part of normal biochemical processes.

In many cases, an inhibitor resembles the substrate structure and binds reversibly at the same site on the enzyme. These are the 'competitive inhibitors', because both the inhibitor and the substrate compete for the same binding site on the enzyme. Competitive inhibition may be prevented if the active site is already occupied by the substrate.

Inhibitors that bind at sites other than the active site of the enzyme but do not compete directly with binding of the substrate are the 'non-competitive inhibitors'. Instead, these act by interfering with the reaction of the enzyme-substrate complex.

Still another possibility is the binding of the inhibitor only to the enzyme-substrate complex and not to the free enzyme, the effect being called 'uncompetitive inhibition'.

Inhibition may be reversible or irreversible, altering the enzyme structure temporarily or permanently.

The kinetic parameters are also modified, as presented in Figure 4.7:

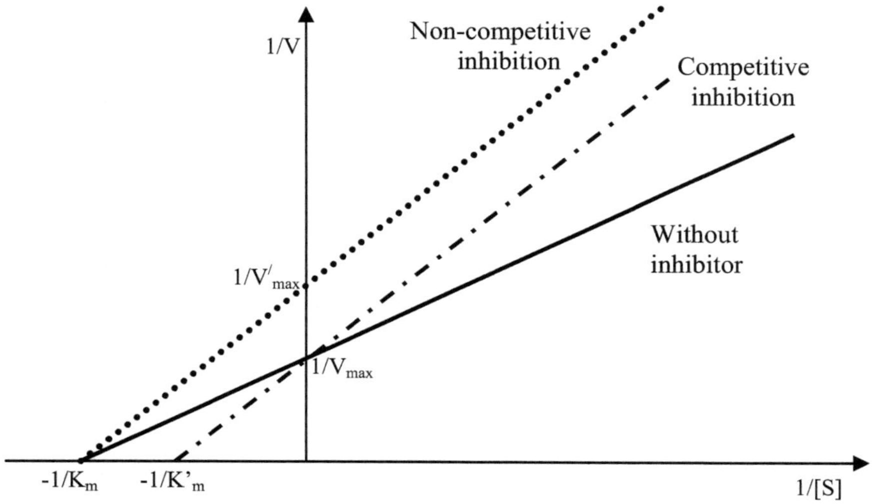

Fig.4.7 The plot of the double reciprocal relation between [S] and the rate of an enzymatic reaction in the absence and presence of some inhibitors

The slopes of the plots and the intercepts on the abscissa are simple, linear functions of $[I]/K_i$, where K_i is the dissociation constant of the inhibitor-enzyme complex.

In the case of a competitive inhibitor, the Michaelis-Menten equation becomes:

$$v = V_m \cdot \frac{[S]}{[S] + K_m\left(1 + \frac{I}{K_i}\right)} \qquad 4.6$$

where v, V_m, [S] and K_m have the same significance as before, and K_i represents the inhibition constant. V_m is constant, while K_m increases.

In the case of a non-competitive inhibitor, the relationship can be written as:

$$v = \frac{V_m}{1 + \frac{[I]}{K_i}} \cdot \frac{[S]}{[S] + K_m} \qquad 4.7$$

K_m remains constant, in contrast to V_m which is much reduced.

Other enzyme effectors act like activators. Indeed, many enzymes require for full activity, the presence of metal activators, inorganic entities that facilitate binding of the substrate to the active site, by forming ionic bridges. The metal is also effective in orientating the substrate so it can attach to the protein at the proper point and in the correct configuration. Some of the metal activators are tightly bound to the enzyme while others are more loosely attached. In the latter case, a supplemental activator must be part of the reaction mixture in order to obtain full enzyme activity.

Mechanisms of enyzme action at the molecular level

Any enzyme binds a molecule of substrate (or in some cases several substrates) into a special region of the enzyme called the active site (catalytic site).

The active site is usually represented by a pocket surrounded by amino acid side chains that help bind the substrate and by other side chains that play a role in catalysis. The pocket fits the substrate quite closely because of the complex tertiary structure of the enzyme. This feature explains the extraordinary specificity of enzyme catalysis.

The general themes frequently occurring in enzymatic reaction mechanisms include:

- proximity effects
- general-acid and general-base catalysis
- electrostatic effects
- nucleophilic or electrophilic catalysis by functional groups, and
- structural flexibility.

The idea underlying proximity effects is that an enzyme can accelerate a reaction between two species simply by holding the two reactants closely together in appropriate mutual orientation.

The general-acid and general-base catalyses avoid the need for extreme pH values. The reactive chemical groups function either as electrophiles or as nucleophiles, their task often being to make a potentially reactive group more reactive by increasing its electrophilic or nucleophilic character, primarily by adding or removing a proton.

The electrostatic interactions can promote the formation of the transition state, by stabilising the distribution of electrical charge in transition states. Electrostatic interactions can be significant even between groups whose net charge is zero.

Nucleophilic catalysis by enzymes involves the formation of an intermediate state in which the substrate is covalently attached to a nucleophilic group of the enzyme. Nucleophilic groups participate in reactions of hydrolysis (of an ester or amide) and addition.

Although precise positioning of the reactants is a fundamental aspect of enzyme catalysis, most enzymes undergo some change in their structure when they bind substrates.

A first hypothesis for enzyme action was proposed by Fischer (1894) and it is the well-known lock-and-key model (Figure 4.8):

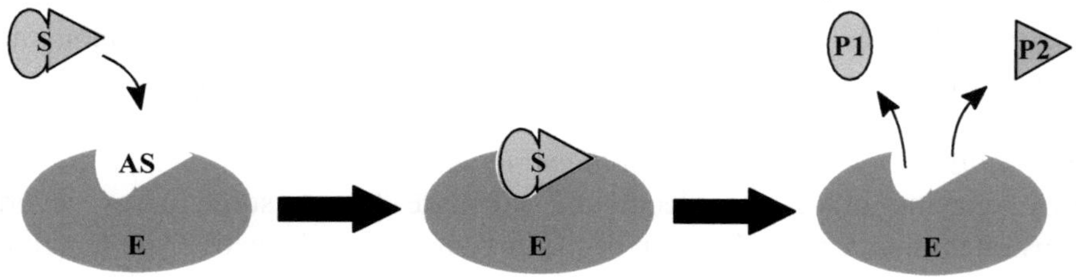

Fig.4.8 The lock-and-key model (the active site of the enzyme fits the substrate as a lock does a key)

According to this model, the enzyme accommodates the specific substrate as a lock does its specific key. However, although this model explained enzyme specificity, it could not explain the catalytic process itself. Thus arose the need for an extension of Fischer's idea: both the enzyme and the substrate must mutually adjust to take up a configuration that stabilises the transition state.

In practice, the enzyme does not simply accept the substrate but instead there is mutual distortion of enzyme and substrate to produce and fit conformations close to the transition state. This model represents Koshland's induced fit hypothesis. As indicated in Figure 4.9, induced fit implies distortion of the enzyme as well as the substrate; this distortion may be local, or it may involve a major change in enzyme conformation.

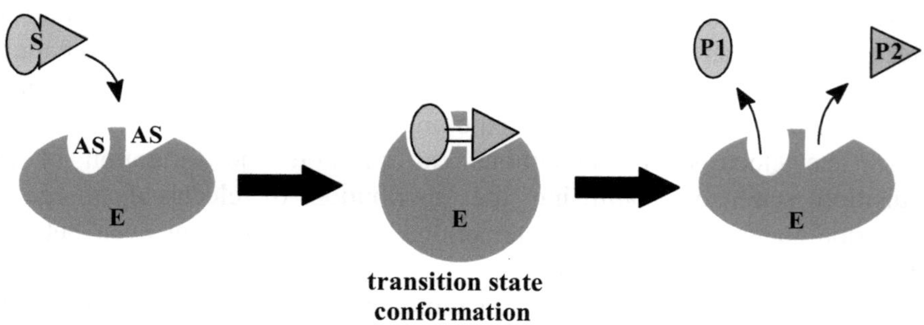

Fig.4.9 The induced fit model, in which both enzyme and substrate are distorted on binding (the enzyme keeps the substrate under stress)

Summarising, we may say that an enzyme:

- binds the substrate(s),
- lowers the energy of the transition state, and
- directly promotes the catalytic event.

When the catalytic process has been completed, the enzyme must be able to release the product(s) and return to its original state, ready for another round of catalysis.

Structural changes also contribute to the high specificity of the enzymatic reactions, concerning either substrate, or reaction specificity.

Recently, the dynamics of these processes, approached through a variety of kinetic methods, were discussed in support of the potential roles of conformational changes in the catalytic process and in terms of dynamic coupling within the enzyme-substrate complex [7].

Specificity of enzymes

Although a fundamental aspect of enzyme catalysis relies on the precise positioning of the reactants, due to their structural flexibility, when binding different substrates, most enzymes undergo some changes in their structure. Commonly such a structural change is referred to as an induced fit, and contributes to the high specificity of some enzymatic reactions. Practically, when an enzyme binds a substrate, its structure changes in a manner that brings together the elements of the active site, the enzyme closing like a net around the substrate. Moreover, it also allows the enzyme to control the electrostatic effects that promote the formation of the transition state; in this way, the substrate is forced to respond to the directed electrostatic fields generated by the enzyme's functional groups, instead of the disordered fields from the solvent.

A particular case of enzyme specificity is stereospecificity, which occurs in so-called prochiral substrates. In some specific situations, even for a symmetrical molecule, when bound by three points to an asymmetric object, two of its previously identical atoms will no longer be equivalent; consequently only one of the two initially equivalent (now, prochiral) atoms would be able to contact the surface properly. An example is the substrate ethanol, CH_3CH_2OH, whose methylene hydrogen atoms become distinguishable when the molecule attaches itself to an asymmetric template.

Regulation of enzyme activities

Cells use two basic strategies for regulating their enzyme activities.

The first strategy refers to adjusting the amount and location of key enzymes, consequently requiring mechanisms for the control of synthesis, degradation, and transport of proteins, while the second strategy, is to regulate the activities of the enzymes. We shall focus in the present material exclusively with the second strategy.

In principle, the activities of many enzymes can be altered by changes in pH. However, this is not a very practical solution especially for intracellular enzymes, because most cells maintain their pH within narrow limits. Therefore, two other strategies are more widely applicable and efficient.

- The first refers to the covalent modification of the enzyme structure, in such a way as to alter either K_m or k_{cat}.
- The second is to use an effector (inhibitor or activator) that binds reversibly to the enzyme and again, alters either the K_m or k_{cat}. Such effectors may bind either at the active site itself, or at some distance from it. In this latter case, we are referring to an 'allosteric effector' (from Greek: *allos* = 'other', and *stereos* = 'space').
- Within the first strategy, the most common regulatory mechanism is phosphorylation. It refers to some specific amino acid residues, such as serine, threonine and tyrosine and uses two separate enzymes: the introduction of the phosphate is catalysed by a protein kinase, while dephosphorylation is effected by a phosphatase (both enzymes themselves usually being under metabolic regulation). Phosphorylation involves consumption of ATP (Figure 4.10):

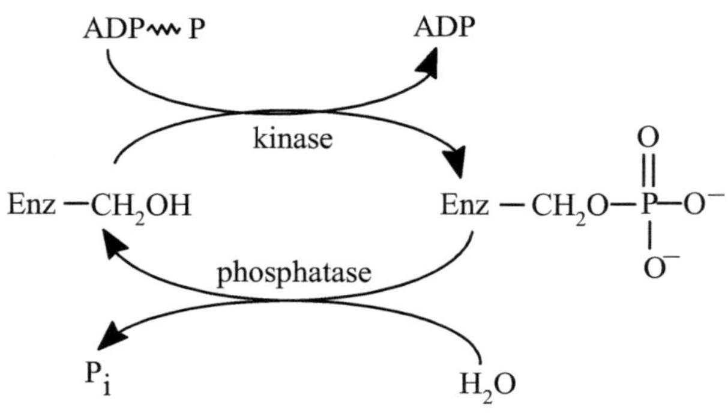

Fig.4.10 Phosphorylation and dephosphorylation mechanism

In eukaryotic organisms, phosphorylation is used to control the activities of hundreds of enzymes, in response to extracellular signals, such as hormones or growth factors. Sometimes phosphorylation can also modify an enzyme's sensitivity to allosteric effectors. For example, phosphorylation of glycogen phosphorylase reduces its sensitivity to the allosteric activator adenosine monophosphate (AMP). Other modifying groups (acting by covalent attachment) include fatty acids, isoprenoid alcohols and carbohydrates. Their regulating enzymatic activity is, however, fragmentary.

Covalent modifications of enzymes allow a cell to regulate its metabolic activities more rapidly and in much more intricate ways than is possible by changing the absolute concentrations of the same enzymes.

• Another mechanism of response to extracellular signals both in eukaryotic and prokaryotic organisms is allosteric regulation. A compound that binds at an allosteric site can serve as either an inhibitor or an activator, depending on the structure of the enzyme and does not need to have any structural relationship to the substrate. Such effectors are, for example, ATP, ADP, AMP, or P_i, often chemically unrelated to the substrate of the enzyme that must be regulated. They usually bind to an allosteric site (rather than to the active site) and their concentrations provide the cell with an indication of the available energy.

However, there also, combined control systems for enzymatic activity exist, a good example being provided by glycogen phosphorylase. This enzyme, which catalyses the removal of a terminal glucose residue from glycogen, exists in two forms, 'a' and 'b', which differ greatly in their catalytic activities. Phosphorylase 'b', virtually inactive, can be activated by low concentrations of AMP; nonetheless, activation may be inhibited competitively by ATP. On the other hand, phosphorylase 'a' becomes fully active at low concentrations of AMP, being also relatively insensitive to inhibition by ATP. The structural basis for the difference between the two forms of phosphorylase is given by a punctiform modification: phosphorylase 'a' has a phosphate residue on serine 14 (which is absent in phosphorylase 'b'). The interconversion of the two forms is catalysed by another enzyme, namely a cAMP-dependent protein kinase, the process being under hormonal control.

In keeping with the complexity of its covalent and allosteric regulation, phosphorylase is a large enzyme, consisting of a dimer of two identical subunits and having the catalytic site buried near the centre of each subunit. Located near the catalytic site is a covalently bound molecule of PALPO – the coenzyme pyridoxal phosphate, derived from vitamin B_6, which probably participates as a general acid in the catalytic mechanism. The binding site for the allosteric effectors (such as AMP, ATP) is about 30 Å from the catalytic site, at one of the interfaces of the two subunits.

As final conclusions, the following aspects should be emphasised:

• cells regulate their metabolic activities by controlling rates of enzyme synthesis and degradation and by adjusting the activities of specific enzymes;

• enzyme activities vary in response to changes in pH, temperature, and the concentrations of substrates and products;

• enzyme activities can also be controlled by covalent modifications of the protein or by interactions with activators or inhibitors;

• the most common type of reversible covalent modification is represented by phosphorylation;

• allosteric effectors (which can act as either activators or inhibitors) bind to enzymes at sites distinct from the active site;

• allosteric regulation allows cells to adjust their enzyme activities rapidly and reversibly, in response to changes in the concentrations of substances that are structurally unrelated to the substrates or products;

• the allosteric enzymes usually have multiple subunits and their kinetics show a sigmoidal dependence on substrate concentration.

Modified enzymes and non-protein catalysts

Despite the variety of enzymatic functions available in nature, modern biotechnology continually faces needs either for substances with new catalytic abilities, or enzymes capable of functioning under unusual conditions, or displaying different specificities. We are addressing here the topic of enzyme design and engineering.

Essentially, three approaches exist: site-directed mutagenesis, hybrid enzymes and catalytic antibodies:

• site-directed mutagenesis is focused especially on developing tolerance to extreme environmental conditions (e.g. enzymes functioning at temperatures as high as 100°C);

• hybrid enzymes are fusioned proteins recombining catalytic and binding sites in novel ways;

• catalytic antibodies (also known as abzymes) are antibodies that can act like enzymes; they are attaining considerable importance, especially in synthetic organic chemistry and are characterized by a remarkable stereospecificity.

Until recently, it was assumed that all biochemical catalysis was carried out by proteins. Recent research has however revealed that some other molecules can also act as enzymes: these are RNA molecules, called ribozymes. Ribonuclease P, an enzyme that cleaves pre-tRNAs (yielding the active, functional RNAs) was known for some time to contain both a proteic portion and a specific RNA co-factor, and it was widely assumed that the active site resided on the protein portion. However, later studies revealed that while the protein alone was wholly inactive, in contrast, the RNA itself, under certain conditions, displayed catalytic abilities. These special conditions referred to either a sufficiently high concentration of magnesium, or a small amount of magnesium plus the presence of a small basic molecule, which was proven to be spermine. Under those circumstances, the RNA is capable of catalysing the specific cleavage of pre-tRNAs, acting like a true enzyme, being unchanged in the process and obeying Michaelis-Menten kinetics [8].

4.3 ENZYME SYSTEMS WITH SPECIFIC ROLES

4.3.1 Phase I enzyme systems

Oxidative enzyme systems

Cytochrome P450-dependent mixed-function oxidation reactions
Oxidation is probably the most common reaction in xenobiotic metabolism. This is catalysed by a group of membrane-bound mixed function oxidases (M.F.O.) located in the smooth endoplasmic reticulum of the liver and other extrahepatic tissues, and called the cytochrome P450 monooxygenase enzyme system or, microsomal hydroxylase [9-11]. The subcellular fraction containing the smooth endoplasmic reticulum is called the microsomal fraction. It should, however, be noted that these membrane-bound enzymes may in fact be present in all membranes and cells [12].
The overall catalytic reaction conforms to the following stoichiometry:

$$\mathrm{NAD(P)}\mathbf{H} + \mathbf{H}^+ + \mathrm{O}_2 + \mathrm{R}\mathbf{H} \xrightarrow[\text{CYT}]{} \mathrm{NAD(P)}^+ + \mathbf{H}_2\mathrm{O} + \mathrm{R}\mathbf{OH} \qquad 4.8.$$

where **RH** represents an oxidisable drug substance,
R**OH**, the corresponding hydroxylated metabolite
NAD(P)**H** + $\mathbf{H}^+$, reducing equivalents.

As evident from the above equation, during the reaction the reducing equivalents are consumed and one atom of molecular oxygen is incorporated into the substrate (as a hydroxyl group) whereas the other is reduced to water [11].

As regards cytochrome P450 and its multiple forms and catalytic cycle, some introductory details have already been given in Chapter 2.

Cytochrome P450 enzymes play critical roles in the biogenesis of sterols and other physiological intermediates, the catabolism of endogenous substrates (such as fatty acids, sterols and prostaglandins), and in exogenous metabolism by catalysing the biotransformation of a wide variety of xenobiotics, including drugs, carcinogens, insecticides, plant toxins, environmental pollutants (such as pesticides, herbicides and aliphatic and aromatic hydrocarbons), and many other foreign chemicals [12]. In practice, these enzymes catalyse the monooxidation of a wide variety of structurally unrelated compounds, whose only common feature appears to be a reasonably high degree of lipophilicity.

Understanding these processes is vital for predicting the reactivity of chemicals and reducing toxic side effects of drugs (see Chapter 8).

Cytochrome P450 is the terminal oxidase component of the electron transfer system present in the smooth endoplasmic reticulum and represents a superfamily of heme-thiolate proteins with molecular weights of approximately 50 000 Da. CYTP450 consists of at least two protein components: a heme protein called cytochrome P450 and a flavoprotein called NADPH-cytochrome P450 reductase, containing both FAD and FMN. A third component essential for electron transport is a lipid, phosphatidylcholine.

The haem-containing component, with iron protoporphyrin IX as the prosthetic group [13] (Figure 4.11) is the substrate and oxygen-binding site of the enzyme system, whereas the reductase serves as an electron carrier.

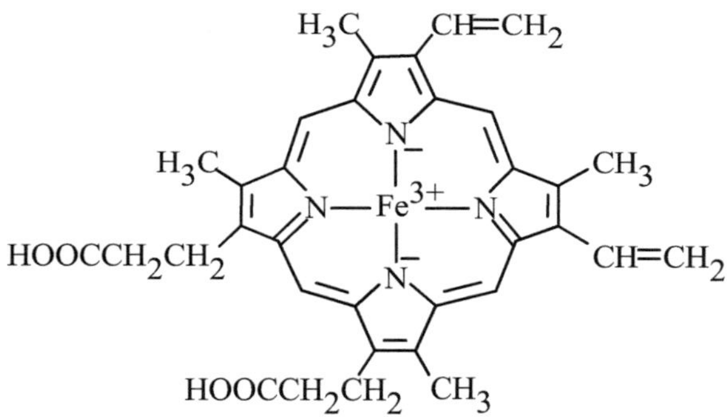

Fig.4.11 Structure of ferric protoporphyrin IX [11]

Crucial to the functioning of this unique superfamily of heme proteins is the coordination of the iron-protoporphyrin to the sulphur atom of the cysteine residue of the apoprotein [14] (Figure 4.12).

The ability of CYTP450 to form a biologically inactive ferrous carbonyl complex with carbon monoxide, with a major absorption band at 450 nm, led both to its discovery and its name.

As already known from the overall oxidation reaction, CYTP450 has an absolute requirement for NADPH and molecular oxygen, the rate at which various compounds are biotransformed being dependent on many factors including species, strain, sex, age, nutritional status, physiological or pathological condition, and so on.

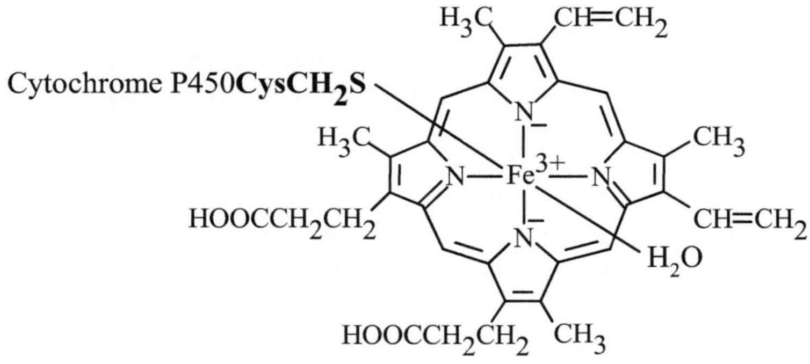

Fig.4.12 Ferric heme thiolate catalytic centre of cytochrome P450

The most important function of CYTP450 is its ability to 'activate' molecular oxygen [15], permitting the incorporation of an oxygen atom into a substrate molecule, with simultaneous reduction of the other oxygen atom, yielding a molecule of water (Figure 4.13):

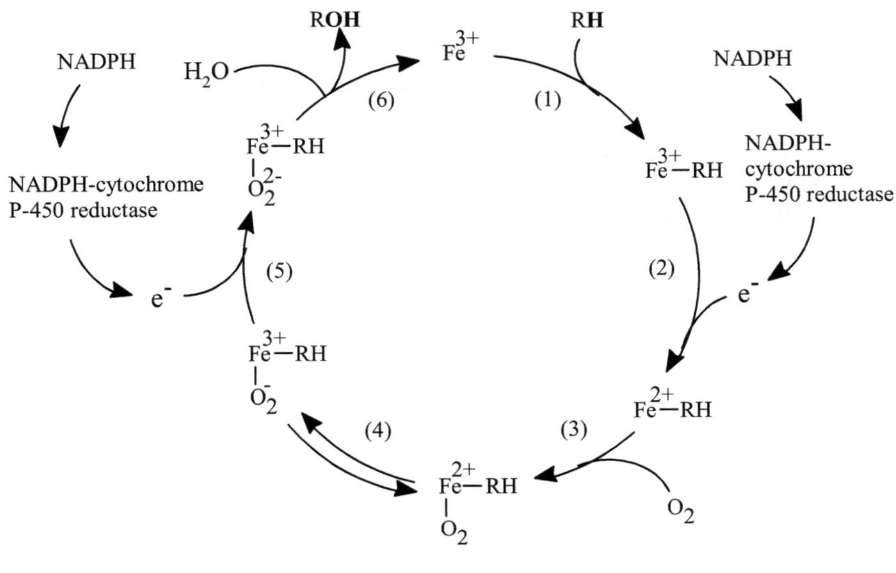

Fig.4.13 Catalytic cycle of CYTP450
(Fig. 2, p.325 from R.E. White and M.J. Coon, 1980. Ann. Rev. Biochem., 49: 315-356. Reprinted by permission of the journal)

From the outset, we should stress that all proteins that directly interact with molecular oxygen have a common characteristic at the most fundamental level, namely the ability to provide either low-energy d-orbitals (e.g. metal ions such as iron and copper) for stabilising unpaired electrons, or extensively delocalized molecular orbitals (e.g. organic co-factors such as flavin, pterin, or porphyrin).

The current view illustrating the cyclic pattern of the reduction and oxygenation of CYTP450 as it interacts with substrate molecules, electron donors, and oxygen (presented in the above figure) can be summarised as follows:

- the ferric cytochrome P450 binds reversibly to a molecule of substrate (RH), with consequent formation of a complex (Fe^{III}-RH) analogous to an enzyme-substrate complex. This binding of the substrate facilitates the first one-electron reduction step;
- the ferric complex formed undergoes reduction to a ferrous complex (Fe^{II}- RH), by an electron originating from NADPH and transferred by the NADPH-cytochrome P450 reductase;
- the reduced cytochrome P450 complex readily binds dioxygen, as the ferrous iron sixth ligand, yielding the oxycytochrome P450 complex (Fe^{II}-O_2-RH);
- further, the oxycytochrome P450 undergoes auto-oxidation, with the subsequent formation of a superoxide anion (Fe^{III}-O_2^-RH);
- by accepting a second electron from the flavoprotein, the ferric superoxide anion undergoes further reduction and forms the equivalent of a two-electron reduced complex, peroxycytochrome P450 (Fe^{III}-O_2^{2-}-RH);
- finally, the ferric peroxycytochrome P450 complex undergoes heterolytic cleavage of peroxide anion, yielding a water molecule and the hydroxylated metabolite [9-11,15-17].

For the first electron reduction, the following mechanism is proposed (Figure 4.14) [9-11]:

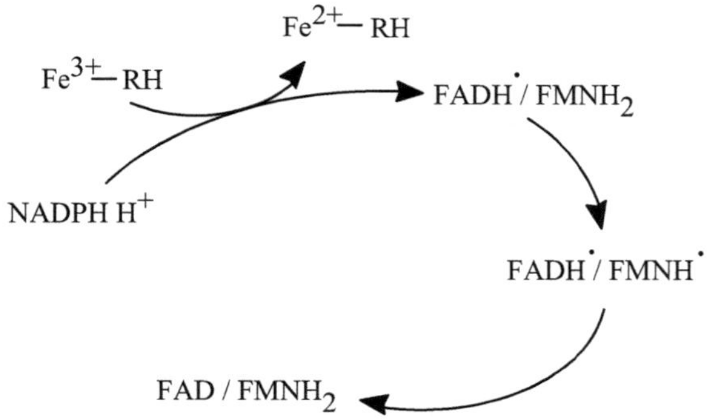

Fig.4.14 Role of flavins in the first electron reduction step

The role of the flavins in the second electron reduction of cytochrome P450 is suggested in Figure 4.15 [9-11]:

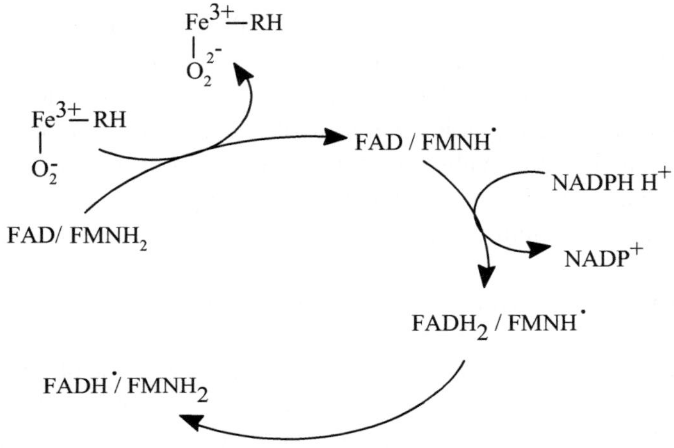

Fig.4.15 Second electron reduction involving binding of molecular oxygen

Cytochrome P450 multiple forms

More then 300 cytochrome P450 isoforms have been characterised to date, with respect to their sequences and corresponding encoding genes. The P450s are grouped into families and subfamilies, according to their probable structural and functional similarities. The family is denoted by a number and the subfamily by a letter. Generally, enzymes with 40% identity of the sequence are assigned to the same family and with more then 55% to the same subfamily [18]. Another number indicates the isoform's order in the subfamily. For example, P4501B5 indicates that this specific isoform is the fifth member of subfamily B of family 1.

The reader is alerted to the current use of an abbreviated nomenclature obtained by replacing cytochrome P450 with the three letters CYP.

The many years of human drug metabolism study have proven that most of the biotransformation reactions occurring are mediated primarily by enzymes of the CYP1, 2, 3 and 4 families, with CYP3A4 as the most abundant isoform (from the spectroscopically detectable CYTP450 in the liver [19], this isoform is assumed to represent about 30% of the total).

Nonetheless, it must be stressed that the relative importance of different isoforms in biotransformation reactions and their resulting products (more reactive, less reactive, with higher toxicity) is strongly dependent on the genetic idiosyncrasies of the individual, as well as on exposure to different environmental factors (including drugs, for example). Moreover, certain P450 isoforms are polymorphically distributed in the human population (details are presented in Chapter 7, Pharmacogenetics).

As mentioned earlier, CytP450 has a characteristic absorption spectrum at about 450 nm, determined by the presence of the ferrous-CO complex (the value being indicative of the thiolate-ligated haemoprotein); on denaturation, a shift of the absorption maximum to approximately 420 nm is observed, characteristic of the ferrous-CO complex in imidazole-ligated proteins (as in myoglobin) [20].

Selected examples of various cytochrome P450 families

Family 1

The CYP1A subfamily plays an integral role in the metabolism of two important classes of environmental carcinogens, polycyclic aromatic hydrocarbons (PAHs) and aryl amines. The PAHs are commonly present in the environment, either as a result of industrial combustion processes, or in tobacco smoke. Several potent carcinogenic aryl amines result from pyrolysis of amino acids in cooked meats. It is an inducible isoenzyme, present mainly in extrahepatic tissues where it is very well expressed [21].

CYP1A2 (also known as phenacetin O-deethylase, caffeine demethylase, or antipyrine N-demethylase) metabolises aryl amines, nitrosoamines, and aromatic hydrocarbons and is considered primarily responsible for the activation of the carcinogen aflatoxin B_1 under ordinary conditions in human exposure [10]. CYP1A2 is also an important determinant in the metabolism of certain drugs, being involved in their general metabolic disposition and possible drug-drug interactions [22,23]. An interesting study has been performed fairly recently on zolmitriptan (a highly selective 5-HT receptor agonist used in the acute oral treatment of migraine) as substrate for CYP1A2, concurrently administered with other drugs, with the aim of establishing the enzyme kinetics, the metabolism of zolmitriptan and possible drug-drug interactions. Investigations were made on rat hepatic microsomes induced with different inducers [24].

Family 2

The CYP2B subfamily is represented by several isoforms including CYP2B6, CYP2D6, CYP2C8, CYP2C9, CYP2C10, CYP2C18 and CYP2C19, metabolising a range of drugs and other compounds [25].

CYP2B6 has been intensively investigated with respect to its extensive genetic polymorphism [26], its role in the biotransformation of certain drugs [26-29], possible inhibitors [30,31] and inducers [32] and their mechanism of action.

Another isoform of particular importance, because it metabolises a wide range of commonly prescribed drugs, is CYP2D6.

Numerous compounds including psychotropic, cardiovascular and many miscellaneous agents, have been demonstrated to undergo CYP2D6-mediated biotransformations, although this may not be the only or the main pathway of their oxidative metabolism [25].

Possible implications of certain family 2 cytochromes in chemical toxicity and oxidative stress have recently been investigated, particularly 2E1, an alcohol-inducible enzyme [33].

Family 3

The CYP3A subfamily includes the most abundantly expressed CYTP450s in humans, about two thirds of the CYP450 in the liver belonging to the CYP3A subfamily [25], but with only one having been recently characterized, namely CYP3A4 [34]. The CYP3A subfamily metabolises a range of clinically important drugs, including nifedipine, cyclosporine, erythromycin, midazolam, diazepam, dextromethorphan, lidocaine, diltiazem, tamoxifen, verapamil, cocaine, dapsone, terfenadine, imipramine, rifampicin, valproic acid, carbamazepine and theophylline.

CYP3A5 has been detected to a greater extent in adolescents than in adults and does not appear to be inducible. In contrast, it is polymorphically expressed. CYP3A4 is glucocorticoid-inducible and CYP3A7, present only in foetal livers, is involved in the hydroxylations of allylic and benzylic carbon atoms [25].

In addition to oxidative reactions, which are quite numerous (see the review in Chapter 2), cytochrome P450 is known to catalyse reductive reactions as well [35]. These reactions are usually most important under anaerobic conditions, but can, in some instances, compete with oxidative reactions under aerobic conditions.

The principal reductive reaction specifically catalysed by CYTP450 is the dehalogenation of alkyl halides, but cytochrome P450, or at least, cytochrome P450 reductase, has been shown to participate in other reactions, such as reduction of azo- and nitro- compounds. In these types of reactions, the important catalytic species is the ferrous deoxy intermediate, in which the iron is not coordinated to oxygen [25].

Microsomal flavin-containing monooxygenase

The microsomal flavin-containing monooxygenase (F.M.O.) system is the second most important monooxygenase system in xenobiotic metabolism.

It is also known as 'Ziegler's' enzyme in the older literature. These enzymes belong to the large class of flavoproteins, polymeric proteins exhibiting an apparent molecular weight in the range 52 000 to 65 000 Da, containing a single molecule of FAD as the prosthetic group and being at the same time NADPH- and oxygen-dependent [36].

The F.M.O. system catalyses the oxygenation of many nucleophilic organic nitrogen and sulphur compounds (including many drugs, such as phenothiazines, ephedrine, N-methylamphetamine and norcocaine) and uses as a source of reducing equivalents either NADH or NADPH (although K_m for NADH is about ten times higher than that for NADPH; the higher the value, the smaller the affinity for substrate).

The proposed mechanism for the F.M.O. system is presented in Figure 4.16:

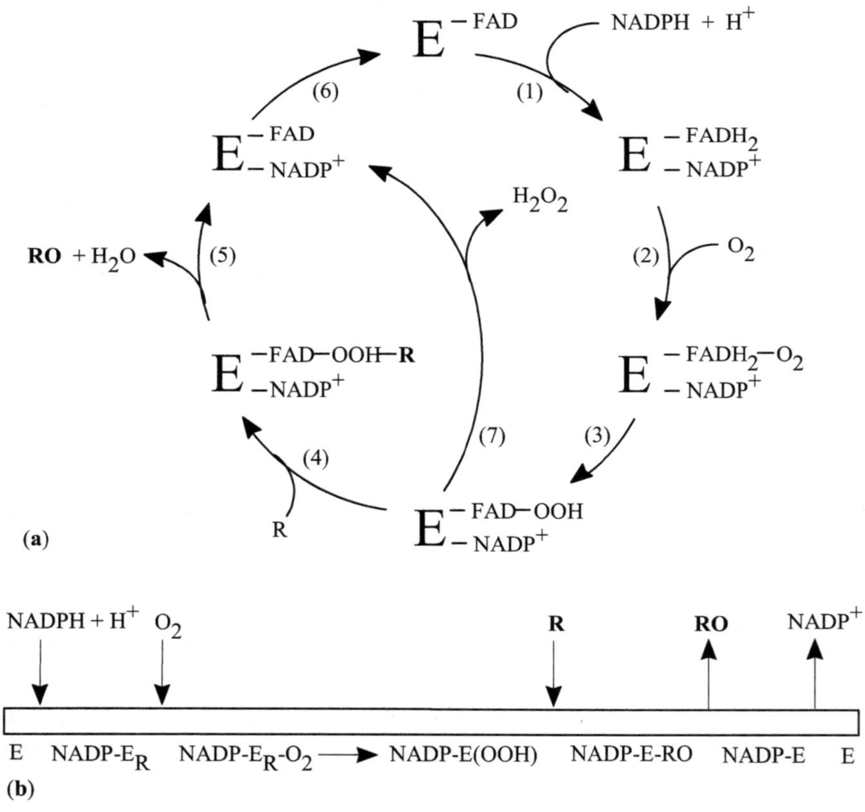

Fig.4.16 Proposed mechanism for the microsomal flavin-containing monooxygenase (Reproduced with the permission of Nelson Thornes from 'Introduction to Drug Metabolism', 2001, 3rd Ed., isbn 0 7487 6011 3 - Gibson & Skett - first published in 1986)

where, in part (a)
E-FAD, represents the oxidized enzyme,
E-$FADH_2$, the reduced enzyme,
R, the oxidisable substrate,
RO, the monooxygenated product,

and in part (b),
E, represents the oxidised enzyme and
E_R, the reduced enzyme.

The reaction mechanism, as presented in the above figure, involves flavin reduction, followed by oxygen binding and, by an internal electron transfer to oxygen, the formation of a peroxy-flavin complex. Substrate nucleophiles attack the distal oxygen of this hydroperoxide, with resultant oxygen transfer to the substrate. Finally, the enzyme complex dissociates, yielding the oxidised enzyme. As shown in the figure, in the absence of an oxidisable substrate, the peroxy-flavin intermediate may slowly decompose, yielding H_2O_2 (step 7). The reaction sequence is summarised in part (b) of the figure [9,11,36].

Extensive experimental studies focused on structure-metabolism relationships, and established that the best substrates were the lipophilic amines [37]. Unlike the CYTP450s, the F.M.O. system is not induced by exogenously administered xenobiotics, being however under dietary and hormonal control [25].

Prostaglandin endoperoxide synthase and prostaglandin-dependent co-oxidation of drugs

Prostaglandin endoperoxide synthetase (PES) is an enzyme present in almost all mammalian cell types and catalyses the oxidation of arachidonic acid to prostaglandin H_2, the precursor of other important prostaglandins, thromboxanes and prostacyclins. The enzyme, involved primarily in endogenous metabolism has two distinct catalytic functions [38], namely fatty acid cyclooxygenase activity, forming prostaglandin G_2, and hydroperoxydase activity, reducing the resulting prostaglandin to prostaglandin H_2 (Figure 4.17).

The enzyme has been proven to exist as a dimer with the amino acids sequence established [39], and also to display enantioselectivity [40].

An important aspect to mention, and one that is described in Figure 4.18, is that several drugs display the ability to undergo co-oxidation. Among them are aminopyrine, benzphetamine, oxyphenbutazone as well as chemical carcinogens, including benzidine, benzo[a]pyrene and derivatives.

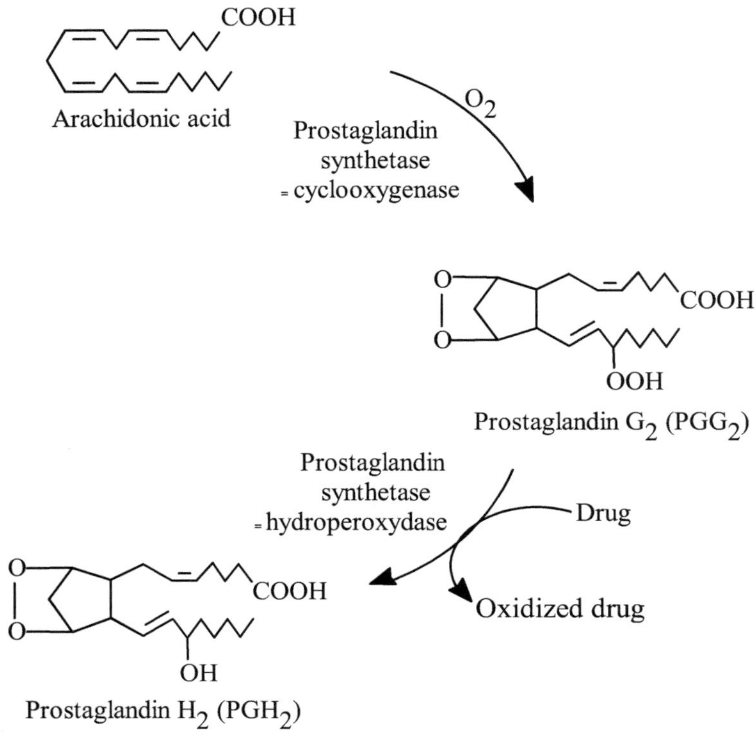

Fig.4.17 Co-oxidation of drugs by the prostaglandin synthetase system (role in drug oxidations) (Reprinted from Biochemical Pharmacology, Vol. 31, P Moldeus et al. "Prostaglandin synthetase catalyzed activation of paracetamol", p.1367, 1982, with permission from Elsevier)

In the case of certain drugs (e.g. paracetamol [11]), the biotransformation process involves a radical-mediated mechanism, resulting in glutathione conjugation of the drug, or reduction back to acetaminophen [41] (Figure 4.18).

The mechanism of acetaminophen oxidation has been the subject of numerous experimental studies [42].

In the first step (i), the reaction most probably involves a one-electron oxidation, resulting in hydrogen abstraction, to yield the phenoxy radical of paracetamol.

The product phenoxy radical may undergo two pathways of further biotransformation: it can either tautomerise, or be reduced with glutathione. In the first case, a carbon-centred quinone radical is formed, which can further react with cellular glutathione, forming the corresponding conjugate of paracetamol (ii).

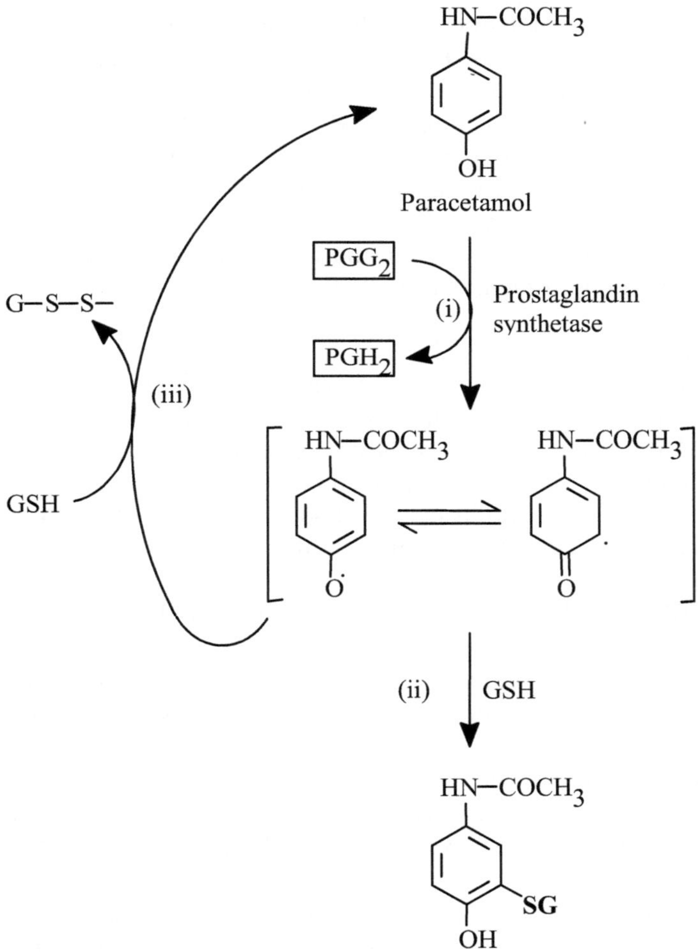

Fig.4.18 Postulated mechanism for the prostaglandin synthetase mediated metabolism of paracetamol (Reprinted from Biochemical Pharmacology, Vol. 31, P Moldeus et al. "Prostaglandin synthetase catalyzed activation of paracetamol", p.1367, 1982, with permission from Elsevier)

Alternatively, the phenoxy radical may be directly reduced with glutathione, reforming paracetamol (iii).

A very interesting aspect that merits highlighting is that acetaminophen oxidation displays a marked isoenzyme selectivity, with the most selective being the CYP1A1 isoform, which binds acetaminophen so that its phenolic group comes into close proximity of the central iron ion [43].

As regards the phenomenon of co-oxidation we should mention that quite a variety of xenobiotics have been demonstrated to act as cofactors for the enzymatic reduction of PGG_2 to PGH_2, therefore being called reducing substrates [44]. In addition, an increasing number of drugs, differing significantly in their structures, have been reported to be co-oxidised by PES, although the resulting metabolites are not known in every case [45].

Therefore, the prostaglandin synthetase-dependent co-oxidation of certain drugs could very well be assumed to play a major role in drug biotransformations, particularly in those tissues low in F.M.O. activity, and naturally, rich in prostaglandin synthetase.

Monoamine oxidase

This is a FAD-containing enzyme widely distributed in most tissues of mammals [46]. MAO is a membrane-bound enzyme, present in two different forms, MAO-A and MAO-B, as protein sequencing, cloning and sequencing cDNA coding for humans have proven [47].

Its most common physiological substrates are primary amines, which are oxidatively deaminated, as follows [48]:

$$RCH_2NR'NR'' + O_2 + H_2O \rightarrow RCHO + NHR'R'' + H_2O_2 \qquad 4.9$$

The detailed, intimate mechanism is partly understood based on studies performed with both substrates and mechanism-based inactivators [49].

MAO is an enzyme of particular medical interest: on the one hand, it represents a target for selective, reversible inhibitors used therapeutically [50] and on the other, displays a considerable capacity to activate exogenous neurotoxins [51].

As an example, we mention the activation of MPTP (a tetrahydropyridine) to a neurotoxin causing Parkinsonism in monkeys and humans. Apparently, the activation of MPTP is mainly due to MAO-B, and follows several steps, with the final formation of a particularly reactive radical intermediate and of $MPDP^+$; finally, $MPDP^+$ is further oxidised (by unidentified membrane-bound enzymes) to MPP^+ (N-methyl-4-phenylpyridinium), which represents the ultimate neurotoxin causing cell death [52].

Xanthine dehydrogenase-xanthine oxidase (XD-XO)

These are two forms of the same homodimeric, cytosolic enzyme, with relatively high levels being localised in the liver and small intestine, tissues known to be implicated in the first-pass metabolism of a variety of agents. Each subunit of XD/XO contains one atom of molybdenum, in the form of a molybdopterin cofactor $[Mo^{VI}(=S)(=O)]^{+2}$, one FAD molecule, and two Fe_2-S_2 centers [53].

The general reaction catalysed by these enzymes can be represented by the following equation:

$$\mathbf{SH} + H_2O \rightarrow \mathbf{SOH} + 2e^- + 2H^+ \qquad 4.10$$

where SH represents the reduced substrate and SOH is the hydroxylated metabolite.

The proposed mechanism of action involves an oxygen insertion step; most probably, addition of the substrate across the Mo==S double bond takes place with simultaneous incorporation of a hydroxide, yielding a three-centre Mo-C-O bond. Finally, electron transfer and rearrangement would then yield the hydroxylated metabolite, and regenerate the molybdopterin cofactor [54].

The molybdenum hydroxylases generally catalyse oxidation of electron-deficient sp^2-hybridized carbon atoms, most commonly found in aromatic heterocycles, aromatic or non-aromatic charged azaheterocycles and aldehydes [55]. In addition, XO plays a role in the oxidation of several chemotherapeutic agents [56] and purine derivatives (6-mercaptopurine, 2, 6-dithiopurine, and 2'-fluoroarabinosyldeoxypurine) [57].

Unfortunately, xanthine oxidase is also implicated in several toxic responses, the most important being the generation of reactive oxygen species, which can cause lipid peroxidation [58].

Aldehyde oxidase (AO)

AO is also a cytosolic molybdozyme, existing only in the oxidised form, displaying a mechanism of action very similar to that described for XO, and fulfilling roles complementary to those of the monooxygenases in the biotransformation of both endogenous and exogenous compounds [59].

Common substrates are represented by nitrogen-containing heterocylic compounds, including several therapeutic agents such as tamoxifen-4-aldehyde [60], pyrimidone [61], and sulindac [62].

Epoxide hydrolase

This is a widely distributed enzyme, that in addition to metabolising epoxides of drugs and xenobiotics, also catalyses the hydration of endogenous epoxides, which suggests a substantial role for this enzyme in endogenous metabolic reactions [63].

In general, epoxide hydrolase is of minor importance in normal drug metabolism, but is significant in the formation of chemical carcinogens from otherwise innocuous xenobiotics and in the metabolism of toxic intermediates formed from certain drugs by cytochrome P450-mediated reactions [64].

Four different isoforms of the enzyme have been demonstrated in humans, two of them displaying specific metabolic roles, and the other two, hydrolysing a range of alkene and arene oxides [65].

Esterases

These enzymes represent a multigene family, involved in the hydrolysis of carboxylesters, carboxyamides, and carboxythioesters (see the review in Chapter 3). Several chemicals have been shown to be detoxified by liver carboxylesterase, including insecticides, herbicides, and drugs in several classes (anaesthetics, analgesics, and antibiotics).

Polymorphism has been detected for cholinesterase, which is important in the hydrolysis of the muscle relaxant succinylcholine and, possibly, diacetylmorphine. From several different allelic variants, most significant to mention are the so-called atypical enzyme, found in 2% of the population and showing defective binding of anionic substrates (such as succinylcholine), and the less common 'silent' variant, for which no enzyme is produced [66].

Dehydrogenases

The most representative is alcohol dehydrogenase, a cytoplasmic NAD^+ dependent zinc metalloenzyme that catalyses the oxidation of an alcohol to an aldehyde; NAD^+ is simultaneously reduced to the corresponding NADH.

It is assumed that the human ADH family consists of seven genes, encoding proteins belonging to one of five classes of ADH isoenzymes based on structural and kinetic features [67]. Although of importance in determining susceptibility to alcoholism and alcohol liver disease, they are not of great importance in the metabolism of commonly prescribed drugs.

4.3.2. Phase II enzymes

UDP-glucuronosyltransferases

The enzyme UDP-glucuronosyltransferase is found in almost all mammalian species, present in many tissues, mostly in the liver, but also in kidney, small intestine, lung, skin, adrenals and spleen. It is mainly localised in the membrane of hepatic endoplasmic reticulum fractions, and therefore ideally positioned to glucuronidate the products of the mixed function oxidase reactions.

This enzyme has no prosthetic group, and its catalytic activity is substantially influenced by the presence of lipids [68].

As already discussed in Chapter 3, this family of enzymes catalyses the transfer of glucuronic acid to a multitude of endobiotic and xenobiotic compounds, including drugs, pesticides, and carcinogens.

Indeed, some key UGTs have evolved to prevent accumulation of potentially toxic endogenous compounds, such as bilirubin (the end product of heme catabolism, excreted from the body as biliary mono- and diglucuronides), bile acids and steroid hormones.

Other UGTs are concerned with maintenance of physiological levels of certain hormones, such as thyroxine and tri-iodothyronine, which are also excreted as glucuronic conjugates in the liver and bile [69].

UGT isoforms in humans have also been reported [70], but the importance of pharmacogenetic variation in the UDP-glucuronosyl transferases is still unclear.

Of interest and relevance in relation to drug metabolism, we should mention an inborn error of metabolism, termed Gilbert's syndrome, which is characterized by mild hyperbilirubinemia affecting an average of 10% of the population. In addition to this impaired bilirubin metabolism, decreased clearance of several drugs (e.g. tolbutamide, acetaminophen, rifampicin) has been reported in patients with this syndrome [71].

Sulphotransferases

Sulphotransferase enzymes conjugate exogenous and endogenous compounds (including neurotransmitters with sulphate derived from PAPS – 3'-phosphoadenosine-5'-phosphosulphate) and play important roles in the metabolism of a range of drugs (phenols, alcohols and hydroxylamines), forming the readily excretable sulphate esters.

The sulphotransferase enzymes are soluble enzymes found in many tissues including liver, kidney, gut and platelets, and apparently they do exist in multiple enzyme forms, with the steroid sulphating enzymes being distinct from the sulphotransferases responsible for drug conjugation reactions [72].

Glutathione-S-transferase

The glutathione-S-transferase family of enzymes comprises soluble proteins predominantly found in hepatocytes that play important roles in the conjugation of a variety of hydrophobic and electrophilic compounds. The latter include epoxides, haloalkanes, nitroalkanes, alkenes, and aromatic halo- and nitro- compounds.

It is, generally, a detoxifying metabolic pathway, with most glutathione conjugates undergoing further metabolism to mercapturic acids before excretion [73].

In addition to their ability to catalyse conjugation reactions, certain glutathione-S-transferases have the ability to bind a variety of endogenous and exogenous substrates without effecting biotransformation. Examples include bilirubin, oestradiol, testosterone, tetracycline and penicillin.

The glutathione-S-transferase enzymes exist in multiple forms (at least 20 isoenzymes) as homo- or heterodimers of two subunits and they are inducible by various xenobiotics, including phenobarbital and polycyclic aromatic hydrocarbons. However, in the case of heterodimers, each subunit may be differently and independently regulated, especially by transcriptional gene activation systems.

4.4 FINAL REMARKS

In describing above the role and nature of enzymes in the most prominent classes, as well as some aspects of their action at the molecular level, the authors' intention has in part been to prepare the way for an appreciation of two extremely important phenomena that govern enzyme activity and that have crucial implications for the pharmacological effects of drugs. These are the occurrence of adverse reactions and drug-drug interactions, both of which can be understood on the basis of enzyme induction and enzyme inhibition, the topics of the next chapter.

References

1. Price NC, Stevens L. 2000. Fundamentals of Enzymology. The Cell and Molecular Biology of Catalytic Proteins, 3rd ed., Oxford Univ Press, pp 118-266; 370-399.
2. Cornish-Bowden AC. 1995. Fundamentals of enzyme kinetics. London: Portland Press, pp 46-111.
3. Metais P, Agneray J, Ferard G, Fruchard J-C, Jardillier J-C, Revol A, Siest G, Stahl A. 1990. Enzymes. In: Biochimie clinique. 1. Biochimie analitique 2e ed. Paris Cedex 06: Simep, pp 144-163.
4. Hennen G. 1995. Cinetique enzimatique. In: Biochimie, 1ercycle. Paris: Dunod, pp 166-176.
5. Zubay GF, Parson WW, Vance DE. 1994. Enzyme Kinetics and How Enzymes Work. In: Sievers EM, editor. Dubuque, Iowa: Wm C Brown Publishers, pp 135-153; 154-174.
6. Matthews CK, van Holde KE, Ahern KG. 1999. Enzymes: biological Catalysts. In: Roberts B, Weber L, Marsch J, editors. Biochemistry, 3rded. San Francisco: An Imprint of Addison Wesley Longman Inc., pp 360-408.
7. Hammes, Gordon G. 2002. Multiple conformational changes in enzyme catalysis. Biochemistry 41:8221-8228.

8. Zubay GF, Parson WW, Vance DE. 1994. How Enzymes Work. In: Sievers EM, editor. Dubuque, Iowa: Wm C Brown Publishers, pp 165-169.

9. Testa B. 1995. The Nature and Functioning of Cytochromes P450 and Flavin-containing monooxygenases. In: Testa B, Caldwell J, editors. The Metabolism of Drugs and Other Xenobiotics: Biochemistry of Redox Reactions. London: Academic Press (Harcourt Brace and Company, Publishers), pp 70- 121.

10. Ortiz de Montellano PR. 1999. The Cytochrome P450 Oxidative System. In: Woolf TF, editor. Handbook of Drug Metabolism. New York: Marcel Dekker Inc., pp 109-130.

11. Gibson GG, Skett P. 1994. Cytochrome P450-dependent mixed function oxidation reactions. In: Introduction to Drug Metabolism, 2nd ed. London: Blackie Academic & Professional, An Imprint of Chapman & Hall, pp 37-49.

12. Coon MJ, Person AV. 1980. Microsomal cytochrome P450: a central catalyst in detoxication reactions. In: Jacobi WB, editor. Enzymatic Basis of detoxication. New York: Academic Press, pp 117-134.

13. Caughey WS, Ibers JA. 1977. Crystal and molecular structure of free base porphyrin, protoporphyrin IX dimethyl ester. J Am Chem Soc 99:6639-6645.

14. Black SD, Coon MJ. 1986. Studies on the identity of the heme-binding cysteinyl residue in rabbit liver microsomal cytochrome P-450 isozyme 2. Biochem Biophys Res Co 128:82-89.

15. Castro CE. 1980. Mechanisms of reaction of hemeproteins with oxygen and hydrogen peroxide in the oxidation of organic substrates. Pharmacol Therapeut 10:171-189.

16. Gerber NC, Sligar SG. 1992. Catalytic mechanism of cytochrome P-450: evidence for a distal charge relay. J Am Chem Soc 114:8743-8743.

17. Ortiz de Montellano PR. 1986. Oxygen activation and transfer. In: Ortiz de Montellano PR, editor. Cytocrome P450: Structure, Mechanism, and Biochemistry, 2nd ed. New York: Plenum, pp 217-271.

18. Nelson D. Koymans L, Kamataki T, Stegeman G, Feyereisen R, Waxman D, Watterman M, Gotoh O, Coon M, Estabrook R, Gunsalus R, Nebert D. 1996. P450 superfamily: Update to new sequences, gene mapping, accession numbers and nomenclature. Pharmacogenetics 6:1-9.

19. Shimada T, Yamazaki H, Mimura M, Inui Y, Guengerich FP. 1994. Interindividual variations in human liver cytochrome P450 enzymes involved in the oxidation of drugs, carcinogens, and toxic chemicals: Studies with liver microsomes of 30 Japanese and 30 caucasians. J Pharmacol Exp Ther 270:414-421.

20. Correia MA. 1995. In: Ortiz de Montellenano PR, editor. Cytocrome P450: Structure, Mechanism, and Biochemistry, 2nd ed. New York: Plenum, pp 607-630.

21. Guengerich FP. 1995. Human cytochrome P450 enzymes. In: Ortiz de Montellenano PR, editor. Cytocrome P450: Structure, Mechanism, and Biochemistry, 2nd ed. New York: Plenum, pp 473-575.

22. Orlando R, Piccolli P, De Martin S, Padrini R, Floreani M, Palatini P. 2004. Cytocrome P450 1A2 is a major determinant of lidocaine metabolism in vivo: effects of liver function. Clin Pharmacol Ther 75:80-88.

23. Peterson TC, Peterson MR, Wornell PA, Blanchard MG, Gonzales FJ. 2004. Role of CYP1A2 and CYP2E1 in the pentoxifylline ciprofloxacin drug interaction. Biochem Pharmacol 68:395-402.

24. Yu L-S, Yao T-W, Yeng Su. 2003. In vitro metabolism of zolmitriptan in rat cytocromes induced with β-naphthoflavone and the interaction between six drugs and zolmitriptan. Chem-Biol Interact 146:263-272.

25. Daly AK. Pharmacogenetics. 1999. In: Woolf TF, editor. Handbook of Drug Metabolism. New York: Marcel Dekker Inc., pp 175-202.

26. Lang T, Klein K, Fischer J, Nussler AK, Neuhaus P, Hofmann U, Eichelbaum M, Schwab M, Zanger UM. 2001. Extensive genetic polymorphism in the human CYP2B6 gene with impact on expression and function in human liver. Pharmacogenetics 11:399-405.

27. Xie H-J, Yasar U, Lundgren S, Griskevicius L, Terelius Y, Hassan M, Rane A. 2003. Role of polymorphic CYP2B6 in cyclophosphamide bioactivation. Pharmacogenomics J 3:53-61.

28. Hesse LM, Venkatakrishnan K, Court MH, von Moltke LL, Duan SX, Shader RI, Greenblatt DJ. 2000. CYP2B6 mediates the in vitro hydroxylation of bupropion: a potential drug interactions with other antidepressants. Drug Metab Dispos 28:1176-1183.

29. Ko JW, Desta Z, Flockart DA. 1998. Human *N*-demethylation of (S)-mephenytoin by cytochromes P450s 2C9 and 2B6. Drug Metab Dispos 26:775-778.

30. Richter T, Schwab M, Eichelbaum M, Zanger UM. 2005. Inhibition of human CYP2B6 by *N,N',N''*,-triethylenethiophosphoramide is irreversible and mechanism-based. Biochem Pharmacol 69:517-524.

31. Richter T, Murdter TE, Heinkele G, Pleiss J, Tatzel S, Schwab M, Eichelbaum M, Zanger UM. 2004. Potent mechanism-based inhibition of human CYP2B6 by clopidrogel and ticlopidine. J Pharmacol Exp Ther 308:189-197.

32. Martin H, Sarsat JP, de Waziers I, Housset C, Balladur P, Beaune P, Albaladejo V. 2003. Induction of cytochromes P4502B6 and 3A4 expression by Phenobarbital and cyclophosphamide in cultured human liver slices. Pharm Res 20:557-568.

33. Gonzales FJ. 2005. Role of cytochromes P450 in chemical toxicity and oxidative stress: studies with CYP2E1. Mutat Res 569:101-110.

34. Scott EE, Halpert JR. 2005. Structures of cytochrome P450 3A4. Trends Biochem Sci 30:5-7.

35. Goeptar AR, Scheerens H, Vermeulen NPE. 1995. Oxygen and xenobiotic reductase activities of cytocrome P450. Crit Rev Toxicol 25:25-31.

36. Ziegler DM. 1988. Flavin-containing monooxygenases: catalytic mechanism and substrate specificities. Drug Metab Rev 19:1-32.

37. Rose J, Castagnoli N Jr. 1983. The metabolism of tertiary amines. Med Res Rev 3:73-88.

38. Eling TE, Thompson DC, Foureman GL, Courtis JF, Hughes MF. 1990. Prostaglandin H synthase and xenobiotic oxidation. Ann Rev Pharmacol Toxicol 30:1-45.

39. DeWitt DL, El-Harith EA, Kraemer SA, Andrews MJ, Yao EF, Armstrong RL, Smith WL. 1990. The aspirin and heme-binding sites of ovine and murine prostaglandin endoperoxide synthases. J Biol Chem 265:5192-5198.

40. Cashman JR, Olsen LD, Bornhaim LM. 1990. Enantioselective S-oxygenation by flavin-containing and cytochrome P-450 monooxygenases. Chem Res Toxicol 3:344-349.

41. Ben-Zvi Z, Weissman-Teitellman B, Katz S, Danon A. 1990. Acetaminophen hepatotoxicity: is there a role for prostaglandin synthesis? Arch Toxicol 64:299-304.

42. Vermeulen NPE, Bessems JGM, Van der Straat R. 1992. Molecular aspects of paracetamol-induces hepatotoxicity and its mechanism-based prevention. Drug Metab Rev 24:367-407.

43. Patten CJ, Thomas PE, Guy RL, Lee M, Gonzales FJ, Guengerich FP, Yang CS. 1993. Cytochrome P450 enzymes involved in acetaminophen activation by rat and human liver microsomes and their kinetics. Chem Res Toxicol 6:511-518.

44. Eling TE, Curtis JF. 1992. Xenobiotic metabolism by prostaglandin synthase. Pharmacol Therap 53:261-273.

45. Marnet LJ. 1983. Cooxidation during prostaglandin biosynthesis: A pathway for the metabolic activation of xenobiotics. In: Hodgson E, Bend JR, Philpot RM, editors. Reviews in Biochemical Toxicology (vol.5). New York: Elsevier Biomedical, pp 135-172.

46. Singer TP. 1991. Monoamine oxidases. In: Mueller F, editor. Chemistry and Biochemistry of Flavoenzymes (vol.2). CRC, Boca Raton, FL, pp 437-470.

47. Wu HF, Chen K, Shih JC. 1993. Site-directed mutagenesis of monoamine oxidase A and B: role of cysteines. Mol Pharmacol 43:888-893.

48. Tripton KF, O'Carroll AM, Mc Crodden JM. 1987. The catalytic behaviour of monoaminoxidase. J Neural Transm 23:25-35.

49. Kyburtz E. 1990. New developments in the field of MAO inhibitors. Drug News Perspect 3:592-599.

50. Youdim MBH, Finberg JPM. 1991. New directions in monoamine oxidase A and B selective inhibitors and substrates. Biochem Pharmacol 41:155-162.

51. Maret G, Testa B, Jenner P, El Tayar N, Carrupt PA. 1990. The MPTP story: MAO activates tetrahydropyridine derivatives to toxins causing parkinsonism. Drug Metab Rev 22:291-332.

52. Chacon JN, Chedekel MR, Land EJ, Truscott TG. 1987. Chemically induced Parkinson's disease: intermediates in the oxidation of 1-methyl-4-phenyl-1,2,3,6-tetrahydropyridine to the 1-methyl-4-phenyl-pyridinium ion. Biochem Biophys Res Co 144:957-964.

53. Beedham C. 1985. Molybdenum hydroxylases as drug-metabolizing enzymes. Drug Metab Rev 16:119-156.

54. Oertling AM, Hille R. 1990. Resonance-enhanced Raman scattering from the molybdenum center of xanthine oxidase. J Biol Chem 265:17446-17450.

55. Krenitsky TA, Neil SM, Elion GB, Hitchings GC. 1972. A comparison of specificities of xanthine oxidase and aldehyde oxidase. Arch Biochem Biophys 150:585-599.

56. Pristos CA, Gustafson DL. 1994. Xanthine dehydrogenase and its role in cancer chemotherapy. Oncol Res 6:477-82.

57. Qing WG, Powell KL, Stoica G, Szumlansky CL, Weinshilboum RM, Macleod MC. 1995. Toxicity and metabolism of 2,6-dithiopurine, a potential chemoprotective agent. Drug Metab Dispos 23:854-860.

58. Nishino T. 1997. The conversion of xanthine dehydrogenase to xanthine oxidase and the role of the enzyme in reperfusion injury, Biochem Soc Trans 25: 783-786.

59. Krenitsky TA. 1978. Aldehyde oxidase and xanthine oxidase – functional and evolutionary relationships. Biochem Pharmacol 27: 2763-2764.

60. Ruenitz PC, Bai X. 1995. Acidic metabolites of tamoxifen: Aspects of formation and fate in the female rat. Drug Metab Dispos 23:993-998.

61. Guo W, Tung-Lerner M, Chen H, Chang C. Zhu J, Pizzorno G., Lin T, Cheng Y. 1995. 5-Fluoro-2-pyrimidone, a liver aldehyde oxidase-activated prodrug of 5-fluorouracil. Biochem Pharmacol 49:1111-1117.

62. Lee S, Renwick AG. 1995. Sulphoxide reduction by rat and rabbit tissues in vitro. Biochem Pharmacol 49:1557-1565.

63. Gibson GG, Skett P. 1994. Epoxide hydrolase. In: Introduction to Drug Metabolism, 2nd ed. London: Blackie Academic & Professional, An Imprint of Chapman &Hall, pp 57-59.

64. Guenthner TM. 1990. Epoxide hydrolases. In: Mulder GJ, editor. Conjugation Reactions in Drug Metabolism: An integrated approach. London: Taylor & Francis, pp 365-367.

65. Oesch F, Timms CW, Walker CH, Guenthner TM, Sparrow A, Watabe T, Wolf CR. 1984. Existence of multiple forms of microsomal epoxide hydrolases with radically different substrate specificities. Carcinogenesis 5:7-9.

66. Lockridge O. 1990. Genetic variants of human serum cholinesterase influence metabolism of the muscle relaxant succinylcholine. Pharmacol Ther 47: 35-39.

67. Lohr JW, Willsky GR, Acara MA. 1989. Renal Drug Metabolism. Pharm Rev 50:107-141.

68. Gibson GG, Skett P. 1994. Epoxide hydrolase. In: Introduction to Drug Metabolism, 2nd ed. London: Blackie Academic & Professional, An Imprint of Chapman & Hall, pp 61-63.

69. Flock EV, Bollman JL, Owens CA, Zollman PE. 1965. Conjugation of thyroid hormones and analogues by the Gunn rat. Endocrinology 77:303-308.

70. Burchell B, Coughtrie MWH. 1989. UDP-glucuronyltransferases: genetic factors influencing the metabolism of foreign compounds. Pharmacol Ther 43:261-289.

71. de Morais SM, Uetrecht JP, Wells PG. 1992. Decreased glucuronidation and increased activation of acetaminophen in Gilbert's syndrome. Gastroenterology 102:577-581.

72. Weinshilboum R. 1990. Sulphotransferase pharmacogenetics. Pharmacol Ther 45:93-102.

73. Mannervick B, Danielson UH. 1988. Glutathione transferases – structure and catalytic activity. Crit Rev Biochem 23:281-288.

Chapter 5

INDUCTION AND INHIBITION OF DRUG-METABOLISING ENZYMES

5.1 INTRODUCTION

The state of enzymatic systems involved in drug biotransformation represents an important factor in pharmacokinetic and/or pharmacodynamic variability. The changes in the state of enzymatic systems may be qualitative and quantitative.

Qualitative changes are commonly due to impairments in the state of the enzymatic systems, genetically pre-determined (enzymopathies).

Quantitative changes may evolve in two directions: either towards the stimulation of enzyme activity (enzyme induction) or the reverse, a reduction in enzyme activity (enzyme inhibition).

The pharmacological consequences of both phenomena are quantitative and refer to the modification of intensity and/or duration of the pharmacological effect of the drug, the modification of $t_{1/2}$ and biotransformation rate, the appearance of adverse reactions of overdosing, and therapeutic inefficacy.

In this chapter, both enzyme induction and enzyme inhibition are examined closely, with an emphasis on the cytochrome P450 system. Many examples are quoted to illustrate these effects and their general impact. An extensive discussion of the role of other factors affecting drug biotransformation follows. Recognizing that there is much interaction between them, these factors are systematically treated under the categories of internal and external factors. The present chapter deals with some of the internal factors that have direct implications for the cytochrome P450 levels/activity, namely the dietary factors, comprising macro- and micro-nutrients, as well as tobacco smoking (considered also as adietary component since it is inhaled deliberately). Under these factors are also discussed the so-called non-dietary factors, such as pyrolysis products (compounds normally formed during cooking) and various food additives.

5.2 INDUCTION

5.2.1 Induction of the Cytochrome P450 system

Induction is defined as an increase in enzyme activity associated with an increase in intracellular enzyme concentration [1-5]. From a genetic point of view, this increase in enzymic protein is usually caused by an increase in transcription of the associated gene. The stimulation of enzyme activity represents a process of temporal, adaptive increase in the concentration of a specific enzyme, due either to an increase in its rate of synthesis, or in a decrease of its degradation rate. The direct consequence is an accelerated rate of biotransformation of both endogenous compounds and xenobiotics, by co-administration of another compound, designated as an inducer. The inducer will modify drug metabolism in different ways (either qualitatively or quantitatively) and it is therefore expected to alter the pharmacological effects of drugs (increase in the metabolism of the drug involved in interaction, and decrease in the quantity of drug available for pharmacological activity). In order for this to take place, 1-2 weeks are usually needed, the process being an adaptive, temporal one, as indicated above. Inductive properties may be displayed by compounds with quite different chemical structures, pharmacological actions, or even different toxicities. Drugs of abuse are also known to induce gene expression [6].

The pharmacodynamic and pharmacotherapeutic consequences are reflected by a decrease in pharmacodynamic activity, and hence inefficacy at the usual therapeutic doses. As a first conclusion, we may therefore emphasise that enzyme induction contributes to inter- and intraindividual variation in drug efficacy and potential toxicity associated with drug-drug interactions. On the other hand, we note that alterations in drug efficacy are directly dependent on several factors, including the extent of enzyme induction (in a particular individual), the relative importance of the enzyme in multiple pathways of metabolism, and on the therapeutic ratio of substrate and metabolite(s). In this context, we may add that potential toxicity will depend on changes in metabolic pathways associated with an alteration in the balance between drug activation and detoxication [7].

In an attempt to localise the site of induction of drug metabolism, significant advances have been made in considering the role of the liver. As is already known, the major organ responsible for drug metabolism in most species (including man) is the liver. It then becomes evident that the main enzymatic system involved will be that of the cytochromes P450, the well-known family of oxidative hemoproteins responsible for a wide variety of oxidative transformations in a variety of organisms (see Chapter 4). The extent of induction of hepatic metabolism can reasonably be expected to be reflected in experimentally accessible indicators such as increased drug clearance, decreased drug plasma half-life, increased plasma bilirubin levels and others.

On the other hand, it should be borne in mind that the wide range of drugs and chemicals that act as inducers has been investigated for the most part on hepatocytes *in vitro*. Kinetic data obtained with isolated hepatocytes *in vitro* were then extrapolated to laboratory animals. However, although enzyme induction commonly results in increased rate of xenobiotic metabolism *in vitro*, the effects of enzyme induction may be dampened by physiological constraints *in vivo*. Furthermore, animal experiments can give only an indication of possible human response, necessitating very cautious extrapolation [8]. One of the most common types of induction is that which is substrate-dependent. A well-known example of this phenomenon is the influence of phenobarbitone on the metabolism and duration of action of several drugs. Drugs affected include oral anticoagulants (anticoagulant effect decreased due to increased metabolism) [9], tricyclic antidepressants (antidepressant effect decreased, by the same mechanism) [10], corticosteroids (corticosteroid effect decreased, by the same mechanism) [11], narcotics (increased CNS depression with meperidine, increased meperidine metabolites) [12], theophyllines (theophylline effect decreased due to increased metabolism) [13] and the muscle relaxant zoxazolamine (substantial metabolism increase, and consequently a significant decrease in the paralysis time) [2].

Pre-treatment with phenobarbitone has also been shown to markedly increase the metabolism of felodipine and its pyridine analogue [14].

Moreover, the inducing action of phenobarbitone may affect the expression of several specific CYP450 isoforms, as revealed in a recent study [15]. The same phenomenon was observed even more recently in pregnant rat and foetal livers and placenta, impacting on different cytochrome P450 isoforms [16].

From the above examples it is obvious that phenobarbitone induces the metabolism of different drugs, thus affecting the intensity and duration of their pharmacological action. The assumed molecular mechanism is a substantial increase in intranuclear RNAs that represent precursors to cytochrome P450 and mRNA. The immediate consequence will be a substantial increase in the hepatic levels of certain P450 isoforms, particularly CYP2B1and CYP2B2, with the former considered as the major phenobarbitone-inducible cytochrome P450 [17]. Accordingly, we may conclude that the major inductive effect of phenobarbitone in the liver is to increase specific mRNA levels by augmenting transcription, rather than translational efficiency, or by stabilising the pre-existing levels of protein precursor.

The other main type of induction response is receptor-mediated, by interactions with important regulatory pathways. For example, many drug-metabolising enzymes that are also involved in metabolism of endogenous cellular regulators (steroids, eicosanoids) have been proven to be inducible by hormones [18]. Moreover, there is evidence now that other hormones, such as the growth hormone, are capable of altering human cytochrome P450 expression [19].

In the case of the CYP1 family, this type of induction response is mediated by a specific cytosolic aryl hydrocarbon (Ah) receptor [20]. The best-known example is that of induction of cytochrome P450s by polycyclic aromatic hydrocarbons (PAHs) [21], which combine with this specific receptor (in a similar manner to hormone response), resulting an inducer-receptor complex. Furthermore, this complex is translocated to the nucleus of the hepatocyte whereupon induction-specific mRNA is transcribed from DNA. In the nucleus, the translocated Ah receptor forms a heterodimer (with a second nuclear protein), which will then bind to a common response element, known as the XRE (xenobiotic responsive element) that functions as a transcriptional enhancer, resulting in stimulation of gene transcription [22]. Large amounts of newly translated, specific cytochrome P450 are then incorporated into the membrane of the hepatic endoplasmic reticulum, resulting in the observed induction of metabolism of certain drugs and xenobiotics including tamoxifen, tacrine, acetaminophen, dietary phytochemicals and carcinogens, such as the aromatic amines produced in cooking and those found in cigarette smoke.

This type of response is common both for phase I and phase II metabolic reactions (details of which appear in the following subsection).

A third type of induction response is that which is inhibitor-mediated interaction with the heme group of the cytochrome P450s, resulting in inhibition of endogenous function and consequent disruption of endogenous pathways catalysed by a specific cytochrome P450 isoform. Well known examples include the induction of CYP2E1 by isoniazid [23] and induction of CYP3A1 by macrolide antibiotics [24].

However, in this context we have to remind the reader that the CYP2E1 isoform is one of the most toxicologically important of the cytochrome P450 enzymes. This is borne out by numerous studies which have revealed that this specific isoform is implicated in the activation of acetaminophen and organic solvents to hepato- and nephrotoxic intermediates, as well as in the activation of nitrosamine procarcinogens and in the etiology of alcohol-induced liver damage [21,23]. Unfortunately, many xenobiotics, as well as some dietary and pathophysiological factors, increase CYP2E1 expression [23]. An interesting fact worth emphasis is that the induction of the CYP2E1 isoenzyme arises through multiple mechanisms, depending on the induction stimulus [2]. Induction has been proven to occur at all regulatory levels, from transcription to mRNA stabilisation, increases in translational efficiency, and post-translational protein stabilisation [23]. Nevertheless, the predominant xenobiotic-dependent induction is assumed to be *via* stabilisation and inhibition of degradation of the CYP2E1 apoprotein. A well-studied case is that of ethanol, which at low concentrations has been proven to stabilise the CYP2E1 apoprotein [25]. A question of great interest in this connection is whether ethanol is an enzyme inducer or an enzyme inhibitor. It has been proven that at low concentrations ethanol acts like an inducer, whereas at high concentrations it acts as an inhibitor [26].

An interesting aspect, related to the variability in the metabolism of narcotic drugs, was reviewed relatively recently [27]. The liver P450s are primarily responsible for their biotransformation, ensuring both their oxidative or reductive metabolism, which is of crucial importance as regards the toxicity of their metabolites. These are commonly reactive intermediates or free radicals, which may either combine with macromolecular cellular components (generating an autoimmune response), or induce peroxidation of membrane lipids. However, what the study cited above revealed is that since different isoforms of CYTP450 are greatly induced by pre- or co-administration of certain drugs, so also may the metabolism of the narcotic drugs vary greatly from one patient to another, depending on previous or concurrent drug treatments.

5.2.2 Induction of other enzyme systems

Although the CYTP450 is by far the dominant enzymatic system involved in drug metabolism, it should be pointed out that other enzyme systems may undergo induction as well. We refer especially to some of the phase II metabolising enzymes, such as UDP-glucuronosyl transferase (UDPGT) and glutathione-S-transferase (GST). Naturally, since the phase II enzyme systems are involved in the major routes of detoxication, there is much interest in the induction of these systems, especially for cancer chemoprevention.

The induction of UDPGT and GST has been proven to be highly dependent on the nature of the inducer under consideration, as well as on the species variability. In experimental animals, both enzymes have been inducible by phenobarbitone-type inducers and Ah receptor ligands [28,29], and an ethanol-inducible UDPGT has been described in rabbit [30]. As far as endogenous compounds are concerned, the bilirubin UDPGT has been reported to be induced by several drugs, including phenobarbitone and rifampicin [31].

GSTs in the alpha class (GST A1 and A2) have been reported to be inducible by several xenobiotics including phenobarbitone, dithiolethiones and oltipraz [32].

5.3 INHIBITION

5.3.1 Inhibition of the Cytochrome P450 system

Inhibition of drug metabolism by pre- or co-administration of other drugs or xenobiotics is a well-recognized phenomenon, with pharmacological and toxicological effects, reflected by exacerbated pharmacodynamic activity and adverse effects of relative overdosing at the usual therapeutic doses. In the context of the common practice of polypharmacy, another topic of great interest arises, namely drug-drug interaction [33-35] (Consequences of this phenomenon are discussed in Chapter 8). The other major interest in enzyme inhibition is based on a very important sector of therapy, namely the selection of enzymes as targets for drug action.

As in the case of induction, inhibition can also take place by different pathways and mechanisms. In principle, inhibition involves either the blocking of enzymatic synthesis, the destruction of pre-formed enzymes, or inactivation of the enzyme by their complexation with drug metabolites.

The direct consequence of enzyme inhibition is the delay in the biotransformation of certain drugs, resulting thus in increased plasma concentrations and potentiation or prolongation of their pharmacological action. The level of drug biotransformation may be decreased to the point of complete inhibition by various compounds which can interfere either before their contact with the MMFOs or, more commonly, through direct action on this enzymatic system.

It is important to note that drug binding at the level of different tissues is also a significant factor in limiting the *in vivo* rate of biotransformation (decreased concentration of free, unbound drug).

Basic mechanisms of enzyme inhibition involve one of the following:

1) competitive inhibition
2) non-competitive inhibition
3) uncompetitive inhibition
4) product inhibition
5) transition-state analogues
6) slow, tight-binding inhibitors
7) mechanism-based enzyme inactivation

8) inhibitors that generate reactive intermediates that can covalently be attached to the enzyme.

Overviews of some special mechanisms will be presented, followed by a few prominent examples involving various enzymes.

Competitive inhibition occurs when the 'normal' substrate and the inhibitor share structural similarities. The inhibitor may or may not be a co-substrate and the intermediate complex [ES] may or may not be present. Screening for inhibition is a very important and routine practice in the pharmaceutical industry today and therefore new approaches have been introduced to handle the ever-increasing numbers of new drug candidates. One of the successful statistical, experimental approaches is commonly designated as virtual kinetics [36]. Competitive inhibition is a relatively common phenomenon in drug metabolism, being of significant relevance especially in the field of drug-drug interactions (see Chapter 8), since we acknowledge that many enzymes have multiple drug substrates that can compete with each other.

In classical non-competitive inhibition, the inhibitor will usually bind to a site distinct from that of the substrate, resulting in a decrease of V_{max} without a change in K_m, according to the Michaelis-Menten equation (or the linearised Lineweaver-Burk representation).

Un-competitive inhibition, although well defined, is seldom observed. It presumes binding of the inhibitor only to the [ES] complex and affects both the V_{max} and the K_m values, which decrease, but still maintaining the V_{max}/K_m ratio constant. In this context, it is evident that the enzyme efficiency would not really change.

'Product inhibition' occurs when the metabolic product generated by the enzyme inhibits the reaction on the substrate ('feedback inhibition'). This usually occurs when the product has physical characteristics very similar to those of the substrate. A well-known example is that of benzene, which is oxidised to phenol by a specific P450 isoform. It has been proven that both the initial substrate (benzene) and the product (phenol) compete with each other [37].

Transition-state analogues are compounds that are non-covalently bound to the enzyme, resembling the transition state of the enzymatic reaction. The complex formed is [EI], leading to inactivation of the enzyme. However, it is important to note that enzyme activity may be restored by removal of the inhibitor using different methods, the most common being gel-filtration and dialysis.

Slow, tight-binding inhibitors bind to the enzyme at a slow rate, inhibit competitively, and practically inactivate the enzyme irreversibly [38]. Possible causes of inactivation are associated with different mechanisms: conformational change of the three-dimensional structure of the enzyme (including therefore alteration of activity), a change in the protonation state of the enzyme, reversible formation of a covalent bond, or displacement of a water molecule at the active site.

Mechanism-based enzyme inactivators are special, unreactive compounds, with structures similar either to the substrate or to the product of the enzyme; due to this similarity these compounds undergo a catalytic transformation by the enzyme. The characteristic feature here is that the resulting species inactivates the enzyme before leaving the active site [39]. Some characteristic patterns of mechanism-based inhibition include first-order kinetics, irreversibility, and covalent binding to either the protein moiety, or the prosthetic group of the enzyme. Inhibitors that generate reactive intermediates that can covalently be attached to the enzyme are not particularly effective themselves, but following their oxidative biotransformation, they bind tightly to the heme of the CYT, preventing in this way further involvement of the iron atom in catalysis.

As far as the CYTP450 enzyme superfamily is concerned, several types of inhibition have been described.

An interesting inhibition mechanism, and one with profound pharmacological implications, is the destruction of hepatic cytochrome P450. Compounds having the ability to effect this include xenobiotics containing either an olefinic, or an acetylenic function, such as acetylene, allobarbital, ethylene, fluoroxene, vinyl chloride and norethindrone. The majority of these compounds are relatively inert *per se*, requiring metabolic activation by cytochrome P450, after which they form substrate-haem adducts, thus destroying the enzyme. Because of this pattern of action, these compounds are commonly designated as 'suicide substrates' of the haemoprotein [40].

The immediate and major consequence of the formation of these adducts (involving haem modifications), will obviously be a significant and sustained decrease in the levels of functional CYTP450, leading to a reduction in drug-metabolising capacity of the liver. A significant point to note is that since the target of the olefinic-induced inhibition is the haem, administration of exogenous haem would be very helpful in restoring both the liver CYTP450 content and the drug-metabolising activity.

Another important group of inhibitors of CYTP450 activity, though acting through other mechanisms, comprises metal ions [41]. It is generally accepted that they do not modify the existing CYTP450-haem, but in contrast, act by modulating both the synthesis and the degradation of the haem prosthetic group of the cytochrome.

Studies in the 1970s established lead as an inhibitor of a number of CYTP450 related oxidations. Further research revealed that the inhibitory effect may be a combined one, involving both the protein synthesis and the haem cofactor [42]. Other metals of interest as regards their inhibitory potential include copper, cobalt, cadmium, arsenic and mercury.

Cobalt, for instance, in the form of cobalt-haem (substituting the iron from the prosthetic group of CYTs) has a pronounced influence on both the biosynthesis and biodegradation of hepatic haem, and consequently on drug metabolism in general. Because of the substantial decrease in both hepatic microsomal content of CYTP450 and total haem, following cobalt pre-treatment, a substantial decrease in the metabolising activity of liver enzymes has often been reported [2].

In the 1980s, Abbas revealed the inhibition of CYTP450 by mercury [43]; several years later, it was established that the molecular mechanism involves loss of cytochrome P450 content and impairment in the formation of the whole cytochrome [44]; in addition, enhanced rates of haem degradation were established [45].

Interestingly, recent studies reported contrary results, namely specific induction of a particular CYTP450 isoform, 1A1, in *in vitro* cultured cells [46].

Cadmium is another toxic metal having a long history of investigation. It has been proven to inhibit drug biotransformation in particular species (rats, for example), most probably by inducing haem oxygenase (like lead), and resulting in decreased levels of the whole cytochrome contents [47]. More recent studies have revealed a broad specificity of inhibitory effects of cadmium on particular CYTP450 isoforms, such as 2E and 3A [48], and CYP 4A11 from human kidney [49].

Two other interesting aspects that warrant mention are the age and sex dependence of cadmium inhibitory effects: inhibition has been shown to increase with age [50], and to differ with sex, being for example greater in male rats than in females [51].

Being responsible for so many, severe toxicological consequences (including renal dysfunction and cardiovascular effects accompanied by hypertension), cadmium continued to be a focus of attention for toxicologists who established a dose-dependent relationship between the effects of cadmium (according to its burden in different tissues) and the expression of several specific CYTP450 isoforms. One of these isoforms has been proven to be the same CYP 4A11 from human kidney; with respect to its enzymatic activity, this isoform is involved in the hydroxylation of unsaturated fatty acids, which in turn are involved in the regulation of the salt balance in the kidney. The decrease in its enzymatic activity, as a consequence of the inhibitory effect of cadmium, was associated with the development of high blood pressure [52].

As in the case of mercury, more recent studies revealed that in some instances, cadmium may display an inductive effect *in vitro* as well, acting on specific CYTP450 isoforms such as CYP1A1 and CYP2C9; no species differences have been observed thus far [53,54]. Unfortunately, 'cadmium poisoning' is relatively frequent, if we acknowledge that food crops grown in cadmium-containing soils are the major source of cadmium exposure to humans. The inevitable consequence of such exposure will definitely be adverse effects on specific organ systems, with the most severe impact on the kidney. Cadmium-induced renal dysfunction, associated with polluted areas, such as those in Japan for example, has recently been proven to increase the risk of mortality [55].

Another well-known and potent toxic metal is arsenic. It exerts its inhibitory effects in a relatively specific manner that differs from those above, in the sense that its action involves two steps: after an initial increase in cytosolic 'free' haem, there follows a dramatic decrease both in cytosolic 'free' haem and the general content of CYTP450. In this situation, it has also been proven that there is a several-fold increase in haem oxygenase, resulting in a significant decrease of CYTs haem content, and consequently, the general content of cytochrome [56]. Numerous other studies revealed both the importance of the involvement of arsenic species in the inhibition of the cytochrome P450 system, as well as the fact that significant decrease in P450 is largely associated with the 1A1 isoform [57]. As far as the various species of arsenicals have been shown to display different effects on the CYTP450 system, studies revealed that the most significant inhibition is associated with the arsenite species, As^{3+} [58]. Very interesting on-going research associates the genetic polymorphisms identified in humans with the role of different isoforms in inducing cancer, in populations exposed to arsenic [59]. As a concluding remark concerning the impact of toxic metals on drug metabolism, we may affirm that it concerns either haem biosynthesis or haem degradation, with resulting changes in the synthesis of the global P450 cytochromes.

An interesting inhibition mechanism, well understood at the molecular level, involves compounds that, though not being inhibitors *per se*, are capable of forming inactive complexes with hepatic cytochrome P450. These compounds require, as do the olefinic or acetylenic compounds, a pre-activation, being in fact common substrates for the P450CYTs. After the production of the metabolic intermediate, the latter will bind tightly to the haemoprotein, forming spectrally detectable, inactive complexes, which are thus prevented from further participation in drug metabolism. This mechanism is supported by experiments on laboratory animals, which, following pre-treatment with such compounds, showed delayed drug biotransformation, resulting for instance in a significant increase in both hexobarbital narcosis and zoxazolamine paralysis times. Compounds with such properties include amphetamine, cimetidine, dapsone, isoniazid, sulphanilamide, piperonal and piperonyl butoxide [2].

A very recent example refers to the inhibitory effect of *N,N',N''*-triethylenethiophosphoramide (thioTEPA) on the CYP2D6 isoform, involved in the biotransformation (by 4-hydroxylation) of cyclophosphamide

[60]. ThioTEPA is frequently used in chemotherapy regimens that include cyclophosphamide, being required for its activation. The detailed mechanism of this inhibition, studied on human liver microsomes using recombinant P450 enzymes and bupropion as a substrate, revealed a time- and concentration-dependence of the inhibition. The inhibitory action of two compounds and the pharmacokinetic consequences of another irreversible inhibition, on the CYP 2B6 isoform have also been investigated [61]. An evaluation of the potential inhibitory or inductive action of daptomycin was recently performed on human hepatocytes [62].

Human P450 cytochromes can also be inhibited by nicotinic acid and nicotinamide [63]. Recent spectrophotometric analysis indicates, as expected for nitrogen-containing heteroatomic molecules, that in this case the inhibition occurs *via* coordination of the pyridine nitrogen atom to the heme iron.

As mentioned earlier, either the inductive or inhibitive action of different compounds may affect both the phase I as well as phase II enzymes. A study was recently performed on some of the most important hepatic and extrahepatic (kidney, lung and gut) phase II enzymes, including UDPGTs, GSTs and NAT, with propofol in various concentrations, as the chosen substrate. The study was performed on microsomal and cytosolic preparations from both human and other species. As regards the human conjugation enzymes involved, the study revealed that propofol displayed a concentration-dependent inhibition, with the activity of UDPGT significantly decreased, that of GST affected insignificantly, and with NAT activity practically unaffected. Inter-species differences have been demonstrated [64].

5.4 CONSEQUENCES OF THE ABOVE PHENOMENA

As final conclusions, we should mention the following:

- the human cytochrome P450 enzyme system, and to some extent phase II metabolising enzymes, are susceptible either to induction or inhibition mechanisms;
- sometimes, because of a too rapid biotransformation (direct consequence of increased enzyme activity), megadoses of drug are required, which may not always be possible (e.g. for drugs with broad therapeutic index); plasma levels are too rapidly achieved and the clearance is also too rapid, so the drug has insufficient time to display its pharmacological action;
- induction, by increasing enzyme activity, will result in decreased metabolism of certain drugs, contributing to significant inter- and intra-individual variations in drug efficacy and potential toxicity, associated with drug-drug interactions;

• as shown in the corresponding subsection, it is important to mention that the clinical effects following alterations in drug efficacy are dependent on several factors, including the extent of enzyme induction, the relative importance of the enzyme in the multiple pathways of metabolism, and on the therapeutic ratio of substrate and metabolite;

• different inducers (drugs, or other compounds from food and the environment) display substrate specificity for CYTP450 isoforms;

• prevalent inducing conditions in humans include smoking, alcohol consumption and diet;

• elevated levels of specific CYP450 isoforms (1A and 2E1) may contribute to increased risk of cancer; in particular, CYP2E1 may contribute to alcohol-induced liver damage and acetaminophen toxicity in alcohol consumers.

• inhibition of drug metabolism represents a subject of great interest for several reasons. It can give rise to a decrease in drug biotransformation, low plasma levels, decreased clearance (possibility of overdosing at common, therapeutic doses) and increased risk in the occurrence of drug-drug interactions. Besides decreasing the therapeutic effect on one drug by concurrent administration of another, it is unfortunately proven that some drug-drug interactions may be even fatal;

• practical aspects of inhibition include an understanding of the phenomenon at the molecular level, especially as it relates both to such drug-drug interactions (prediction, avoidance or minimisation of the risks), as well as the utilisation of enzymes as therapeutic targets.

In this context we recommend that the reader consult available general reviews dealing with the clinical implications and modern procedures involved in metabolic screening *in vitro* [65,66].

5.5 DIETARY AND NON-DIETARY FACTORS IN ENZYME INDUCTION AND INHIBITION

Numerous studies and clinical observations have revealed that drug biotransformation can also be significantly increased by some exogenous compounds present in the diet [67-70] or in the environment [71].

Before discussing the main dietary factors, we should emphasise from the outset that generally, food intake has been proven to have a considerable influence on the bioavailability of drugs with extensive pre-systemic metabolic clearance. This phenomenon commonly occurs with lipophilic bases, but very rarely, if ever, with lipophilic acids. Thus, concurrent food intake with compounds acting like lipophilic bases will significantly reduce their pre-systemic clearance, consequently enhancing their bioavailability. In contrast, pre-systemic clearance of acidic drugs is commonly unaffected by food. In addition, studies have revealed that even among lipophilic bases, concurrent food intake will act variably, in direct correlation with the type of biotransformation involved; this is usually a decrease in pre-systemic clearance in drugs undergoing hydroxylations, glucuronidations and acetylations, while in contrast, the bioavailability of lipophilic bases that undergo pre-systemic dealkylation usually remains unaffected [69].

Another general consideration is that nutritional deficiencies usually result in decreased rates of drug metabolism, with some notable exceptions of certain vitamins (B_1 and B_2) that enhance the rates of metabolism of some xenobiotics. At the same time, a deficient diet may, in certain instances, favour occurrence of drug-induced toxicity and carcinogenity [70].

In discussing the dietary factors, two distinct major groups have to be considered: the macro- and the micro-nutrients. It should be emphasised as well, that under dietary factors we shall discuss also alcohol consumption and the components of tobacco smoke.

The group of macronutrients includes proteins, lipids and carbohydrates.

The impact of a correct protein diet is obvious, if we consider that all enzymes are proteins [72]. Consequently, we may affirm that if there is a general decrease in overall microsomal protein, the extent of drug metabolism will decrease as well. This may affect not only the pharmacological response, by decreasing it, but also result in toxicological effects, generally because of delayed clearance. Protein deficiency in rats, for example, decreased the total liver cytochrome P450 content, with concomitant alterations in its composition, namely increased Ile and Leu levels, decreased Asp, Glu and Phe levels [73].

In constrast, in some specific cases, low protein diet can be beneficial; one example is that of aflatoxin-induced hepatotoxicity, which may be reduced either by low protein diet (acting like an inhibitor of phase I metabolism), or by phenobarbitone treatment (induction of phase I metabolism), in both cases the production of the epoxide intermediate being reduced.

An interesting study focused on the effects on theophylline metabolism accompanying a change from a high-protein to a high-carbohydrate diet [74]. The results indicated a decrease in drug biotransformation by almost one-third (very similar to the effect exerted by cimetidine).

Lipids are necessary as well for normal drug metabolism for several reasons: they are required by the drug metabolising enzymes as membrane components and possibly also for specific interactions. Experiments showed that the quantity and quality of dietary fat affect lipid composition, and consequently, physical characteristics of biological membranes, including their stability and drug passage into the membrane [75]. In addition, they may affect the enzymatic activity of several phospholipid-dependent enzymes associated with these membranes. These changes will accordingly affect the inducibility of these enzymes significantly, resulting in alterations of the pharmacological response to certain drugs.

The most important components of a correct lipid diet are assumed to be linoleic acid and arachidonic acid. Therefore, treatment with polyunsaturated fatty acids is considered to be beneficial, because it increases the microsomal content of these fatty acids and consequently increases drug metabolising capacity. Essential fatty acids are known to be required for the interaction of different substrates with the active site of cytochrome P450. (See Chapter 4 for a discussion of the importance of the lipid component in some enzymatic systems).

The content of dietary lipids on the activities of different hepatic microsomal drug-metabolising enzymes such as demethylases, hydroxylases and cytochrome P450 was proven to be of utmost significance. For example, experimental studies on rats revealed that in some instances, a high-fat diet might produce more hepatotoxic effects [76].

From the class of macronutrients, the carbohydrates seem to have less significant effects. However, a well-known example is that of glucose, which, at high intake levels, can inhibit certain drugs (e.g. phenobarbital), resulting in specific secondary effects and lengthening the sleeping time caused by the drug. At the same time, excess of glucose has been shown to decrease the general hepatic cytochrome P450 content, and consequently, the enzymatic activity [2].

Interestingly, a high-carbohydrate diet, in a comparative study with a high-protein diet, has been shown to display quite the opposite effect. While the increased protein content, as expected, increased the hepatic biotransformation rate by increasing the total content of CYTP450, experiments based on pharmacokinetic studies with a specific substrate revealed that increased carbohydrate content in the diet produced the opposite effect on the activity of the same enzymatic system [77].

Similar effects were evident in a comparative study of long-term feeding with high-fat (FAT) diet versus high-carbohydrate (CHO) diet. The study was performed on rats, and the control substrate used was pentobarbital. The results suggested that the FAT-diet increased the activity of hepatic metabolising enzymes, while the CHO did not. The results also revealed sex differences, only the females being affected in this way [78].

On the other hand, lack of carbohydrate in the diet was associated with a two- to threefold increase in the biotransformation of various xenobiotics in rats [79]. The experiments proved that, contrary to general belief, the CHO play an important role in regulating hepatic drug-metabolising enzyme activity, acting like enhancers when in low concentrations and as repressors when in excess. The control substrate was ethanol and the results showed that a decrease in CHO intake may significantly increase the action of ethanol, while an opposite effect was observed at high-CHO content in the diet. Similar effects on rats were evident from studies that employed orally administered or intraperitoneally injected phenobarbital, polychlorinated biphenyl and 3-methylcholanthrene as control substrates [80].

Having a much greater impact on the drug-metabolising capacity of certain enzymatic systems by far, is a special group of micro-nutrients, namely, the vitamins.

Vitamins are biochemical effectors indispensable for life and are essential constituents (or at least, should be) of a normal, correct diet. Apart from other functions (specific, or as enzyme cofactors), vitamins have been proven to be required also for the biosynthesis of proteins and lipids. We have already presented the role of the macro-nutrients as vital components of drug-metabolising enzymatic systems; thus it is obvious that changes in vitamin levels (particularly, deficiencies) will have an important impact on drug-metabolising activity in general. Vitamins influence enzymatic activity predominantly as inhibitors.

In different vitamin deficiencies, the enzymatic activity is generally decreased through various mechanisms, involving either (and more commonly) a direct decrease in cytochrome P450 levels, or a reduction in other CYTP450 system components, such as NADPH-cytochrome P450 reductase. Sometimes the inhibitory action is very specific; for example, vitamin A deficiency has enzyme-selective effects on drug metabolism. Studies have proven that a vitamin A-free diet (in rats for example) will result in lower levels of some specific enzymes relative to control animals. A relatively recent study showed that after four days of retinol administration to BALB/c mice, the activity of only CYP3A was decreased, while the catalytic activity of other enzymes (both phase I and phase II enzymes) remained relatively unchanged [81]. The control substrate was paracetamol and further observations of the study were that vitamin A potentiates paracetamol-induced hepatotoxicty, without involving changes in the corresponding biotransformation enzymes.

Other interesting aspects have been revealed experimentally. For example, the potentiation of paracetamol-induced hepatotoxicty was proven not to occur in the kidney; the suggested conclusion indicated an organ-specific response [82].

An interesting study on the impact of vitamin A-deficient, or supplemented diet, on the expression of different CYTP450 isoforms was performed on Syrian hamsters [83]. After a six-week observation period, the vitamin A-deficient diet resulted in a decrease in the total CYTP450 content, implicit in the catalytic activities of different CYT isoenzymes. The opposite effect was observed with the vitamin A-supplemented diet, which resulted in a marked increase in the activity of testosterone 7α-hydroxylase. The authors thus suggested that dietary vitamin A deficiency or supplementation displays not only organ-specific response, but enzyme specificity as well.

A point worth highlighting is that in certain instances, vitamin A supplementation can display an inhibitory effect on drug-induced hepatocarcinogenesis in the rat [84]. A dietary supplementation with β-carotene (the most common vitamin A precursor) proved to be effective in increasing certain microsomal and cytosolic enzymes such as cytochrome b5, NADPH cyt c reductase, and aryl hydrocarbon hydroxylase. It was therefore suggested that β-carotene is particularly protective in limiting the initiation phase of the toxic process.

Another important vitamin displaying opposite effects in deficiency and excess is vitamin B_1 (thiamine). The mechanisms by which thiamine deficiency can affect hepatic microsomal enzyme activities have been elucidated by investigating the activities of two constitutive cytochrome P450 isoforms, namely P 450IIE1 and P450IIC11 [85]. The experiment used male Sprague-Dawley rats and was performed for a period of three weeks. The results showed an increase in the IIE1 isoform, while the activity of the other enzyme remained unchanged. In addition, an elevation in the activity of cytosolic glutathione S-transferase was also observed. The overall conclusion of the study was that thiamine deficiency displays enzyme specificity.

Supplementation of vitamin B_1, either in the diet, or by direct parenteral administration has been proven to result in significant effects on the hydroxylation function of rat liver [86]. Experiments on rats showed an increase in the general CYTP450 content, as well as in some other specific microsomal enzymes, namely demethylases and hydroxylases. An interesting aspect revealed during the same experiment was that when given in excess (e.g. by repeating parenteral doses), thiamine caused an opposite effect.

A diet deficient in vitamin B_1 results in organ-specific effects, the tendency being to increase some specific hepatic and pulmonary microsomal enzyme activities, while for renal drug metabolism the consequences are quite the opposite [87].

Commonly, these effects – similarly to vitamin A - are attributed to changes in the microsomal content of cytochrome P450 or its NADPH-reductase component. However, more recent studies have found that thiamine can also act by changing the type of cytochrome present [2].

Vitamin B_2 (riboflavine) likewise displays specific effects on drug-metabolising enzymes, initial studies having been made about 30 years ago [88]. They refer to the induced decrease in certain enzyme activities (particularly demethylation and hydroxylation enzymes) by riboflavine deficiency. The experiments revealed decreased levels of both hepatic cytochrome P450 and cytochrome b5.

Further experimental studies have proven that the activities of both drug-metabolising enzymes and lipid peroxidation were decreased in low- or deficient-riboflavine groups. Experiments involving supplementary administration of vitamin B_2 resulted in increased activities of drug enzymes [89].

Other experiments proved that NADPH-dependent lipid peroxidation was markedly decreased in the liver microsomes of groups with riboflavin-deficient diet [90].

Experiments comparing the effects of a standard diet and a riboflavin-deficient diet proved that concurrent consumption of ethanol enhanced the intestinal phospholipid concentration in the deficient diet group, whilst no significant effects on the concentration of proteins or phospholipids was observed in the standard diet group. Riboflavin deficiency decreased both intestinal phase I and phase II enzyme levels [91].

A more recent, related study followed variations of the drug-metabolising enzyme activities mediated by vitamin B_2 deficiency. In this study, the substrates were polychlorinated biphenyls (PCBs), known to induce liver lipid peroxide formation in rats. These components are well-known inducers of the liver microsomal cytochrome P450, and vitamin B_2 deficiency has been proven to promote this induction [92].

Vitamin B_2 being a component of NADPH-cytochrome P450 reductase (which is itself a component of the MMFO system), it is not surprising that a deficiency in riboflavin will result in a general decrease in activity of the enzymatic system.

Vitamin C (ascorbic acid) has special status among the vitamins, being the only one that cannot be synthesised in some organisms, including man, monkey and guinea-pig, due to an inherited enzymopathy. Therefore, these species in particular will show a special requirement for this vitamin.

Studies on guinea-pig species proved that individuals with vitamin C deficiency are more sensitive to the effects of particular drugs (e.g. pentobarbitone, procaine) [93]. This increased sensitivity was attributed to a marked decrease in drug-metabolising capacity. Studies on experimental animals suggest that the vitamin C deficiency may interfere with heme biosynthesis, consequently directly affecting cytochrome P450 levels.

An interesting experiment concerning the effects of vitamin C on hepatic drug metabolising function in hypoxia/re-oxygenation was relatively recently performed [94]. Results revealed that increase in total and oxidised glutathione levels were attenuated by vitamin C by itself, or in combination with vitamin E. Total and oxidised glutathione levels were markedly increased during hypoxia, but markedly decreased in the presence of vitamin C, E, or their combination. By increased oxidation and glucuronidation, vitamins C and E synergistically improve the hypoxia/reoxygenation hepatocellular damage as indicated by abnormalities in drug metabolising function. This protection appears to be mediated by decrease in oxidative stress.

Vitamin E has been under the scrutiny of researchers since 1976, when attempts were made to correlate drug metabolism and hepatic heme proteins in a vitamin E-deficient diet in rats [95]. Hepatic homogenates have been analysed for CYTP450 content and specific drug metabolizing enzyme activities. While no difference could be detected between the vitamin E-deficient population and the control group, decreased levels have been noticed for some microsomal hydroxylases and demethylases. No relevant association could be made between heme protein synthesis and a vitamin E-deficient diet. Interesting studies performed in the same period revealed another important aspect for a drug to display its pharmacological action, through examination of the effect of vitamin E deficiency on intestinal transport of passively absorbed drugs. Sodium barbital was used as the control compound. In vitamin E-deficient animals, its absorption rate was enhanced compared with the control group. This increase in the absorption rate was attributed to vitamin E-deficiency-induced alterations of the intestinal membrane structure, and was confirmed by using other control compounds passively absorbed through the intestinal membrane. On the other hand, it was observed that the transport rate for drugs normally rapidly transported (e.g. salicylates) was not modified by the deficiency state [96].

As far as hepatic drug metabolism is directly involved, a comparative study tried to elucidate whether there are significant differences in normal and vitamin E-deficient animals [97]. The general conclusion of the study, which monitored several specific enzymes as marker parameters, was that vitamin E-deficiency did not influence these parameters significantly, only a slight increase in NADPH oxidase activity being noted.

Another study proved that in the presence of polychlorinated biphenyls, vitamin E displayed an inductive effect on various microsomal enzymes [98]. This observed activity enhancement was associated with increased liver weight and the amount of liver microsomal protein after application of Delor 103 (a specific polychlorinated biphenyl preparation). Vitamin E has been proven to stimulate catalytic activity of certain microsomal hydroxylases and demethylases.

Concerning the microsomal hydroxylations of certain drugs, after oral doses of vitamin E, the result was a clear stimulatory effect [99]. However, important observations were that this inductive effect is relatively specific and can be reversed (or prevented) by pre-administration of actinomycin D.

It is well recognized that by manipulating dietary levels of vitamin E, lipid peroxidisability (especially of bio-membranes) can be altered. Experimental studies tried to associate the increasing lipid peroxidation induced by a vitamin E-deficient diet and the adriamycin-induced inhibition of hepatic drug metabolism [100]. The results proved that vitamin E deficiency produced a significant elevation in hepatic lipid peroxidation, but without any considerable alterations in the activities of specific microsomal hydroxylases and demethylases. In contrast, pre-treatment with adriamycin displayed a significant inhibitory effect, decreasing the same enzyme activities by up to 63%. Nonetheless, the final conclusion of the study was that decreases in drug metabolism were independent of dietary vitamin E and did not correlate with lipid peroxidisability. It was thus suggested that adriamycin-induced depression of hepatic drug-metabolising enzymes was not mediated by elevated lipid peroxidation.

In vitamin E-supplemented diets, studies revealed a significant increase both in the total cytochrome P450 content and NADPH-cytochrome P 450 reductase activity, in rat liver. On the other hand, further experiments proved that vitamin E protected protein sulphhydryl groups and lipids against peroxidation, which can induce apoprotein loss. Interestingly, observations established that the protective effect against –SH and lipid peroxidation extend to protection of the CYTP450 apoprotein, but not to the enzyme activity, which was only partially protected. The most significant conclusion was that, at least *in vitro* peroxidation-dependent loss of P450 is not directly related to lipid/SH oxidation, but is instead mediated by heme degradation from the P450 holoenzyme [101].

The same protective effect was observed in experimental animals infected with influenza virus, which resulted in depression of different monooxygenase enzyme activities. In a dose-dependent relation, vitamin E was demonstrated to both decrease lipid peroxidation enhanced by the viral infection, and to increase the enzyme activities depressed by the same cause [102].

Based on its protective anti-peroxidation effects, an association between a retinoic acid metabolism blocking agent (RAMBA) and vitamin E was relative recently proposed [103]. Examples of compounds in this category include 2-benzothiazolamine derivatives. Combined formulations with vitamin E are administered as capsules or injectable solutions.

Vitamin E is metabolised by CYTP450-mediated side-chain oxidations. Often, these enzymes are regulated by their substrates themselves. However, tocopherols are able to activate gene expression of different CYP450 isoforms *via* the pregnane X receptor (PXR), a receptor with nuclear localisation capable of regulating a variety of drug metabolising enzymes [104].

Decreased activity in a number of enzymatic systems involved in the biotransformation of drugs has also been reported for vitamin K deficiency. A vitamin K-poor diet, as well as vitamin deficiency, have been proven to be accompanied by a decreased activity of various microsomal enzymes, including demethylases, hydroxylases, NADH- and NADPH-reductases. A very interesting hypothesis aimed at explaining the above effects correlates changes in the enzymatic activity with the weakening of both hydrophobic and polar interactions in the microsomal membranes [105].

A very recent study focused on possible effects of a synthetic vitamin K analogue (menadione) on enzyme activity [106]. It was demonstrated that depending on dose and duration of treatment, menadione displays an inductive effect on both phase I and phase II drug metabolising enzymes.

Dietary vitamin K, among other factors such as age or genetic polymorphism in the CYTP4502C9 isoform, also plays an important role in the inter-individual variability in responses to warfarin. The impact of these findings in clinical practice is still being assessed [107].

Based on the discussion in the above examples, it is evident that we can highlight the fact that vitamins are essential not only for good health, but also for maintaining normal levels of drug metabolism. The impact of deficiencies is reflected in decreased enzyme activity, with the consequences described in the previous subsection.

Minerals are likewise required in very small amounts in the diet, both for maintenance of good health and for normal physiological function, including normal functioning of the enzymatic systems involved in xenobiotic metabolism. Some of the most important minerals which have been proven to influence drug metabolism in one way or another by affecting enzyme activity, include iron, magnesium, calcium, zinc, copper and selenium. Depending on the enzyme affected, the effects may manifest as an increase, or instead as a decrease, in enzyme activity, with consequent impact on drug biotransformations. In addition, it should be mentioned that in some instances, no changes have been reflected as effects on drug metabolism, depending also on the enzyme or enzymatic system involved.

Usually, we refer to mineral deficiencies, which as expected, will generally result in decreased metabolism.

The only mineral deficiency resulting in increased metabolism involves iron. This seems anomalous, given that iron is essential in the haem group of cytochrome P450s. Experimental studies on rats revealed that iron deficiency is sex-dependent, occurring only in the male, in which a significant increase in the CYP3A2 isoform (a male-specific isoenzyme) was observed. The activities of drug-metabolising enzymes in female rats were not increased by iron deficiency (because of lack of the CYP3A2 isoform) [108]. However, the opposite effect was observed during a long-term study focused on the effects of iron-deficiency on different drug metabolising enzymes in both hepatic and extra-hepatic tissues including lung, kidneys and intestinal mucosa [109]. An important aspect involving iron-deficiency is the observed marked decrease in intestinal drug metabolism, which may result in significant toxicological consequences, if we refer particularly to the protective role of the intestinal enzymes against certain procarcinogenics, such as PAHs, for example.

Another experimental system was employed to study the effect of iron-deficiency on both phase I (activating) and phase II (conjugating) xenobiotic metabolising enzymes [110]. Microsomes and cytosolic preparations were made from various tissues including liver, lungs, kidneys and intestinal mucosa at the end of the experiment. Activities of many phase I and phase II enzymes were investigated and the results showed that in most cases, their activities were significantly decreased by iron deficiency; it was concluded that this may result in the persistence of some ingested compounds in the body, without appropriate elimination, which might prove to be harmful to the host.

The effects of copper deficiency are variable, but generally result in decreasing the metabolism of certain drugs by affecting the corresponding enzyme activity. However, important consequences of copper deficiency have been observed during pregnancy, resulting in both structural and biochemical abnormalities of the foetus. The aim of one study was to establish the mechanisms of copper deficiency-induced teratogenesis, available data suggesting that more mechanisms may be involved in the associated dysmorphogenesis, including especially the free radical defence mechanism [111].

An interesting aspect to mention is that copper excess has the same effect as copper deficiency, most commonly a decreased ability of enzymatic systems to metabolise drugs or other xenobiotics (-see also the discussion on copper enzymes in Chapter 4). As an example we may mention a significant decrease in the metabolism of aniline in rats pre-treated over one month with an excess of copper [112].

Another important aspect to highlight is that copper excess (as well as an excess of other transition metals such as molybdenum and zinc) can be toxic, so most organisms have developed defence mechanisms to form detoxification pathways. These mechanisms commonly act by reducing uptake, sequestration or enhancing elimination, and are controlled at different levels (transcriptional, translational and enzymatic) by inducing specific conformational changes which will affect the metal binding [113].

In zinc deficiency, both phase I and phase II biotransformation reactions are influenced, namely decreased, the effects being related to reduced levels of cytochrome P450s and reduced activities of certain phase II enzymes, namely UDPGTs and GSTs [114]. Recent studies have revealed that zinc deficiency may result in a decrease in the activities of specific demethylases and hydroxylases and even in the inhibition of the synthesis of a specific P450 isoform (CYP2D11) [115]. However there are also examples of enzyme activities that remain unchanged in zinc deficiency, for example that of the microsomal epoxide hydrolase [116]. As for the phase II enzymes, the activity of glutathione-S-transferase (important in conjugation reactions and involved in detoxication) has been proven to be significantly decreased in zinc deficiency [116].

On the other hand, as in the case of copper, zinc excess has also been proven to display an inhibitory action. Studies on experimental animals showed that high intake of zinc for a longer period inhibited NADPH-cytochrome C reductase, benzphetamine-N-demethylase and glutathione S-transferase activity, while the cytochrome P450 and cytochrome b5 activities were not obviously changed [117].

Magnesium deficiency, often found to correlate with calcium deficiency, has been proven to influence in the same manner the action of certain enzymatic systems, particularly that of the cytochrome P450s. However, experiments on rats proved that in magnesium deficient ('MgD') diet, no significant alterations were observed as far as different phase I enzymes were studied, with only aniline metabolism reduced by 30%. A fourfold decrease in the MgD group was identified for a phase II enzyme, a UDP-glucuronosyltransferase [118].

An interesting explanation to support especially the action of magnesium is based on the interaction of this mineral with the phospholipids and thyroid hormones. In magnesium deficiency, both thyroid hormone levels and microsomal content of phospholipids are depleted, thus resulting in decreased drug-metabolising capacity [2].

Calcium deficiency was shown to decrease the rates of metabolism of various drugs (prolonging their activity, e.g. for hexobarbital), by both oxidative and reductive pathways in liver microsomes, decreasing specific enzyme activities [119].

An essential trace element, closely related in biochemical action to vitamin E, is selenium. This element is particularly important because it was demonstrated to be a novel regulator of cellular heme metabolism, displaying an enhancement of two essential microsomal and mitochondrial enzymes involved in heme synthesis. However, at high concentrations *in vitro* selenium acted as an inhibitor [120].

As with copper, excessive selenium levels can also be inhibitory, as can be the deficiency. A role of selenium in the biosynthesis of microsomal components and phase II enzyme activities has been suggested [121].

From the above considerations we may conclude that the impact of dietary components and their interaction on drug metabolism can be extremely complex; on the other hand, the effects tend to occur predominantly under conditions of deficiencies. Malnutrition, unfortunately still prevalent in Third World countries, must be addressed in order to prevent or counteract the effects mentioned, since they often have complex and sometimes unpredictable consequences.

However, when discussing effects of diet on enzyme activities and their impact on drug metabolism, we have to consider also the so-called non-nutritional factors as still being part of diet; these would include ingestion of pyrolysis products (formed during cooking) and tobacco smoking.

Considering the pharmaco- and toxicological consequences, particular attention should be paid to the products formed in meat (or fish) when fried. These so-called pyrolysis products, commonly breakdown products of tryptophan, are enzyme inducers, displaying specificity for the P4501A1 isoform. At the same time, as they resemble closely the pyrolysis products from tobacco, they are likewise potential carcinogens/mutagens. The inducing effect will be reflected by increasing rate of biotransformation and consequently, reduced bioavailability of the drug. It is assumed that a polycyclic hydrocarbon inducer of CYTP450 is responsible for this effect. It is formed as a pyrolysis product in fried or charcoal-broiled meat. Interesting studies revealed not only a species difference, but also an organ and an enzyme specificity, in the action exerted by compounds of this type [122-123]. For example, in the case of the pyrolysis product from fried or charcoal-broiled meat, the target organ is commonly the liver, the main detoxication organ in the body. In animals treated with masheri (a form of roasted tobacco paste) the activities of enzymes occurring mostly in extrahepatic tissues was determined. The GI tract has been proved to be principally affected and so could become a predispositional factor in determining susceptibility to carcinogen exposure. As for enzyme specificity, an increase in the activity of phase I and a marked decrease in activity of phase II detoxication enzymes were observed.

The species difference was determined on mice, rats and hamsters [122]. The activity of specific enzymes, including CYTP450, benzo[α]pyrene hydroxylase and glutathione-S-transferase, decreased in the order: hamsters, rats and mice.

Because of chemical similarity with the pyrolysis products of tryptophan, some other groups of components should be considered in this context as well. We refer particularly to the indole type group of compounds found in cabbages and Brussels sprouts. The inductive effect has been proven to be species-dependent. This is supported by the different metabolisms affected in various species. If, for example in rats, these compounds induce the biotransformation of barbiturates, in humans they increase caffeine metabolism [2].

Administered to healthy volunteers, both Brussels sprouts and cabbage displayed stimulatory effects on antipyrine and phenacetin metabolism, by decreasing mean plasma half-life, increasing clearance rate and enhancing phase II conjugation reaction [124].

On other drugs, the effects were different, with more influence on phase II conjugative reactions. The most frequently quoted example is that of acetaminophen; its glucuronidation is enhanced, as is the amount of urinary recovery of the corresponding glucuronide, for which a mean increase of 8% was observed. In contrast, no comparable changes have been noticed in the biotansformation of acetaminophen to its sulphate conjugate. Also, no changes in the plasma glucuronide/oxazepam ratio were observed, suggesting a substrate specificity of the inductive effect exerted by cabbage or Brussels sprout [125].

An interesting and worthwhile aspect to stress is the potential of cabbage (and other Brassica species) as potent dietary cancer-inhibitors. The idea is supported by the fact that dietary cabbage has been reported to increase the aromatic hydrocarbon hydroxylase (AHH) microsomal enzyme system, and consequently to enhance the rate of metabolism of certain procarcinogenic drugs and carcinogens. Bacterial studies have also suggested that cabbage may additionally display a demutagenic activity [126].

Other non-nutrient components (but still categorised in the class of dietary factors) are food additives, flavourings, colourings and the like. These have usually been shown to act either as inducers or inhibitors of the metabolism of particular drugs. As an example, we should mention that di-*t*-butyl hydroxytoluene significantly increases the activity of some enzymes, including demethylases and hydroxylases in rat microsomes. Because of increased metabolism, duration of pentobarbital narcosis is significantly decreased. Final observations suggested that lipid-soluble compounds that are metabolised in liver microsomes, such as di-*t*-butyl hydroxytoluene, may generally increase the activities of drug-metabolising enzymes in liver microsomes [127].

A commonly used colouring agent for foods (as well as for some pharmaceutical preparations) is erythrosine. Examining its action in rat liver homogenates on labelled T4 and T3, experiments proved that in a dose-dependent manner, erythrosine inhibited the de-iodination of T4, and consequently the formation of T3. Further experiments revealed that other pathways of T4 metabolism were inhibited as well [128].

Being inhaled deliberately, tobacco smoke is still considered a 'dietary' component, which can affect drug therapy by both pharmacokinetic and pharmacodynamic mechanisms.

It usually displays intense inducing effects [129], in a way quite similar to that observed for ingestion of charcoal-broiled meat. The related factor is identified as the polycyclic hydrocarbon benzo[*a*]pyrene. The inducing effect will be reflected in low plasma levels of certain drugs, due to their increased biotransformation. A well-known and commonly quoted example is that of phenacetin metabolism in smokers and non-smokers [130]. Another example, considered as a marker for drug metabolism is antipyrine; tobacco smoke (which contains at least 3000 components) was found to increase the drug's clearance, lowering its bioavailability [130].

At the same time other drugs have shown no alterations in their biotransformation, thus indicating that tobacco smoke acts as a selective inducer [131].

A significant aspect is that enzymes induced by tobacco smoking may also increase the risk of cancer by enhancing the metabolic activation of carcinogens. Compounds believed to be implicated here are the polycyclic aromatic hydrocarbons, which are potent inducers of various CYTP450 isoforms. The most significantly affected has been proven to be an extrahepatic enzyme, the CYP1A1 isoform present in the lung. There is some evidence that high inducibility of this enzyme is more frequent in patients with lung cancer.

Drugs for which induced metabolism due to cigarette smoking may have clinical consequence include theophylline, caffeine, tacrine, imipramine, haloperidol, pentazocine, propranolol, flecainide and estradiol.

At the same time, clinical trials suggested that cigarette smoking leads to other pharmacological consequences, such as a faster clearance of heparin, a decrease in the rate of insulin absorption (after s.c. administration, due to the cutaneous vasoconstriction produced), heart-rate lowering during treatment with β-blockers, less analgesia from some opioids and less sedation from benzodiazepines. All these associated effects are attributed to the stimulant action of nicotine.

However, some of the tobacco smoke components have proven to act like enzyme inhibitors; examples include cadmium and carbon monoxide. As studies have thus far been confined to animal studies and the *in vitro* situation, the relevance for human drug metabolism has not been established. In all cases, from all actual data, it is the inductive effects of tobacco smoke that are prevalent [132].

Another extremely important aspect of tobacco smoking is that it can affect drug therapy *via* different mechanisms [133]. In an extended experimental study, both pharmacokinetic and pharmacodynamic drug interactions are described. As a direct consequence, cigarette smoking can reduce the efficacy of certain drugs or make drug therapy quite unpredictable. Pharmacokinetic interactions are presented for various drugs including theophylline, diazepam, propranolol and flecainide. This type of interaction causes enhanced plasma clearance, decrease in absorption, and induction of CYTP450 enzymes. Therefore, patients who are smokers would be in the situation of requiring larger doses of a certain drug for obtaining the desired therapeutic effect. The pharmacodynamic interactions, described in the study for antianginal and antihypertensive agents, oral contraceptives and histamine-2-receptor antagonists have an important impact in increasing the risk of adverse reactions, especially in smoker patients with cardiovascular or peptic ulcer disease, or in women smokers using oral contraceptives.

In the next chapter, we encounter other internal factors that influence drug biotransformation through their direct impact on the cytochrome P450 system. These include species, sex, age, disease state, hormonal control, as well as some external, environmental factors (excepting the heavy metals treated in the present chapter). A separate chapter discusses the impact of genetic factors.

References

1. Ronis MJJ, Ingelman-Sundberg M. 1999. Induction of Human Drug-Metabolising Enzymes: Mechanisms and Implications. In: Woolf TF, editor. Handbook of Drug Metabolism. New York, Marcel Dekker Inc, pp 239-262.
2. Gibson GG, Skett P. 1994. In: London Blackie Academic & Professional, An Imprint of Chapman & Hall, pp 77-95.
3. Ionescu C. 2001. In: Romania, Cluj-Napoca, "I. Haţieganu" Universitary Medical Printing House. Biotransformarea medicamentelor, pp 138-154.
4. Bock KW, Lipp HP, Bock-Hennig BS. 1990. Induction of drug-metabolising enzymes by xenobiotics. Xenobiotica 20:1101-1111.
5. Park BK, Kitteringham NR. 1990. Assessment of enzyme induction and enzyme inhibition in humans: toxicological implications. Xenobiotica 20:1171-1185.
6. Rhodes JS, Crabbe JC. 2005. Gene expression induced by drugs of abuse. Curr Opin Pharm 5:26-33.
7. Park BK, Kirtteringham NR, Pirmohamed M. 1995. In: Avan G, Balant LP, Bechtel PR, editors. Relevance of induction of human drug-metabolizing enzymes: Pharmacological and toxicological Implications, Specificity and Variability on Drug Metabolism. Luxembourg: Office for Official Publications of the European Communities EU, pp 169-190.
8. Kedderis GL. 1997. Extrapolation on *in vitro* enzyme induction data to humans *in vivo*. Chem-Biol Interact 107:109-121.
9. MacDonald MG, Robinson MG. 1968. Clinical observations of possible barbiturate interference with anticoagulation. JAMA-J Am Med Assoc 204:97-103.
10. Alexanderson B, Evans DA, Sjoqvist F. 1969. Steady-state plasma levels of nortriptyline in twins: influence of genetic factors and drug therapy. Brit Med J 4:764-768.
11. Brooks PM, Buchanan WW, Grove M, Downie WW. 1976. Effects of enzyme induction on metabolism of prednisolone. Ann Rheum Dis 35:339-343.
12. Stambaugh JE; Hemphill DM; Wainer IW; Schwartz I. 1977. A potentially toxic drug interaction between pethidine (meperidine) and phenobarbitone. Lancet 1: 398-399.
13. Kandrotas RJ, Cranfield TL, Gal P, Ransom JL, Weaver RL. 1990. Effect of phenobarbital administration on theophylline clearance in premature neonates. Ther Drug Monit 12:139-143.

14. Baarnhielm C, Skanberg I, Borg KO. 1984. Cytochtome P-450-dependent oxidation of felodipine a 1,4-dihydropyridine to the corresponding pyridine. Xenobiotica 14:719-726.

15. Martin H, Sarsat J-P, de Waziers I, Housset C, Balladur P, Beaune P, Albaladejo V, Lerche-Langrand, C. 2003. Induction of cytochtome P450s 2B6 and 3A4 expression by phenobarbital and cyclophosphamide in culture human liver slices. Pharm Res 20: 557-568.

16. Noriko E, Kei-ichi K, Kunio D. 2005. Induction of cytochrome P450 isozymes by phenobarbital in pregnant rat and fetal livers and placenta. Exp Mol Pathol 78:150-155.

17. Waxman DJ, Azaroff L. 1992. Review: phenobarbital induction of cytochrome P450 gene expression. Biochem J 281:577-592.

18. Waxman DJ, Chang TKH. 1995. Hormonal regulation of liver cytochrome P450 enzymes. In: Ortiz de Montellano PR, editor. Cytochrome P450: Structure, Mechanism, and Biochemistry. New York, Plenum Press, pp 391-417.

19. Cheung NW, Liddle C, Coverdale S, Lou JC, Boyages SC. 1996. Growth hormone treatment increases cytochrome P450-mediated antipyrine clearance in man. J Clin Endocr Metab 81:1999-2001.

20. Safe S, Krishnan V. 1995. Cellular and molecular biology of aryl hydrocarbon (Ah) receptor-mediated gene expression. Arch Toxicol 17:S99-115.

21. Guengerich FP. 1995. Human cytochrome P450 enzymes. In: Ortiz de Montellano PR, editor. Cytochrome P450: Structure, Mechanism, and Biochemistry. New York, Plenum Press, pp 473-575.

22. Hankinson O. 1994. A genetic analysis of processes regulating cytochrome P450 1A1 expression. Adv Enzyme Regul 34:159-171.

23. Ronis MJJ, Lindros KO, Ingelman-Sundberg M. 1996. The CYP2E family. In: Ioanides C, editor. Cytochromes P450: Metabolic and Toxicological Aspects. Boca Raton: CRC Press, pp 211-239.

24. Watkins PB, Wrighton SA, Schuetz EG, Maurel P, Guzelian PS. 1986. Macrolide antibiotics inhibit the degradation of the glucocorticoid-responsive cytochrome P450p in rat hepatocytes *in vivo* and in primary monolayer culture. J Biol Chem 261:6264-6271.

25. McGree RE, Ronis MJJ Jr, Cowherd RM, Ingelman-Sundberg M, Badger TM. 1994. Characterization of cytochrome P450 2el induction in a rat hepatoma FGC-4 cell model by ethanol. Biochem Pharmacol 48:1823-1833.

26. Hoensch H. 1987. Ethanol as enzyme inducer and inhibitor. Pharm Therapeut 33:121-128.

27. Omura T. 1994. Induction of drug metabolizing enzymes. Masui to Sosei 30:69-71.

28. Morel F, Fardel O, Meyer DJ, Langouet S, Gilmore KS, Meunier B, Tu CP, Kensler TW, Ketterer B, Guillouzo A. 1993. Preferential increase of glutathione S-transferase class alpha transcripts in cultured human hepatocytes by phenobarbital, 3-methylcholanthrene and dithiolethiones. Cancer Res 53:231-234.

29. Schuetz EG, Beck WT, Schuetz JD. 1996. Modulators and substrates of P-glycoprotein and cytochrome P450 3A co-ordinately up-regulate these proteins in human colon carcinoma cells. Mol Pharmacol 49:311-318.

30. Hutabarat RM, Yost GS. 1989. Purification and characterization of an ethanol-induced UDP-glucuronosyltransferase. Arch Biochem Biophys 273:16-25.

31. Doostdar H, Grant MH, Melvin WT, Wolf CR, Burke MD. 1993. The effects of inducing agents on cytochrome P450 and UDP-glucuronosyltransferase activities in human HepG2 hepatoma cells. Biochem Pharmacol 46:629-635.

32. Egner PA, Kensler TW, Prestera T, Talalay P, Libby AH, Joyner HH, Curphey TJ. 1994. Regulation of phase 2 enzyme induction by oltipraz and other dithioethiones. Carcinogenesis 15:177-181.

33. Guengerich FP. 1999. Inhibition of Drug Metabolizing Enzymes: Molecular and Biochemical Aspects. In: Woolf TF, editor. Handbook of Drug Metabolism. New York, Marcel Dekker Inc, pp 203-209.

34. Gibson GG, Skett P. 1994. In: London Blackie Academic & Professional, An Imprint of Chapman & Hall, pp 96-106.

35. Ionescu C. 2001. In: Romania, Cluj-Napoca, "I. Hațieganu" Universitary Medical Printing House. Biotransformarea medicamentelor, pp 155-157.

36. Bronson DD, Daniels DM, Dixon JT, Redick CC, Haaland PD. 1995. Virtual kinetics: Using statistical experimental design for rapid analysis of enzyme inhibitor mechanisms. Biochem Pharmacol 50:823-831.

37. Guengerich FP, Kim D-H, Iwasaki M. 1991. Role of human cytochrome P-450 IIE1 in the oxidation of many low molecular weight cancer suspects. Chem Res Toxicol 4: 168-176.

38. Tian GC, Mook RA, Loss ML, Frye SV. 1995. Mechanism of time-dependent inhibition of 5alpha-reductase by delta(1)-4-azasteroids: Toward perfection of rates of time-dependent inhibition by using ligand-binding energies. Biochemistry US 34:1343-1349.

39. Silverman RB. 1995. Mechanism-based enzyme inactivators. Method Enzymol 249: 240-254.

40. Ortiz de Montellano PR. 1988. Suicide substrates for drug metabolizing enzymes: mechanisms and biological consequences. In: Gibson GG, editor. Progress in Drug Metabolism (vol.11). London: Taylor and Francis, pp 99-148.

41. Moore MR. 2004. A commentary on the impacts of metals and metalloids in the environment upon the metabolism of drugs and chemicals. Toxicol Lett 148:153-158.

42. Meredith PA, Moore MR.1979.The influence of lead on haem biosynthesis and biodegradation in the rat. Proc Biochem Soc T 7:637-639.

43. Abbas-Ali B.1980. Effect of mercuric chloride on microsomal enzyme system in mouse liver. Pharmacology 21:59-63.

44. Veltman JC, Maines MD. 1986. Alterations of heme cytochrome P450, and steroid metabolism by mercury in rat adrenal. Arch Biochem Biophys 248:467-478.

45. Xiao GH, Wu JL, Liu YG. 1989. The effects of cadmium, mercury and lead in vitro on hepatic microsomal mixed function oxidase and lipid peroxidation. J Tongji Med Univ 9:81-85.

46. Ke Q, Yang Y, Ratner M, Zeind J, Jiang C, Forrest JN, Xiao YF. 2002. Intracellular accumulation of mercury enhances P4501A1 expression and Cl-currents in cultured shark rectal gland cells. Life Sci 70:2547-2566.

47. Rosemberg DW, Kappas A. 1991. Induction of heme oxygenase in the small intestinal epithelium: a response to oral cadmium exposure. Toxicol 67:199-210.

48. Alexidis AN, Rekka EA, Kourounakis PN. 1994. Influence of mercury and cadmium intoxication on hepatic microsomal CYP2E and CYP3A subfamilies Res Commun Mol Pathol Pharmacol 85:67-72.

49. Baker JR, Satarug S, Edwards RJ, Moore MR, Williams DJ, Reily PEB. 2003. Potential for early involvement of CYP isoforms in aspects of human cadmium toxicity. Toxicol Lett 137:85-93.

50. Jahn R, Klinger W. 1982. Influence of age on in vitro effect of cadmium on rat liver cytochrome P450 concentration and monooxygenases activity. Acta Pharmacol Toxicol 50:85-88.

51. Schnell RC, Pence DH. 1981. Differential effects of cadmium on the hepatic microsomal cytochrome P450 system in male and female rats. Toxicol Lett 9:19-25.

52. Holla VJ, Adas F, Imig JD, Zhao X, Price E, Olsen N, Kovacs WJ, Magnuson MA, Keeney DS, Breyer MD, Falck JR, Waterman MR, Capdevila JH. 2001. Alterations in the regulation of androgen-sensitive CYP 4a monooxygenases cause hypertension. Proc Nat Acad Sci 98:5211-5216.

53. Tully DB, Collins BJ, Overstreet JD, Smith CS, Dinse GE, Mumtaz MM, Chapin RE. 2000. Effects of arsenic, cadmium, chromium, and lead on gene expression regulated by a battery of 13 different promoters in recombinant Hep G2 cells. Toxicol Appl Pharmacol 168:79-90.

54. Satarug S, Ujjin P, Reily PEB. 2000. Changes in CYPIA specific monooxygenase in rat liver caused by cadmium and lipopolysaccharide administration. The 13th International Symposium on Microsomes and drug Oxidations.

55. Satarug S, Baker JR, Urbenjapol S, Haswell-Elkins MR, Reilly PEB, Williams DJ, Moore MR. 2003. A global perspective on cadmium pollution and toxicity in non-occupationally exposed population. Toxicol Lett 137:65-83.

56. Albores A, Cebrian ME, Bach PH, Connelly JC, Hinton RH, Bridges JW. 1989. Sodium arsenite induced alterations in bilirubin excretion and heme metabolism. J Biochem Toxicol 4:73-78.

57. Vernhet L, Allain N, Le Vee M, Morel F, Guillouzo A, Fardel O. 2003. Blockage of multidrug resistance-associated protein potentiates the inhibitory effects of arsenic trioxide on CYP1A1 induction by aromatic hydrocarbons. J Pharmacol Exp Ther 304:145-155.

58. Brown J, Kitchin KT, George M. 1997. Dimethylarsinic acid treatment alters six different rat biochemical parameters: relevance to arsenic carcinogenesis. Teratog Carcinog Mutagen 17:71-84.

59. Katiyar SK, Matsui MS, Muktar H. 2000. Ultraviolet-B exposure of human skin induces cytochromes P450 1A1 and 1B1. J Invest Dermatol 114:328-333.

60. Richter T, Schwab M, Eichelbaum M, Zanger UM. 2005. Inhibition of human CYP2D6 by N,N',N''-triethylenethiophosphoramide is irreversible and mechanism-based. Biochem Pharmacol 69:527-524.

61. Richter T, Muerdter TE, Heinkele G, Pleiss J, Tatzel S, Schwab M. 2004. Potent mechanism-based inhibition of human CYP2B6 by clopidogrel and ticlopidine. J Pharmacol Exp Ther 308:189-197.

62. Oleson FB, Berman CL, Li AP. 2004. An evaluation of the P450 inhibition and induction potential of daptomycin in primary human hepatocytes. Biochim Biophys Acta 1675:120-129.

63. Gaudineau C, Auclair K. 2004. Inhibition of human P450 enzymes by nicotinic acid and nicotinamide. Biochem Bioph Res Co 317:950-956.

64. Chen TL, Wu CH, Chen TG, Tai YT, Chang HC, Lin CJ. 2000. Effects of propofol on functional activities of hepatic and extrahepatic conjugation enzyme systems. Brit J Anaesth 84:771-776.

65. Lin HJ, Lu AY. 1998. Inhibition and induction of cytochrome P450 and the clinical implications. Clin Pharmacokinet 35:361-390.

66. Riley RJ, Grime K. 2004. Metabolic screening in vitro: metabolic stability, CYP inhibition and induction. Drug Discov Today: Technologies 1:365-372.

67. Anderson KE, Conney AH, Kappas A. 1982. Nutritional influences on chemical biotransformations in humans. Nutr Rev 40:161-171.

68. Kanke Y, Iwama M. 2000. Xenobiotic metabolism and dietary-nutritional factors. Shokuhin Eiseigaku Zasshi 41:95-108.

69. Melander A, McLean A. Influence of food intake on presystemic clearance of drugs. Clin Pharmacokinet 8:286-296.

70. Yang CS, Brady JF, Hong JY. 1992. Dietary effects on cytochromes P450, xenobiotic metabolism, and toxicity. FASEB J 6:737-744.

71. Alvares AP, Pantuck EJ, Anderson KE, Kappas A, Conney AH. 1979. Regulation of drug metabolism in man by environmental factors. Drug Metab Rev 9:185-206.

72. Campbell TC. 1978. Effects of dietary protein on drug metabolism. Nutr Drug Interrelat [Pap Int Symp] 409-422.

73. Amelizad Z, Narbonne JF. 1982. Interaction of xenobiotics and nutritional factors with drug metabolizing systems. Dev Biochem 23:63-66.

74. Anderson KE, McCleery RB, Vesell ES, Vickers FF, Kappas A. 1991. Diet and cimetidine induce comparable changes in theophylline metabolism in normal subjects. Hepatology 13:941-946.

75. Wade AE. 1986. Effects of dietary fat on drug metabolism. J Environ Pathol Tox 6:3-4.

76. Watanabe A, Nakatsukasa H, Kobayashi M, Nagashima H. 1985. Effect of dietary lipid on hepatic microsomal drug-metabolizing enzyme activities in sake-ingested rats. Res Commun Substance 6:243-246.

77. Janus K, Muszczynski Z, Skrzypczak WF. 1996. Hepatic biotransformation rate in calves fed food rations with protein or carbohydrate supplements. Acta Vet Brno 65:123-131.

78. Satoh A, Shimomura Y, Suzuki M. 1990. Effects of long-term feeding of a high-carbohydrate diet or a high-fat diet on hepatic drug metabolism in rats. Taiiku Kagakukei Kiyo 13:161-167.

79. Sato A, Nakajima T. 1984. Dietary carbohydrate-and ethanol-induced alteration of the metabolism and toxicity of chemical substances. Nutr Cancer 6:121-132.

80. Nakajima T, Sata A. 1982. Effect of low carbohydrate diet on the induction of drug-metabolizing enzymes by phenobarbital, polychlorinated biphenyl and 3-methylcholanthene. Igaku no Ayumi 123:1003-1006.

81. Bray B, Inder RE, Rosengren RJ. 2000. Retinol-mediated effects on the activation and detoxification pathways of paracetamol. Australas J Ecotoxicol 6:75-59.

82. Bray BJ, Goodin MG, Inder RE, Rosengren RJ. 2001. The effect of retinol on hepatic and renal drug-metabolizing enzymes. Food Chem Toxicol 39:1-9.

83. Ushio F, Fukuhara M, Bani M-H, Narbonne J-F. 1996. Expression of cytochrome P450 isoenzymes in Syrian hamster after dietary vitamin A supplementation or deficiency. Int J Vitam Nutr Res 66:197-202.

84. Sarkar A, Mukherjee B, Chatterjee M. 1994. Inhibitory effect of β-carotene on chronic 2-acetylaminofluorene induced hepatocarcinogenesis in rat: reflection in hepatic drug metabolism. Carcinogenesis 15:1055-1060.

85. Yoo J, Sook H, Park HS, Ning SM, Lee MJ, Mao J, Yang CS. 1990. Effects of thiamin deficiency on hepatic cytochromes P450 and drug-metabolizing enzyme activities. Biochem Pharmacol 39:519-525.

86. Sushko LI, Lukienko PI. 1981. Effect of vitamin B1 on the hydroxylating function of rat liver. Vestsi Akademii Navuk BSSR, Seryya Biyalagichnykh Navuk 4:74-77.

87. Omaye ST, Green MD, Dong MH. 1981. Influence of dietary thiamin on pulmonary, renal, and hepatic drug metabolism in the mouse. J Toxicol Environ Health 7:317-326.

88. Patel JM, Pawar SS. 1974. Riboflavine and drug metabolism in adult male and female rats. Biochem Pharmacol 23:1467-1477.

89. Galdhar NR, Pawar SS. 1975. Effects of dietary riboflavine levels and phenobarbital pretreatment on hepatic drug metabolizing enzymes and lipid peroxidation in young male rats. Indian J Med Res 63:507-517.

90. Taniguchi M. 1980. Effects of riboflavin deficiency on lipid peroxidation of rat liver microsomes. J Nutr Sci Vitaminol 26:401-413.

91. Hietanen E, Koivusaari U, Norling A. 1980. Influence of riboflavin deficiency on intestinal drug metabolizing enzyme activities in rat. Adv Physiol Sci 12:89-93.

92. Saito M, Nagayama S, Ikegani S, Innami S. 1990. Influence of vitamin B_2 deficiency on polychlorinated biphenyls-induced liver lipid peroxides formation in rats. Int J Vitam Nutr Res 60:255-260.

93. Omaye ST, Turnbull JD. 1980. Effect of ascorbic acid on heme metabolism in hepatic microsomes. Life Sci 27:441-449.

94. Yoon Ki-W, Lee Sang-H, Lee Sun-M. 2000. Effects of vitamins C and E on hepatic drug metabolizing function in hypoxia/reoxygenation. Yakhak Hoechi 44:237-244.

95. Horn LR, Machlin LJ, Barker MO, Brin M. 1976. Drug metabolism and hepatic heme proteins in the vitamin E-deficient rat. Arch Biochem Biophys 172:270-277.

96. Meshali MM, Nightingale CH. 1976. Effect of alpha tocopherol (vitamin E) deficiency on intestinal transport of passively absorbed drugs. J Pharm Sci 65:344-349.

97. Gourlay GK, Savage JK, Stock BH. 1977. Hepatic drug metabolism in normal and vitamin E-deficient female Merino sheep. Toxicol Appl Pharmacol 39:365-375.

98. Kosinova A, Hudecova A, Madaric A. 1978. Study of combined effect of polychlorinated biphenyls and vitamin E on liver [enzyme] induction in rat. Cesk Hygiena 23:369-376.

99. Carpenter MP. 1972. Vitamin E and microsomal drug hydroxylations. Ann NY Acad Sci 203:81-92.

100. Atkinson JE, Gairola CC, Lubawy WC. 1983. Increasing lipid peroxidation by vitamin E deficiency does not augment adriamycin-induced inhibition of hepatic drug metabolism. Toxicology 29:121-129.

101. Murray M. 1991. *In vitro* and *in vivo* studies on the effect of vitamin E on microsomal cytocrome P450 in rat liver. Biochem Pharmacol 42:2107-2114.

102. Stoeva E, Tantcheva L, Mileva M, Savov V, Galabov AS, Braykova A. 2001. Preventive effect of vitamin E on the processes of free radical lipid peroxidation and monooxygenase enzyme activity in experimental influenza virus infection. Adv Exp Med Biol 500:257-260.

103. De Porre PM-Z R, Bruynseels JPJM, Wouters WBL. 1999. Combination of a RAMBA and a tocopherol. Janssen Pharmaceutica N.V., Belg PCT Int Appl. 24 pp.

104. Landes N, Pfluger P, Kluth D, Birringer M, Ruhl R, Bol G-F, Glatt H, Brigelius-Flohe R. 2003. Vitamin E activates gene expression via the pregnane X receptor. Biochem Pharmacol 65:269-273.

105. Pentyuk AA, Bogdanov NG, Khadur R, Lutsyuk NB, Borisenko BA. 1989. Activity of xenobiotic-metabolizing enzymes and the state of rat liver microsomal membranes in vitamin K deficiency. Biochemistry (Moscow) 54:1700-1708.

106. Sidorova Yu A, Grishanova A Yu. 2004. Dose-and Time-Dependent Effects of Menadione on Enzymes of Xenobiotic Metabolism in Rat Liver. Bull Exp Biol Med 137:231-234.

107. Loebstein R, Yonath H, Peleg D, Almog S, Rotenberg M, Lubetsky A, Roitelman J, Harats D, Halkin H, Ezra D. 2001. Interindividual variability in sensitivity to warfarin-nature or nurture? Clin Pharmacol Ther 70:159-164.

108. Takiguchi M, Arizono K, Ariyoshi T. 1996. Effects of different levels of iron deficiency on cytochrome P 450 isoenzyme in rats. Biomedical Research on Trace elements 7:217-218.

109. Rao NJ, Jagadeesan V. 1995. Effect of long term iron deficiency on the activities of hepatic and extra-hepatic drug metabolising enzymes in Fischer rats. Biochem Mol Biol 110:167-173.

110. Rao N J, Jagadeesan V. 1994. Activities of carcinogen metabolising enzymes in hepatic and extra-hepatic tissues of iron-deficient rats. J Clin Biochem Nutr 16:177-185.

111. Keen CL, Uriu-Hare JY, Hawk SN, Jankowski MA, Daston GP, Kwik-Uribe, Catherine L, Rucker RB. 1998. Effect of copper deficiency on prenatal and pregnancy outcome. Am J Clin Nutr 67:1003S-1011S.

112. Moffit AE Jr, Murphy SD. 1973. Effect of excess and deficient copper intake on hepatic microsomal metabolism and toxicity of foreign chemicals. Trace Substances in Environmental Health 7:213-223.

113. Dameron CT, Harrison MD. 1998. Mechanisms for protection against copper toxicity. Am J Clin Nutr 67:1091S-1097S.

114. Jagadeesan V, Oesch F. 1988. Effects of dietary zinc deficiency on the activity of enzymes associated with phase I and II of drug metabolism in Fisher-344 rats: activities of drug metabolising enzymes in zinc deficiency. Drug Nutr Interact 5:403-413.

115. Iizuka Y, Sakurai E, Tanaka Y. 2001. Effects of trace element deficiency on drug metabolizing enzymes in rats. RIKEN Rev 35:3-4.

116. Jagadeesan V. 1989. Study of activating and conjugating enzymes of drug metabolism in zinc deficiency. J Exp Biol 27:799-801.

117. Ding H, Peng R, Chen J. 1998. Effects of high dietary zinc intake on liver function, hepatic drug metabolism enzymes and membrane fluidity in mice. Weisheng Yanjiu 27:180-182.

118. Brown RC, Wang W, Meskin MS, Bidlack WR. 1997. Effect of dietary magnesium deficiency on rat hepatic drug metabolism and glucuronidation. Environ Nutr Interact 1:253-271.

119. Dingell JV, Joiner PD, Hurwitz L. 1966. Impairment of hepatic drug metabolism in calcium deficiency. Biochem Pharmacol 15:971-976.

120. Maines MD, Kappas A. 1976. Selenium regulation of hepatic heme metabolism: induction of δ-aminolevulinate synthase and heme oxygenase. P Natl Acad Sci USA 73:4428-4431.

121. Obol'skii OL, Kravchenko LV, Avren'eva LI, Tutel'ian VA. 1998. Effect of dietary selenium on the activity of UDP-glucuronosyltransferases and metabolism of mycotoxin deoxynivalenol in rats. Vop pitan 4:18-23.

122. Nair UJ, Ammingan N, Kulkarni JR, Bhide SV. 1991. Species difference in intestinal drug metabolizing enzymes in mouse, rat and hamster and their inducibility by masheri, a pyrolysed tobacco product. Indian J Exp Biol 29:256-258.

123. Nair UJ, Ammigan N, Kayal JJ, Bhide SV. 1991. Species differences in hepatic, pulmonary and upper gastrointestinal tract biotransformation enzymes on long-term feeding of masheri – a pyrolyzed tobacco product. Digest Dis Sci 36:293-288.

124. Pantuck EJ, Pantuck CB, Garland WA, Min BH, Wattenberg LW, Anderson KE, Kappas A, Conney AH. 1979. Stimulatory effect of Brussels sprouts and cabbage on human drug metabolism. Clin Pharmacol Ther 25:88-95.

125. Pantuck EJ, Pantuck CB, Anderson KE, Wattenberg LW, Conney AH, Kappas A. 1984. Effects of Brussels sprouts and cabbage on drug conjugation. Clin Pharmacol Ther 35:161-169.

126. Alberto-Puleo M. 1983. Physiological effects of cabbage with reference to its potential as a dietary cancer-inhibitor and its use in ancient medicine. J Ethnopharmacol 9:261-272.

127. Takanaka A, Kato R, Omori Y. Effect of food additives and colors on microsomal drug-metabolizing enzymes of rat liver. Shokuhin Eiseigaku Zasshi 10:260-265.

128. Ruiz M, Ingbar SH, Charles A. 1982. Effect of erythrosine (2',4',5', 7'-tetraiodofluorescein) on the metabolism of thyroxine in rat liver. Endocrinology 110:1613-1217.

129. Jusko JW. 1979. Influence of cigarette smoking on drug metabolism in Man. Drug Metab Rev 9:221-236.

130. Scavone JM, Joseph M, Greenblatt DJ, LeDuc BW, Blyden GT, Luna BG, Harmatz, Herold S. 1990. Differential effect of cigarette smoking on antipyrine oxidation and acetaminophen conjugation. Pharmacology 40:77-84.

131. Barau M, Flotats I, Massot M. 2001. Interactions between tobacco and drugs. Circ Farm 59:10-16.

132. Zevin S, Benowitz NL. 1999. Drug interactions with tobacco smoking: an update. Clin Pharmacokinet 36:425-438.

133. Schein JR. 1995. Cigarette smoking and clinically significant drug interactions. Ann Pharmacother 29:1139-1148.

Chapter 6

FACTORS THAT INFLUENCE DRUG BIOTRANSFORMATION

6.1 INTRODUCTION

An account of the influence of species, sex, age, hormonal status and disease state on drug biotransformation forms the major part of this chapter, these parameters collectively being referred to as 'intrinsic factors'. Reported results of several recent studies of these factors are reviewed, with the aim of indicating their variable nature as well as their interdependence. Thus, for example, we will encounter cases where for certain drugs, there are only subtle differences in the biotransformation routes in different species, while for others, dramatically different pathways are adopted, leading to the formation of vastly different metabolic products. Interdependence of most of these factors is a natural expectation, given that the status of an individual's metabolising activity and pathological status vary over a lifetime. Thus, for example, the effects of natural attrition of the metabolising activity in an aged patient and a specific disease state can interact in a way that results in a unique mode of metabolic clearance of a drug.

One significant point emerges from the discussion, namely the variability in the outcomes of drug metabolism observed for different species. This unpredictable element reminds us of the caution that should be exercised in extrapolating from animal studies to humans and the implications that this has in the evaluation of new drugs in the pharmaceutical industry.

The final section of this chapter summarises the effects of external (environmental) factors on drug biotransformation, examples of which were encountered in earlier chapters. Continual introduction of new chemical substances into the environment through waste production and industrial activity remains a major international issue, necessitating *inter alia* on-going studies of their effects on human drug metabolism.

6.2 INTRINSIC FACTORS

Drugs, as well as other xenobiotics are metabolised *via* various pathways, including phase I and phase II reactions, which involve participation of numerous enzyme systems. Therefore, it is reasonable to assume that there are many factors that can determine or influence along which pathway a particular drug will undergo biotransformation and the extent to which this will proceed.

These factors are usually arbitrarily divided into internal and external factors, with nevertheless considerable interaction between them [1,2].

6.2.1 Species

Examples of species differences in drug biotransformation are numerous, continuously investigated, and encountered in both phases of biotransformation [3,4]. An interesting observation is that they may involve the same route, but differ in the rate along that particular pathway (i.e. quantitatively different) or they may adopt different pathways (i.e. differing qualitatively) [5,6]. It should be noted as well that there is not always a direct relationship between metabolism, half-life and action of a drug [7].

Selected examples

An interesting quantitative species difference in phase I metabolism is known for caffeine, both in terms of total metabolism and metabolite production [8]. Thus, the total metabolism is highest in humans, decreasing in the order - monkey, rat and rabbit. While there are no significant differences in the formation of theobromine, marked differences have been recorded for the other two metabolites, paraxanthine and theophylline, with paraxanthine formation highest in humans and lowest in monkey, whereas the reverse obtains for theophylline [8].

An interesting aspect is the way caffeine biotransformation reactions proceed in higher plants, the variability of caffeine catabolism again being dependent on species and to a greater extent, on the age of different tissues investigated. As an example, it was reported that in young tea leaves, theophylline is re-utilised for caffeine biosynthesis, while in aged leaves of Coffea arabica, it undergoes further metabolism resulting in 7-methylxanthine accumulation. Other species of Coffea have been proven to convert caffeine to methyluric acids. Obviously, these cases exemplify qualitative differences, as well as species- and age-dependence [9].

A well-known quantitative example is that of species variation in hexobarbitone metabolism, affecting half-life and sleeping time. Investigations have been made on man, dog, mice and the rat [10]. The longest half-time was registered for man (~360 min). The sleeping time increased in the following order: mice, rats, dogs and man. The main conclusion of the experiment, apart from demonstrating that the oxidative metabolism of hexobarbitone is markedly influenced by species, was that the biotransformation is inversely related to the half-time and duration of action of the investigated drug, the highest metabolism being registered for mice and decreasing in the opposite order as for the sleeping time for example.

A recent example refers to the variation in the metabolism of selegiline (structure in Figure 6.1) ((-)-form of deprenyl) in seven different species [11]. From literature data, it is known that selegiline undergoes N-dealkylation, yielding several metabolites, namely N-desmethylselegiline, methamphetamine and amphetamine.

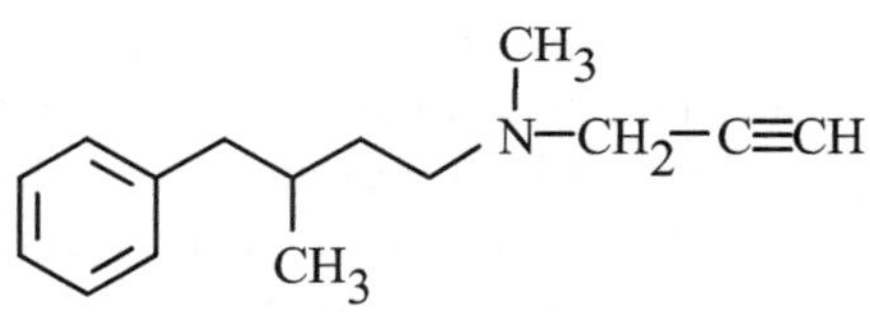

Fig.6.1 Selegiline

The investigations made during the study referred to, and performed on liver microsomes of different species, in addition to characterizing the potential metabolic variations, also proved the existence of another metabolite, the N-oxide. The rate and extent of formation of this metabolite was found to be markedly influenced by species, the highest rate of production occurring in dog and hamster, being much lower in humans, and zero in the rat.

Another example of quantitative variation was revealed from experimental studies investigating the metabolic profile of a relatively novel diuretic. A comparative approach was adopted, aimed at demonstrating its metabolism in experimental animals and human liver microsomes [12]. Increased rates of metabolism were observed in rats and monkeys, and six metabolites, designated RU1, RU2, RU3 and MU1, MU2, MU3 for the respective species, were identified in their urine. Quantitatively, only three of these were considered to be major metabolites in rat and monkey urine, namely RU3, RU1 and MU3 respectively, whereas in the dog, the unchanged drug was observed as the major urinary component. This indicated a net difference between the rat and the monkey, both displaying extensive biotransformation, and the dog, in which only little metabolism occurred. In contrast with dogs, humans showed similarities with rats, suggesting a common metabolic pathway.

Six species have been investigated in connection with the psychoactive drug of abuse 4-bromo-2,5-dimethoxyphenethylamine (2C-B) (street names 'Venus', 'Bromo', 'Erox', 'XTC' or 'Nexus') (Figure 6.2).

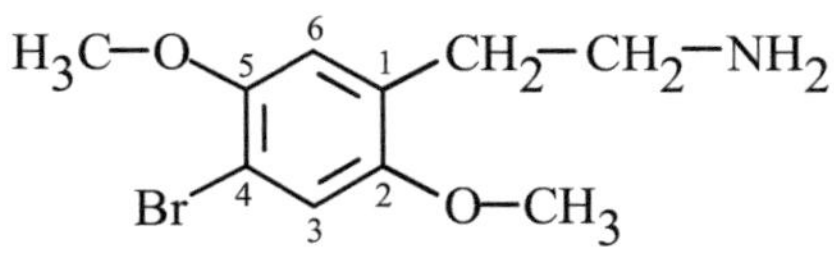

Fig.6.2 Structure of the psychoactive drug of abuse, 4-bromo-2,5-dimethoxyphenethylamine

Hepatocytes from human, monkey, dog, rabbit, rat and mouse were incubated with 2C-B in an attempt to identify the resulting metabolites and to monitor possible toxic effects [13]. Investigations established that the drug under study undergoes oxidative deamination with successive formation of two metabolites, which may or not undergo further metabolism by demethylation. Marked differences were noticed with two other, less common metabolites identified, one of these occurring only in mouse hepatocytes, the other in human, monkey and rabbit, but not in dog, rat and mouse, supporting the idea of qualitative interspecies variations. Another aim of the study, as mentioned above, was to compare the toxic effects exerted by 2C-B on hepatocytes of the six investigated species: the differences observed were only minor. However, another important aspect was revealed, namely that large differences in susceptibility of hepatocytes may occur between different individuals.

The biotransformation pathways of a relatively novel drug used as an acute oral treatment for migraine, namely zolmitriptan (Figure 6.3), were comparatively investigated in human and rat liver microsomes [14].

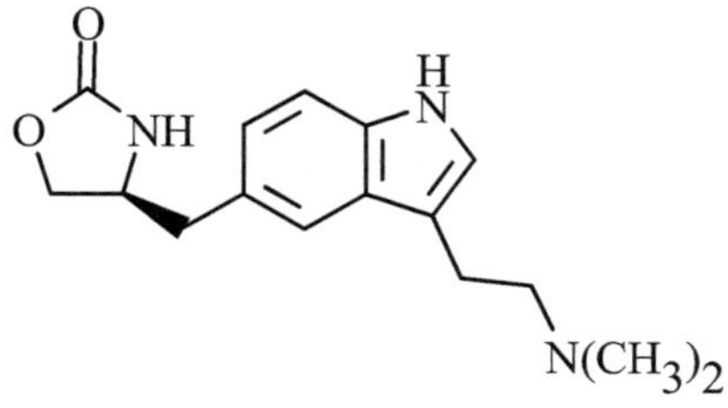

Fig.6.3 Structure of zolmitriptan

Although the reports indicated that the drug was metabolised by the same CYP isoform in both types of microsome, the numbers of metabolites nevertheless differed. This suggests that the report presents a reasonable and economical *in vitro* model for comprehensive studies of zolmitriptan metabolism, including biotransformation pathways, enzyme kinetics, induction and inhibition phenomena, interspecies differences and the possible occurrence of drug interactions.

An interesting study, involving both phase I and phase II biotransformations, has been performed in an approach using comparative interspecies data for both prospective design and extrapolations from animal findings to humans [15]. The aim was to reduce the potential for human risk and increase therapeutic benefit. For paclitaxel (Figure 6.4) for instance, markedly different metabolites were observed to occur in rats and humans, which renders metabolic drug-drug interaction investigations in rats practically irrelevant for humans (thus, qualitative differences). In contrast, for zidovudine (AZT), the variations were quantitative, with a high rate of glucuronidation in humans, resulting in a much shorter half-life than that observed in animals, which display negligible glucuronidation. This study revealed more significant features: qualitative differences in phase I biotransformation and quantitative variations in phase II, with no relevant similarities to allow extrapolations and drug-drug interaction predictions from animals to humans.

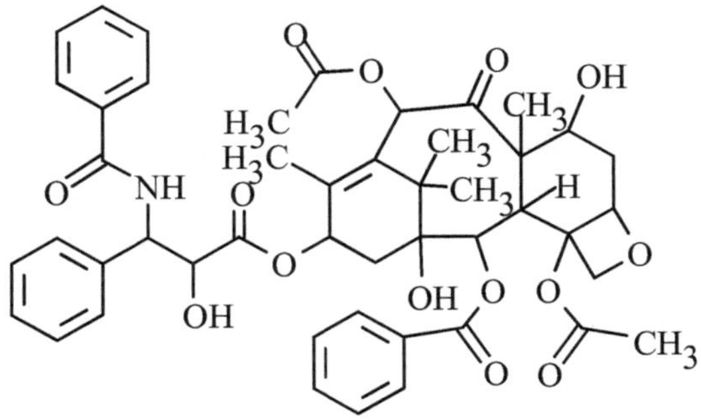

Fig.6.4 Paclitaxel

Advanced analytical procedures (e.g. LC/MS, high field NMR spectroscopy) have been used to examine the potential differences in the biotransformation of efavirenz [16], a potent and specific inhibitor of reverse transcriptase commonly recommended in the treatment of HIV infections. Metabolites produced by humans, rats, guinea pigs, hamsters and monkeys were investigated. Observations confirmed that efavirenz (Figure 6.5) is extensively metabolised by all species, with marked species differences in the metabolites isolated and structurally determined. Although the major metabolite, namely the O-glucuronide conjugate, proved to be common to all five species studied, other metabolites displayed species specificities as follows: the sulphate conjugate was found in rats' and monkeys' urine, but not in that of humans, while GSH-related metabolites were identified only in the urine of rats and guinea pigs.

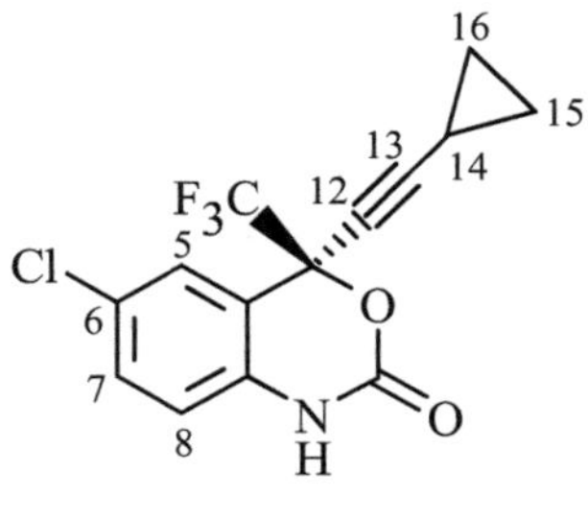

Fig.6.5 Efavirenz

Differences in the production of reactive metabolites may sometimes result in species-selective nephrotoxicity. For example, efavirenz was reported to produce renal injury (necrosis of the renal tubular epithelial cells) in rats, but not in monkeys or humans. Here, a species-specific glutathione adduct, produced only by rats, was deemed responsible for this nephrotoxic effect [17].

Species differences involve, as mentioned above, both phases of biotransformation. An interesting study was performed to investigate the maintenance of drug-metabolising capacities in collagen gel sandwich and immobilisation cultures of human and rat hepatocytes [18]. L-proline was added to the medium to improve albumin secretion. As far as most important phase I enzyme systems are concerned, namely the cytochrome P450-dependent monooxygenase (CYP) and microsomal epoxide hydrase (mEH) systems, comparative measurements of enzyme activities in the absence and presence of L-proline, revealed that their biotransformation enzyme activities were not affected by the addition of L-proline. Instead, the activity of an important phase II enzyme, GST, was decreased in rat hepatocytes, whereas in humans it remained almost unchanged. As human hepatocytes showed a better maintenance of GST activities than the rats in the presence of L-proline, species differences were again demonstrated.

Another study investigated whether there are also species variations in maintaining certain phase I and phase II enzyme activities after cryopreservation of liver slices prepared from five different species, namely mouse, rat, dog, monkey and human [19]. The conclusion of the study was that although the metabolic patterns and rates of biotransformation varied among these species, the phase I and phase II metabolic capacities of the liver slices were well maintained after cryopreservation.

For certain drugs, and depending upon the species investigated, variations have not proven to be very significant. For example, an experiment concerning orbifloxacin metabolism in two species, pigs and calves, aimed at establishing possible species differences, proved that in both species the metabolic pathway of the drug was the same, differing only in the amount of the excreted metabolite [20]. Indeed the final, common metabolite was the glucuronide, excreted in average amounts of 3% and 1% in pigs and calves respectively. In addition, the remainder of the drug was excreted unchanged in both species. However, a qualitative difference was noted, namely that calf urine contained also a product of oxidative metabolism.

Quantitative species differences were established for the immunosuppressive drug cyclosporine A (CSA) (Figure 6.6). [21]. Investigations were performed on liver and small intestinal microsomes from rat, hamster, rabbit, dog, baboon and man.

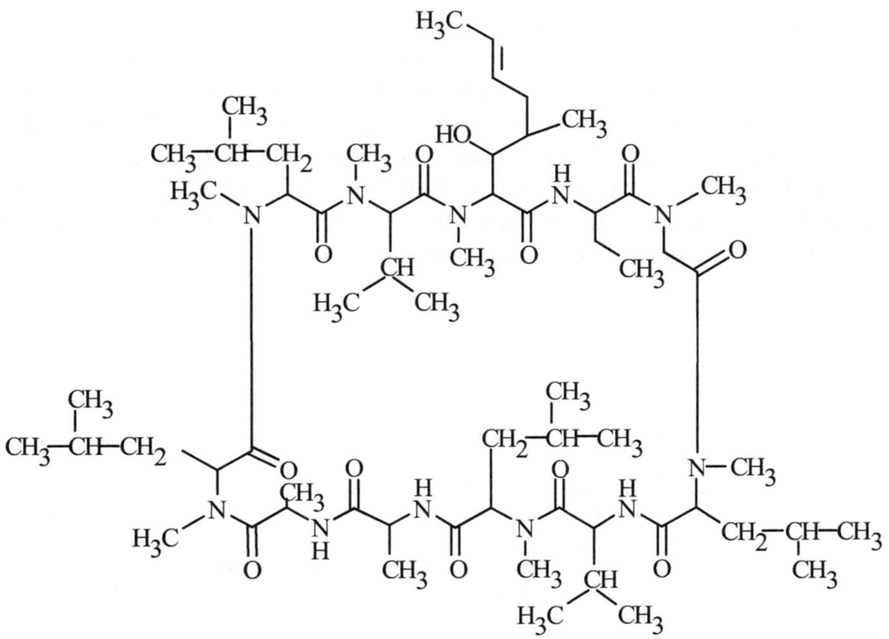

Fig.6.6 The structure of cyclosporine A

The metabolic pathways of CSA are known to result in two principal metabolites, the hydroxylated and N-demethylated CSA, which accounted for most of the CSA metabolised in all tested species. However, marked variations occurred in the biotransformation rate, measuring only 2-8% over 30 min in rats, in contrast to dogs, whose liver microsomes proved to be very efficient, yielding a 70-100% change in the same period. Investigations having been performed on both liver and small intestinal microsomes, another objective of the study was to determine possible differences determined by different tissues of the same organism. Measurements of the formation of the principal metabolites in the two investigated organs indicated a similar metabolic profile, but with differences in the rate of metabolism, that in the small intestine being slightly slower.

Differences in the metabolic profiles were the subject of investigation for panomifene (Figure 6.7), an analogue of tamoxifen, an anti-estrogen for hormone-dependent tumors [22]. Liver microsomes from mouse, rat, dog and human were used. The observed routes of biotransformation were hydroxylation and side chain modifications. Although seven metabolites were detected in the incubated mixtures, there was only one produced by all species that had lost the side chain. Interspecies differences concerned the metabolites with the truncated side chain, as follows: in the case of rodents, the microsomal system led to loss of the hydroxyethylamino group, while for incubated mixtures containing microsomes of all three other investigated species, only the loss of the hydroxyethyl group was detected. Other important observations made during the experiment were (a) that of the seven metabolites detected, three were produced exclusively by the dog and (b) that human liver microsomes produced an oxidised form of the metabolite containing a double bond in the side chain, this compound not being detectable in the other species investigated.

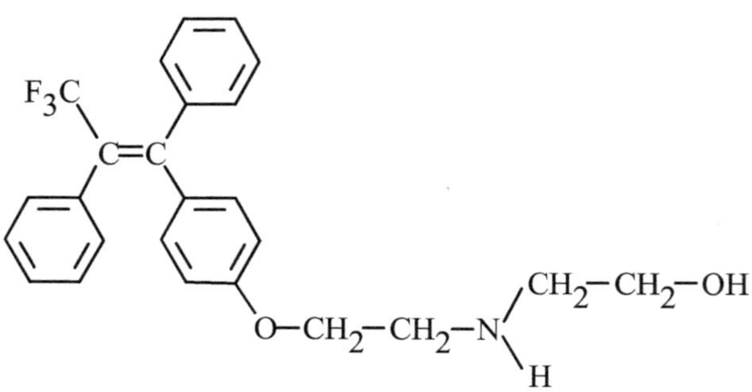

Fig.6.7 Panomifene (analogue of tamoxifen)

Different profiles, as well as quantitative species differences, were observed in the metabolism of L-775,606, a selective 5-HT1D receptor agonist, developed for the acute treatment of migraine [23]. Species investigated included human, dog, monkey and rat. For three of these (human, monkey and rat), the main metabolites were the hydroxylated M1 and the N-dealkylated M2. In contrast, in the dog the N-oxide metabolite (M3) was prevalent, representing an average of about 40%, whereas in the other investigated species, its formation represented a minor pathway, with the excreted metabolite corresponding to less then 5%.

In an interesting experiment accomplished both *in vitro* and *in vivo*, the metabolic fate and the toxicity of dapsone (Figure 6.8) were comparatively investigated in rat, mouse and man [24]. The metabolites were determined by HPLC/MS and metHb formation was used as toxic endpoint. The investigations focused especially on the toxic aspects and possible consequences during dapsone administration. As for the *in vitro* investigations, the results revealed that the greatest toxicity occurred in rats, with a significant difference between sexes: ~36% metHb formed in males and only 8.2% in females. In humans; the metHb toxic metabolite was found in an amount of ~11%, while in the mouse, only 4% under the same conditions. The rank order of toxicity was in direct relation to the formation of the hydroxylamine metabolite *in vitro*. However, experiments proved that the microsomes from all tested species were able to reverse the reaction, reducing the hydroxylamine back to dapsone. In contrast, under *in vivo* conditions the species most susceptible to dapsone toxicity proved to be the human, the sensitivity to toxic effects decreasing in the order: human, mouse, rat. Interspecies and sex differences also occurred in the biotransformation of the drug, in that the hydroxylamine and its glucuronide were detected only in male rats and humans, but not in female rats or mice.

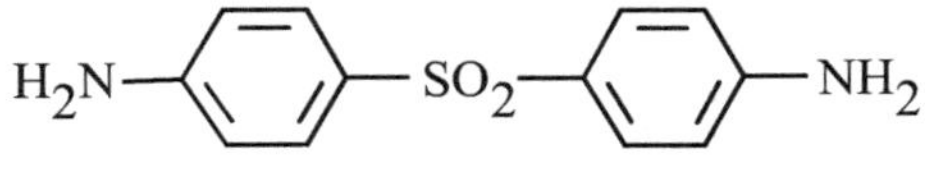

Fig.6.8 Dapsone

Species differences may also account for stereoselective reactions. Experiments were performed with fifteen O-acyl propranolol (PL) prodrugs, using rat and dog plasma and liver subfractions [25]. The aim of the study was to investigate both species differences and substrate specificities for the stereoselective hydrolysis of the tested prodrugs. As far as species was concerned, significant differences in the hydrolytic activities of prodrugs were established, in rat plasma being in the range of 5-119-fold greater than those in dog plasma. In contrast, dogs displayed a higher hepatic hydrolytic activity, especially in cytosolic fractions. The significant differences in the hydrolytic rates therefore represent quantitative species differences. As for stereoselectivity, the study also revealed important interspecies differences: hydrolysis in dogs generally showed a preference for the (R)-enantiomer, whereas in the rat, for all of the prodrugs containing substituents of low carbon number, the (S)-enantiomer was preferentially hydrolysed.

Following a previous report of species differences in the tolerability of rhein (a constituent of rhubarb), with rabbits displaying the highest susceptibility to kidney disturbances, a complex phase I and phase II metabolic investigation was performed in an attempt to elucidate species differences in the biotransformation of this compound [26]. Experiments were performed *in vivo*, with ^{14}C-labelled rhein; tested species included the rat, rabbit, dog and man. The common major metabolites determined in all tested species were the phenolic monoglucuronide and monosulphate. The urine samples of rabbits showed an additional hydrophilic metabolite fraction. The *in vitro* experiments performed on subcellular liver fractions of rabbits revealed the presence of several metabolites, including three monohydroxylated metabolites, their corresponding quinoid oxidation products and a bis-hydroxylated derivative. The hydroxylated phase I metabolites were further detected as glucuronides in all tested species, whereas the quinoid product was found only in rabbit urine. It is assumed that this metabolite displays a potential reactivity with endogenous macromolecules and generates that species-dependent susceptibility.

Species differences can also impact on inhibition phenomena. Investigations of the inhibition of pentobarbital biotransformation in the presence of empenthrin (Figure 6.9) support this idea [27]. Empenthrin (a synthetic pyrethroid) has been reported to display an inhibitory effect on pentobarbital metabolism, resulting in prolongation of the sleeping time.

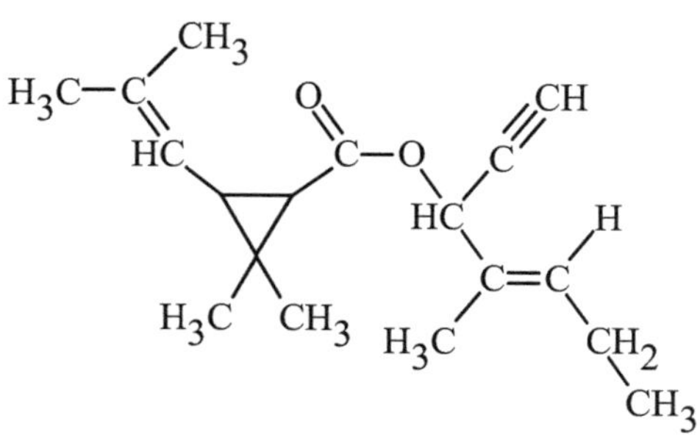

Fig.6.9 Empenthrin (a synthetic pyrethroid)

This phenomenon was observed for mice (the inhibitory effect being determined as dose-dependent), but not for other species investigated, namely rats, dogs, guinea pigs or hamsters. Further experiments using microsomal fractions expressing human CYPs were performed to determine the possible inhibitory effect of empenthrin on pentobarbital metabolism in humans. The final results revealed that the inhibition of pentobarbital by empenthrin occurred only in mice and not in any other of the other species investigated, including humans.

As previously mentioned, species differences may be implicated in several aspects, including qualitative and quantitative differences in drug biotransformation route, influences on stereoselective biotransformations and even on inhibition phenomena. An interesting and relatively recent study revealed the impact of species differences also on the distribution of drug metabolising enzymes. Complex investigations followed the expression of nine CYP450 isoenzymes and three GSTs in the pancreas of several species including humans [28]. The seven species tested in comparison to humans, were mice, hamsters, rabbits, rats, dogs, pigs and monkeys. A first finding was the large variation in the cellular localisation of the enzymes among the eight investigated species, with most of the enzymes expressed only in the pancreas of hamster, mouse, monkey and man. The other tested species were lacking several enzyme isoforms. However, in human tissue, four enzymes were lacking in almost half the cases. All of these observations concerning interspecies differences in the distribution of some of the most important drug-metabolising enzymes support the notion that great caution needs to be exercised when attempting to predict or extrapolate from animal data to humans.

This last observation confirms the importance of species differences, especially for drug-design in the pharmaceutical industry, where a suitable model reflecting human patterns of biotransformation and toxicity is desirable.

6.2.2 Sex

As already indicated in some of the above examples, qualitative and quantitative differences in both phases of drug metabolism are related to sex as well [29]. Initial observations of this feature were made in the early 1930s, when researchers noticed that female rats required only half the dose of a barbiturate compared to male rats to induce sleep. Later investigations indicated that this was due to the reduced capacity of the female to metabolise the barbiturates [30].

Sex differences have been intensively studied, not only in relation to sex-dependent metabolism of various xenobiotics [31], but also with the aim of correlating sex-dependent pharmacokinetics, pharmacodynamics, efficacy, and the possible occurrence of adverse reactions [32].

Sex differences, sometimes related to species or age, are now being observed for a wide range of substrates, including commonly prescribed drugs or even endogenous compounds, including steroid sex compounds [30]. Like other factors that influence drug metabolism, sex differences are considered to determine also biotransformation variations. Therefore, before introducing a new drug into therapy, combined studies investigate both species and sex differences on the metabolic profile of the candidate.

As an example, we refer to such a combined study for the *in vitro* investigation of sex and species differences in the metabolism of BOF-4272, a drug intended for the treatment of hyperuricaemia [33]. Rats, mice and monkeys of both sexes were used in the study. The results of the investigations made on various incubation mixtures revealed that both the pathways involved (i.e. types of metabolites resulting) as well as the rates of biotransformation of the tested drug were significantly influenced by both sex and species differences. On the other hand, results of other investigations examining the influence of sex and age on different enzyme activities showed no significant differences [34].

6.2.3 Age

It has long been recognized that the newborn, young and elderly display marked differences in drug biotransformation and are more susceptible to drug action. These differences are chiefly due to the enzymatic systems involved in drug biotransformation and the development of their metabolising capacity. Thus, the increased sensitivity of neonates may be related to their very low, undeveloped metabolising capacity, until adult levels of enzyme activity are achieved. On the other hand, in the elderly, the decrease in drug-metabolising capacity also appears to be dependent on these factors, important changes in the overall metabolism occurring with ageing.

An important aspect to be borne in mind is that the factors influencing drug metabolism are split arbitrarily and that they are interrelated. Examples have been given so far regarding species, sex and age. We should also highlight the fact that the status of enzymatic systems and their metabolising capacity may develop in many different ways, the patterns varying and being dependent on the species and sex [35-41]. Thus, a very recent study in fact reviewed the influence of age and sex on CYP enzymes in relation to drug bioequivalence [42].

The concern for controlling drug therapy, especially in the elderly to provide desired pharmaceutical effects at lower risks, continues to be a principal aim of research. Specific aims include efforts to try and prevent adverse reactions and to optimise therapy for the individual patient [43].

Unfortunately, as already mentioned, important changes in drug metabolism do indeed occur with ageing. For example, the significant reduction in liver volume accompanying ageing will be reflected in a reduction in the total amount of cytochrome P450 produced, and this could be associated with reduced ability of these enzymes to function. Other problems occurring with ageing, still not very well understood and needing to be revisited in view of recent advances, include the following: the effect of age on extrahepatic enzymes (especially CYPs), the impact of induction and inhibition phenomena on enzymatic systems in the elderly, the effect of the environment on drug metabolism in the aged given the increasing complexity of the CYPs involved in human metabolism, pharmacology and function of transporters, the decline in general metabolic capacity, and general frailty of older people [44].

Taking cognisance of the above, it is understandable and expected that all these conditions will result in altered drug handling and especially, altered pharmacodynamic responses. Recognizing the central role of the liver in the general metabolism of both drugs and other xenobiotics, we should also mention, besides the reduction of hepatocyte mass (with corresponding effects on the hepatic enzyme system activity), the reduction of hepatic blood flow and changes in sinusoidal endothelium. These changes will affect drug transfer and oxygen delivery, resulting in reduced hepatic drug clearance. Another current problem in the elderly is related to renal clearance reduction, which is generally disease-related. Altered pharmacokinetics and pharmacodynamics are expected also in patients with cardiovascular diseases. Also worth remembering is the effect of age on pancreatic secretion [45]. But perhaps one of the major problems resulting in adverse reactions and drug-drug interactions is the still very common practice of polypharmacy, responsible for increased morbidity and mortality in the elderly. This is another aspect that is peculiar to elderly patients, who consume a disproportionate amount of prescription and non-prescription medications. Such practice can obviously lead to many negative consequences, primarily placing the elderly at risk of developing significant drug-drug interactions, which often go unrecognized clinically and which are responsible for increased morbidity in this sector of the population. Drugs can interact to mutually alter absorption, distribution, metabolism or excretion characteristics, or interact in a synergistic or antagonistic fashion altering their pharmacodynamics. In addition, one must be aware that co-administered drugs, foods and nutritional supplements can also alter the pharmacological actions of a medication. These alterations may cause the action of a drug to be diminished or enhanced. Another major issue is that drugs may also interact with diseases, potentially worsening disease symptoms. Therefore, prudent use of medications and vigilant monitoring are essential for preventing the elderly from the high risk of adverse reactions and drug-drug interactions, whose unfortunate consequences have been noted above [46-52].

Considering the physiological changes in main organ functions in the elderly as well as the pharmacokinetic parameters of various drugs, accumulations of drug metabolites presents another important problem. In this context, particular attention should be paid to an adequate treatment scheme designed to ensure the optimum therapeutic effect with a minimum risk of toxic effects. In fact, a starting dose which is 30-40% less than the average dose used in adults is generally recommended, not only for renally excreted drugs, but also for compounds metabolised and excreted by the liver [53-54].

Ageing is directly related to ovarian hormonal activity, and progesterone metabolites, specifically, have been proven to affect the response to various psychotherapeutic agents, resulting in increased risk of adverse effects. Studies on benzodiazepines, for example, demonstrated that their metabolism is altered, either resulting in a decrease in their clearance or an alteration of the effect-concentration relationship. These effects may result in increased risk of adverse reactions, particularly in older patients with anxiety disorders. Therefore, establishing the appropriate low dose for optimal treatment will minimise adverse effects. The intimate mechanisms involved are not completely understood, but it has been suggested that they could be related to modulation of the GABA-antagonist receptor by neurosteroids [55].

Other drugs that were investigated with respect to the role of drug-metabolising enzymes and the effects of age included different alkylphenoxazone derivatives, benzodiazepines and neuroleptics, bisphosphonates (BPs) as therapeutic drugs for osteoporosis, anxiolytics and others [56-59].

A special category includes 'ultra-aged' patients. Aspects concerning decreased drug absorption, metabolism and excretion, decline of protein binding, lower blood flow, disturbance of blood brain barrier, adverse reactions and drug interactions for this category of patients have been reviewed, with the purpose of establishing proper therapeutic management [60].

Two other important aspects of age-related changes are sensitivity to environmental factors and nutritional effects on hepatic drug metabolism in the elderly [61,62]. The cited works review pharmacodynamic and toxicokinetic changes in absorption, distribution, metabolism, excretion and sensitivity, as well as age-associated differences in hepatic drug metabolism, and the effects of nutrition on drug bioavailability, distribution and hepatic metabolism.

An important issue in improving the quality of life of the elderly has recently been reviewed and concerns CoQ10 implications in energetic metabolism, a well-known anti-oxidant effect with relevance to health food and medical drugs [63].

At the other end of the scale, special attention is paid to neonates and children, as regards the development of their enzymatic systems. Unpredictable developmental changes in drug biotransformation have been proven to play a role not only in the pharmacokinetic profile, but also in the pathogenesis of adverse drug reactions in children. Most of these developmental changes have a genetic determinant, which causes variations in different metabolising enzymes, whereby normal, therapeutic drug doses can result in functional overdoses due to drug accumulation. This relative overdosing is determined by inefficient elimination *via* the affected pathways. Furthermore, idiosyncratic forms of toxicity may occur when a relative increase in reactive metabolite formation is due to imbalances in bioactivation and detoxification processes. Phenotyping and genotyping would be very helpful under such circumstances to prevent these effects [64].

Extra-hepatic metabolism has to be considered as well, the renal clearance and volume of distribution being at least as important as hepatic metabolism [65].

Typically, drug metabolism is significantly reduced in the neonatal period because of lack of enzymatic activity. A recent investigation reviewed the effect of age on the biotransformation of four drugs [66]. The subjects were infants and children, and the tested drugs included caffeine, midazolam, morphine and paracetamol. The first observation was that in the neonatal period, for all four tested drugs, clearance was markedly reduced. Further observations confirmed that (with the exception of paracetamol) this reduced clearance is maintained in infants and children under the age of two years, and that there is considerable inter-individual variation in clearance values for all ages and for all tested drugs, appearing to be the greatest for midazolam. The third important observation suggests that for children older than two years, the mean plasma clearance values for all four drugs are more or less similar to those in adolescents and even adults.

6.2.4 Pathological status

The way in which the body clears drugs is affected by many disease states. Among them, those of primary concern are considered to be diseases affecting the liver: cirrhosis, alcoholic liver disease, cholestatic jaundice, and liver carcinoma [67].
Other factors responsible for variation in drug metabolism are the endocrine disorders, such as diabetes mellitus [68], hypo-and hyperthyroidism [69], pituitary disorders [70], and various types of infections (bacterial, viral, malaria) [71].

In cirrhosis for example, replacement of parts of the liver by fibrous tissue leads to a reduction in the number of functional hepatocytes. In this situation, it seems absolutely reasonable that drug metabolism should be impaired. It is known for example that human cytochromes P450, particularly the CYP2A6 isoform, catalyse the bioactivation of various drugs and even carcinogens. Recent studies proved that in cases of liver disease, including cirrhosis (but also viral hepatitis or parasitic infestation), this isoform is over-expressed, and as such may therefore be considered a major liver catalyst in pathological conditions [72].

An important consequence of such liver disease (or other organ impairment) arises during transplantation processes; it is well known that prior to transplantation, organ dysfunction may occur because of stress and anxiety, and this may result in altered pharmacokinetic behaviour of some psychotropic agents. In case of cirrhotic patients, an increased drug bioavailability due to portosystemic shunting was noted, which therefore therefore required drug dosage adjustment. Studies on different psychotropic agents suggest that a selection of these, concurrently administered with an appropriate dosage adjustment, could ensure the lowering of risk of drug accumulation [73].

Another recent article reviews the implications of oxidative stress and the role of cytochrome P450s and cytokines in drug-induced liver diseases, which according to some recent studies can be also induced by immunological mechanisms [74,75].

In this context, we should mention that especially in the last few years, great importance has been attributed to antioxidants in the treatment of drug-induced liver oxidative stress, due to the central role of this organ in the general metabolism. Effects of natural antioxidants have been investigated *in vitro* on liver redox status by biochemical, analytical and histological methods, in order to assess the overall free radical-antioxidant balance.

Studies have also been performed in animal models and in humans with Gilbert's disease and alcohol liver disease. The results confirmed the role of free radicals in alcoholic patients, stressing the greater vulnerability of women to alcohol toxicity. As regards Gilbert's disease, investigations found no alterations of free radical-antioxidant balance, but in contrast, an improvement in the non-enzymatic antioxidant defense system [76].

The impact and consequences of drug-induced liver diseases on drug pharmacokinetics and toxicity in the case of pathogenesis are continuously investigated. Recently, the role of polymorphism of drug metabolising enzyme systems has been reviewed [77].

A comparative study was performed on normal mice to investigate the effects of drug-induced liver injury using prednisolone (PSL) versus Angelica sinensis Polysaccharides (ASP), on hepatic metabolising enzyme activities of both phases. ASP was shown to increase content and catalytic activity of several enzymes *viz.* CYTP450, different demethylases and hydroxylases, and GSH-related enzymes. In contrast, PSL significantly decreased the liver mitochondrial glutathione content, whereas all other enzyme activities were increased. An important observation was that treatment with ASP could restore the GSH content, which is important for detoxication (by glutathione conjugation) of certain xenobiotics, including drugs [78].

An interesting aspect recently investigated concerns the CYTP450 superfamily. The multiple CYP450 isoforms (CYPs) are well known as being involved in the biotransformation of numerous drugs, other chemicals, as well as endogenous substrates. Unfortunately, the hepatic CYPs may also be involved in the pathogenesis of several liver diseases, due to their catalytic activity mediating activation of certain drugs to toxic metabolites (see Chapter 8). Incidences of drug-induced hepatotoxicity, as well as nephrotoxicity and cardiac failure are well known and unfortunately relatively frequent. The most frequently cited examples of hepatotoxicity refer to halothane and acetaminophen (see Chapter 8). There are usually several mechanisms involved in drug-induced liver disease. One of them is an immunological one (see ref. 75), presumably determined by the covalent binding of the metabolite to CYP, which will result in formation of anti-CYP antibodies, leading to so-called 'immune-mediated hepatotoxicity'. Another mechanism, related to the CYP2E1 isoform, is associated with lipid peroxidation and production of reactive oxygen species, resulting in damage to hepatocytes and mitochondrial membranes. The explanation for involvement of this particular CYP isoform relies on the observation that in alcoholic patients, its levels are significantly increased. Thus, it was first associated with alcohol-liver disease and non-alcoholic steatohepatitis.

However, due to its ability to activate carcinogens, investigations also suggested a possible role of this isoform in hepatocellular carcinoma.

Considering the liver as the main location for the most important enzymatic systems, it is expected on the other hand that in patients with liver diseases, drug metabolism should be impaired. Particularly vulnerable isoforms have been proven to be different CYPs such as 1A, 2C19 and 3A, while others (2D6, 2C9, 2E1) appeared to be affected to a lesser extent. An interesting feature is that the pattern of CYP isoenzyme alterations differs with the etiology of the liver disease, with the most severe modifications occurring in cirrhosis [79].

Other liver diseases have also been proven to alter drug metabolism by altering the activities of metabolising enzymes. A prime example is alcohol-induced disease, unfortunately the most common type of chronic liver disease in many countries. An important aspect revealed by one study [80] is that alcohol can interact with other factors of risk for hepatic disease, especially hepatitis C infection and also concurrent consumption of hepatotoxic drugs (acetaminophen, for example), resulting in more severe disease and increased risk of adverse reactions and drug-drug interaction occurrence, than occurs when alcohol alone is the risk factor present.

Another interesting aspect to mention, demonstrated in a recent investigation on rats, is that hepatic and extrahepatic (e.g. intestinal) metabolic activities involving especially the cytochrome P450 system are influenced by surgery and/or drug-induced renal dysfunction [81]. The most marked decreases (of about 66%) were observed for the hepatic CYP3A metabolic activities, in the case of nephrectomy. Less marked, but nonetheless significant decreases were observed also in drug-induced renal dysfunction following i.m. injection of glycerol (about 60%), and i.p. injection of cisplatin (about 49%) (Figure 6.10). In contrast, the intestinal metabolic CYP3A activity was weakly increased in rats injected with glycerol, and remained practically unchanged in the case of injected cisplatin or surgery (nephrectomy).

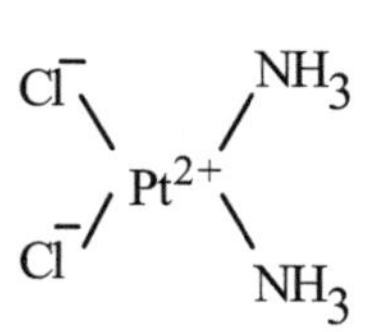

Fig.6.10 Cisplatin

These results suggest a dependence of the extent of lowering of hepatic P450 activities on the etiology of renal failure. In addition, the experimental observations led to the conclusion that alteration of the same enzyme activity in extrahepatic tissues (particularly in intestine, where this tissue was examined experimentally) cannot always be correlated with that in the liver.

6.2.5 Hormonal control of drug metabolism – selected examples

Hormones, known to play a major role in the general metabolism, have similarly been proven to control the biotransformation of drugs, in direct connection with other factors such as age, sex, or in particular physiological states, such as pregnancy.

An example is the apparent connection between certain sex-specific drug- and steroid-metabolising enzyme activities in rats and the sex-dependent expression of those specific enzymes, under gonadal steroid and growth hormone control [82].

Another sex and age connection with the control of the growth hormone (GH) was the focus of interesting cDNA cloning investigations [83,84]. The study examined especially cytochrome P450, it being established that GH is involved in the control of rat hepatic drug- and steroid-metabolism, particularly through the action of this enzymatic system. The results showed low levels of CYTP450 in neonates, and an increase after one month, both in male and female rats. At adult stage, important sex differences were recorded, in female rats the content being about three times higher than in male rats.

Thyroid status contributes to differences for several drugs administered in equi-active doses on several forms of UDPGTs [85]. As experimental animals, rats having different thyroid hormonal status were employed, namely normal (control group), hypothyroid and hyperthyroid. The drugs tested were ciprofibrate, bezafibrate, fenofibrate and clofibrate (Figure 6.11). The responses were markedly modulated by the thyroid status, with an average increase of about 5% in hyperthyroid animals. The results confirmed the role of hormonal control upon the enzyme induction displayed by certain drugs (or other xenobiotics).

The hypothalamo-pituitary-liver axis has also been proven to function as a hormonal control system in the metabolism of drugs and endogenous compounds [86].

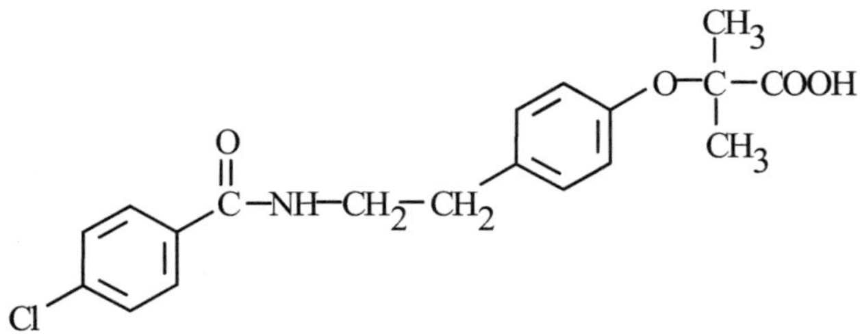

benzafibrate

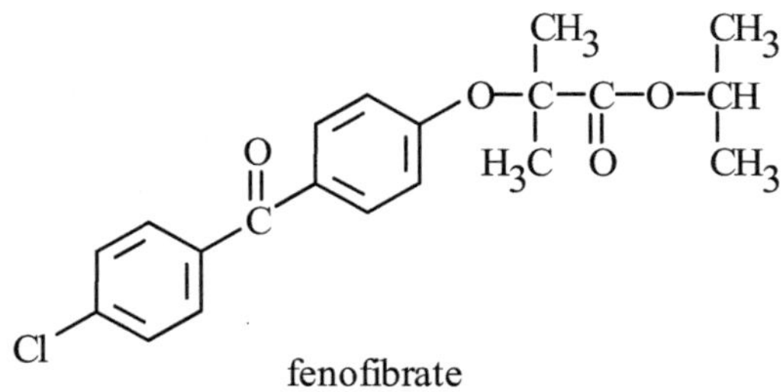

fenofibrate

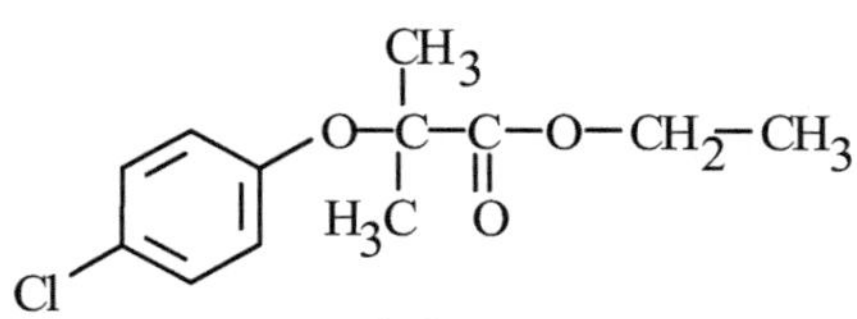

clofibrate

Fig.6.11 Structures of some of the cited fibrates

6.3 ENVIRONMENTAL FACTORS

These are usually considered to be those influences in our surroundings that can affect (sometimes markedly) drug metabolism. Of course, there are a large number of environmental chemicals that potentially could affect drug biotransformations, usually grouped into heavy metals (already discussed, see previous chapter), industrial pollutants and pesticides.

The most important industrial pollutants are typically aromatic or aromatic polycyclic compounds and polychlorinated biphenyls (Figure 6.12). Many of these have been already discussed under different circumstances (inductive enzyme effects, procarcinogenic effects).

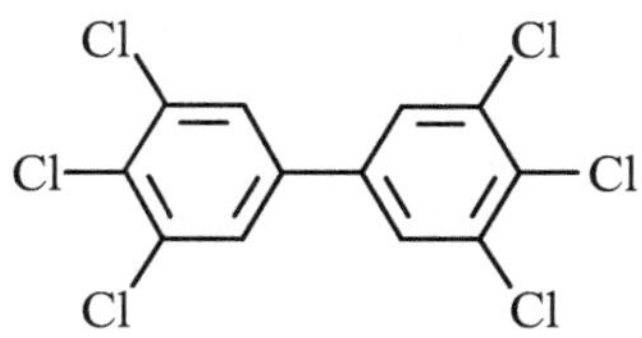

3,3',4,4',5,5'-hexachlorobiphenyl

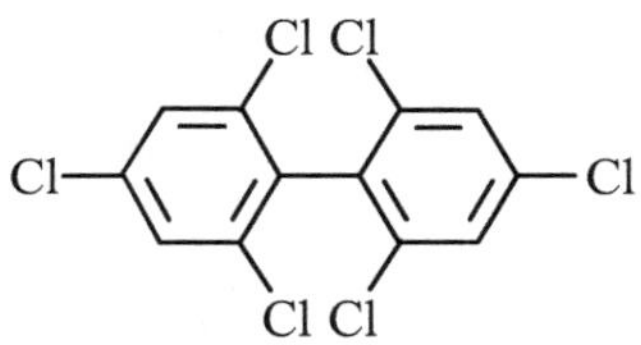

2,2',4,4',6,6'-hexachlorobiphenyl

Fig.6.12 Polychlorinated biphenyls (common industrial pollutants)

Pesticides are also of various types (insecticides, herbicides), and are considered environmental contaminants in air, soil, water and food. They will not be discussed further in the present monograph.

6.4 FURTHER OBSERVATIONS

As has been discussed in the last two chapters, there are numerous factors (some of them interactive) that can affect drug metabolism, therefore making its control an extremely complex problem. With the exception of genetic factors, all the rest are considered variable during a lifetime, so predictions are made with reservation. Also, since most of the studies are performed either *in vitro* or on experimental animals, extrapolations from the *in vitro* to the *in vivo* situation, or from animals to humans must be approached with extreme caution.

References

1. Gibson GG, Skett P. 1994. Factors affecting drug metabolism: internal factors. In: Introduction to Drug Metabolism. London: Blackie Academic & Professional, An Imprint of Chapman & Hall, pp 107-132.
2. Gibson GG, Skett P. 1994. Factors affecting drug metabolism: external factors. In: Introduction to Drug Metabolism. London: Blackie Academic & Professional, An Imprint of Chapman & Hall, pp 133-156.
3. Walker CH. 1980. Species variations in some hepatic microsomal enzymes that metabolise xenobiotics. Prog Drug Metab 5:113-164.
4. Caldwell J. 1982. Conjugation reactions in foreign-compound metabolism: definition, consequences and species variations. Drug Metab Rev 13:745-778.
5. Hucker HB. 1970. Species differences in drug metabolism. Annu Rev Pharmacol 10: 99-118.
6. Smith DA. 1991. Species differences in metabolism and pharmacokinetics: are we close to an understanding? Drug Metab Rev 23:355-373.
7. Walker CH. 1978. Species differences in microsomal mono-oxygenase activity and their relationship to biological half-lives. Drug Metab Rev 7:295-324.
8. Berthou F, Guillos B, Riche C, Dreano Y, Jacqz-Algrain B, Beunes PH. 1992. Interspecies variation in caffeine metabolism related to cytochrome P4501A enzymes. Xenobiotica 22:671-680.
9. Ashihara H, Crozier A. 1999. Biosynthesis and metabolism of caffeine and related purine alkaloids in plants. Adv Bot Res 30:117-205.
10. Quinn GP, Axelrod J, Brodie BB. 1958. Species, strain and sex differences in metabolism of hexobarbitone, amidopyrine, antipyrine and aniline. Biochem Pharmacol 1:152-159.
11. Levay F, Fejer E, Szeleczky G, Szabo A, Eroes-Takacsy T, Hajdu F, Szebeny G, Szatmary I, Hermecz I. 2004. In vitro formation of selegiline-N-oxide as a metabolite of selegiline in human, hamster, mouse, rat, guinea-pig, rabbit and dog. Eur J Drug Metab Ph 29:169-178.
12. Nakajima H, Nakanishi T, Nakai K, Matsumoto S, Ida K, Ogihara T, Ohzawa N. 2002. Studies on the metabolic fate of M17055, a novel diuretic (4): Species difference in metabolic pathway and identification of human CYP isoform responsible for the metabolism of M17055. Drug Metab Pharmacokin 17:60-74.
13. Carmo H, Hengstler JG, de Boer D, Ringel M, Remiao F, Carvalho F, Fernandes E, dos Reys LA, Oesch F, Bastos MdeL. 2005. Metabolic pathways of 4-bromo-2, 5-dimethoxyphenethylamine (2C-B): analysis of phase I metabolism with hepatocytes of six species including human. Toxicology 206:75-89.
14. Yu L-S, Yao T-W, Zeng S. 2003. In vitro metabolism of zolmitriptan in rat cytochromes induced with β-naphthoflavone and the interaction between six drugs and zolmitriptan. Chem-Biol Interact 146:263-272.
15. Collins JM. 2001. Inter-species differences in drug properties. Chem-Biol Interact 134:237-242.

16. Mutlib AE, Chen H, Nemeth GA, Markwalder JA, Seitz SP, Gan LS, Christ DD. 1999. Identification and characterization of efavirenz metabolites by liquid chromatography/mass spectrometry and high field NMR: species differences in the metabolism of efavirenz. Drug Metab Dispos 27:1319-1333.

17. Mutlib AE, Gerson RJ, Meunier PC, Haley PJ, Chen H, Gan LS, Davies MH, Gemzik BC, David D, Krahn DF, Markwalder JA, Seitz SP, Robertson RT, Miwa GT. 2000. The Species-Dependent Metabolism of Efavirenz Produces a Nephrotoxic Glutathione Conjugate in Rats. Toxicol Appl Pharmacol 169:102-113.

18. Beken S, De Smet K, Depreter M, Roels F, Vercruysse A, Rogiers V. 2001. Effects of L-proline on phase I and phase II xenobiotic biotransformation capacities of rat and human hepatocytes in long-term collagen gel cultures. Alternatives to laboratory animals. ATLA 29:35-53.

19. Martignoni M, Monshouwer M, de Kanter R, Pezzetta D, Moscone A, Grossi P. 2003. Phase I and phase II metabolic activities are retained in liver slices from mouse, rat, dog, monkey and human after cryopreservation. Toxicol In Vitro 18:121-128.

20. Matsumoto S, Nakai M, Yoshida M, Katae H. 1999. A study of metabolites isolated from urine samples of pigs and calves administered orbifloxacin. J Vet Pharmacol Ther 22:286-289.

21. Whalen RD, Tata PNV, Burckart GJ, Venkataramanan R. 1999. Species differences in the hepatic and intestinal metabolism of cyclosporine. Xenobiotica 29:3-9.

22. Monostory K, Jemnitz K, Vereczkey L, Czira G. 1997. Species differences in metabolism of panomifene, an analog of tamoxifen. Drug Metab Disposit 25:1370-1378.

23. Prueksaritanont T, Lu P, Gorham L, Sternfeld F, Vyas KP. 2000. Interspecies comparison and role of human cytochrome P450 and flavin-containing monooxygenase in hepatic metabolism of L-775,606, a potent 5-HT1D receptor agonist. Xenobiotica 30:47-59.

24. Tingle MD, Mahmud R, Maggs JL, Pirmohamed M, Park BK. 1997. Comparison of the metabolism and toxicity of dapsone in rat, mouse and man. J Pharmacol Exp Ther 283:817-823.

25. Yoshigae Y, Imai T, Horita A, Otagiri M. 1997. Species differences for stereoselective hydrolysis of propranolol prodrugs in plasma and liver. Chirality 9:661-666.

26. Dahms M, Lotz R, Lang W, Renner U, Bayer E, Spahn-Langguth H. 1997. Elucidation of phase I and phase II metabolic pathways of rhein: species differences and their potential relevance. Drug Metab Disposit 25:442-452.

27. Tsuji R, Isobe N, Kurita Y, Hanai K, Yabusaki Y, Kawasaki H. 1996. Species differences in the inhibition of pentobarbital metabolism by empenthrin. Environ Health Sci 2:331-337.

28. Ulrich AB, Standop J, Schmied BM, Schneider MB, Lawson TA, Pour PM. 2002. Species differences in the distribution of drug –metabolizing enzymes in the pancreas. Toxicol Pathol 30:247-253.

29. Skett P. 1988. Biochemical basis of sex differences in drug metabolism. Pharmacol Therapeut 38:269-304.

30. Gibson GG, Skett P. 1994. Factors affecting drug metabolism: internal factors. In: Introduction to Drug Metabolism. London: Blackie Academic & Professional, An Imprint of Chapman & Hall, p121.

31. Mugford CA, Kederis GL. 1998. Sex-dependent metabolism of xenobiotics. Drug Metab Rev 30:441-498.

32. Ueno K, Negishi E. 2004. Sex differences and EBM in drug therapy. Nippon Yakuzaishikai Zassi 56:1155-1160.

33. Naito S, Nishimura M, Yoshitsugu H. 1999. *In-vitro* biotransformation of BOF-4272, a sufphoxide-containing drug, in rats, mice and cynomolgus monkeys: sex and species differences. Pharm Pharmacol Commun 5:645-651.

34. Tateishi T, Watanabe M, Nakura H, Tanaka M, Kumai T, Kobayashi S. 1997. Sex-or age-related differences were not detected in the activity of dihydropyrimidine dehydrogenase from rat liver. Pharmacol Res 35:103-106.

35. Dawling S, Crome P. 1989. Clinical pharmacokinetic considerations in elderly. Clin Pharmacokinet 17:236-263.

36. Horbach GJMJ, Van Asten JG, Rietjens IMCM, Kremers P, Van Bezooijen CFA. 1992. The effects of age on inducibility of various type of rat liver cytochrome P450. Xenobiotica 22:515-522.

37. Ladona MG, Lindstrom B, Thyr C, Dun-Ren P, Rane A. 1991. Differential foetal development of the *O*- and *N*-demethylation of codeine and dextromethorphan in man. Brit J Clin Pharmacol 32:295-302.

38. Rikans LE. 1989. Hepatic drug metabolism in female Fischer rats as a function of age. Drug Metab Disp 17:114-116.

39. Schmucker DL, Woodhouse KW, Wang RK, Wynne H, James OF, McManus M, Kremers P. 1990. Effect of age and gender on *in vitro* properties of human liver microsomal monooxygenases. Clin Pharmacol Ther 48:365-374.

40. Vestal RE. 1989. Aging and determinants of hepatic drug clearance. Hepatology 9:331-334.

41. Woodhouse K. 1992. Drugs and the Liver. III. Ageing of the liver and the metabolism of drugs. Biopharm Drug Dispos 13:311-320.

42. Bebia Z, Buch SC, Wilson JW, Frye RF, Romkes M, Cecchetti A, Chaves-Gnecco D, Branch RA. 2004. Bioequivalence revisited: Influence of age and sex on CYP enzymes. Clin Pharmacol Ther 76:618-627.

43. Turnheim K. 2004. Drug therapy in the elderly. Exp Gerontol 39:1731-1738.

44. Kinirons MT, O'Mahony MS. 2004. Drug metabolism and ageing. Br J Clin Pharmacol 57:540-544.

45. Annoni G, Gagliano N. 2003. Effect of aging on the liver and pancreas. Interdiscipl Top Gerontol 32:65-73.

46. McLean AJ, Le Couteur DG. 2004. Aging biology and geriatric clinical pharmacology. Pharmacol Rev 56:163-184.

47. Turnheim K. 2003. When drug therapy gets old: pharmacokinetics and pharmacodynamics in the elderly. Exp Gerontol 38:843-853.

48. Delafuente JC. 2003. Understanding and preventing drug interactions in elderly patients. Crit Rev Oncol Hemat 48:133-143.

49. Rane A. 2001. Development and ageing as sourses of variability in drug metabolism. In: Pacifici GM, Pelkonen O, editors. Interindividual Variability in Human drug Metabolism. Basingstoke (UK): Taylor& Francis Ltd., pp 75-84.

50. Miura H, Iguchi A. 1998. Management and drug therapy in the elderly. General comments. Sogo Rinsho 47:93-97.

51. Najma S, Saeed AM. 1993. Aging and drug metabolism. Pakistan J Pharmacol 10:79-92.

52. Meydani M. 1994. Impact of aging on detoxification mechanisms. In: Eaton DL, Groopman JG, editors. Nutrition Toxicology. San Diego: Publisher Academic, pp 49-66.

53. Sawada Y, Ohtani H. 2002. Change of pharmacokinetics by aging. Nippon Yakuzaishikai Zasshi 54:1611-1620.

54. Zeeh J. 2001. The aging liver: consequences for drug treatment in old age. Arch Gerontol Geriat 32:255-263.

55. Kroboth PD, McAuley JW. 2000. The influence of progesterone on the pharmacokinetics and pharmacodynamics of gamma-aminobutyric acid-active drugs. In: Morrison MF, editor. Hormones, Gender and the Aging Brain. Cambridge (UK): Cambridge University Press, pp 334-349.

56. Barnett CR, Ioannides C. 2000. Xenobiotic-metabolizing enzyme systems and aging. Method Mol Med 38:119-130.

57. Kitani K. 1996. Pharmacokinetics and dynamics in the elderly with special emphasis on drug metabolism in the liver. Shinkei Seishin Yakuri 18:293-298.

58. Yamashi K. 1997. Pathogenic mechanism of primary osteoporosis and bisphosphonates. Clin Calcium 7:1954-1958.

59. Yamaoka K, Nomura S. 1995. Interindividual variations in action of antianxiety drugs. Shinkei Seishin Yakuri 17:865-871.

60. Shin K, Okada T, Takasaki M. 1995. View of dosage for ultra-aged patients. Rinsho to Yakubutsu Chiryo 14:695-698.

61. Birnbaum LS. 1990. Age-related changes in sensitivity to environmental chemicals. Eisei Kagaku 36:461-479.

62. Anderson KE. 1990. Nutritional effects on hepatic drug metabolism in the elderly. Prog Clin Biol Res 326:263-277.

63. Yoshimura I. 2003. Supplement in the 21st Century 'CoQ10'. Fragrance Journal 31:76-80.

64. Leeder JS. 2001. Ontogeny of drug-metabolizing enzymes and its influence on the pathogenesis of adverse drug reactions in children. Curr Ther Res 62:900-912.

65. Toshimi K. 2001. Characteristics of renal function development and antibiotic pharmacokinetics in neonates. Farumashia 37:750-751.

66. De Wildt SN, Johnson TN, Choonara I. 2003. The effect of age on drug metabolism. Paediatr Perinat Drug Ther 5:101-106.

67. Hoyumpa AM, Schenker S. 1982. Major drug interactions: effect of liver disease, alcohol and malnutrition. Ann Rev Med 33:113-150.

68. Bourbon J, Jost A. 1982. Control of glycogen metabolism in the developing fetal lung. Paediatr Res 16:50-56.

69. Batlle D, Itsarayoungyuen K, Hays S, Arruda Hose AL, Kurtzman NA. 1982. Parathyroid hormone is not anticalciuric during chronic metabolic acidosis. Kidney Int 22:264-271.

70. Shapiro BH, Albucher RC, MacLeod JN, Bitar MS. 1986. Normal levels of hepatic drug-metabolizing enzymes in neonatally induced, growth hormone-deficient adult male and female rats. Drug Metab Disposit 14:585-589.

71. Azri S, Renton KW. 1991. Factors involved in the depression of hepatic mixed function oxidase during infection with Listeria monocytogenesis. Int J Immunopharmacol 13: 197-204.

72. Christian K, Lang M, Maurel P, Raffalli-Mathieu F. 2004. Interaction of heterogeneous nuclear ribonucleoprotein A1 with cytochrome P450 2A6 mRNA: Implications for post-translational regulation of the CYP 2A6 gene. Pharm BioSci 65:1405-1414.

73. Crone CC, Gabriel GM. 2004. Treatment of anxiety and depression in transplant patients. Clin Pharmacokinet 43:361-394.

74. Saibara T, Nemoto S, Ohnishi S. 2004. Oxidative stress in drug-induced liver injury. Kan, Tan, Sui 48:691-696.

75. Shimizu Y, Murata H, Tajiri K, Yasuyama S, Higuchi K. 2004. Immunological mechanism of drug-induced liver injury. Kan Tan Sui 48:683-690.

76. Hagymasi K, Blazovics A, Lengyel G, Lugasi A, Feher J. 2004. Investigation of redox homeostasis of liver in experimental and human studies. Acta Pharm Hung 74:51-63.

77. Iwasa M, Adachi Y. 2004. Drug metabolism and pathogenesis of drug-induced liver disease. Kan Tan Sui 48:677-681.

78. Xueyan X, Renxiu P, Rui K, Zhegiong Y, Xiao C. 2003. Effects of Angelica sinensis polysaccharides on hepatic drug metabolism enzymes activities in mice. Zhongguo Zhongyao Zazhi 28:149-152.

79. Villeneuve J-P, Pichette V. 2004. Cytochrome P450 and liver disease. Curr Drug Metab 5:273-282.

80. Saunders JB, Devereaux BM. 2002. Epidemiology and comparative incidence of alcohol-induced liver disease. In: Sherman DIN, Preedy VP, Watson RR, editors. Ethanol and the Liver. London: Taylor & Francis Ltd, pp 389-410.

81. Okabe H, Hasunuma M, Hashimoto Y. 2003. The Hepatic and Intestinal Metabolic Activities of P450 in Rats with Surgery- and Drug-Induced Renal Dysfunction. Pharm Res 20:1591-1594.

82. Morgan ET, Gustafsson JA. Sex-specific isoenzymes of P-450. Steroids 49:213:245.

83. Sasamura H, Nagata K, Yamazoe Y, Shimada M, Saruta T, Kato R. 1990. Effect of growth hormone on rat hepatic cytocrome P-450 f mRNA: a new mode of regulation. Mol Cell Endocrinol 68:53-60.

84. Schuetz EG, Schuetz JD, May B, Guzelian PS. 1990. Regulation of cytochrome P 450b/e and P 450 p gene expression by growth hormone in adult rat hepatocytes cultured on a reconstituted basement membrane. J Biol Chem 265:1188-1192.

85. Goudonnet H, Mounie J, Magdalou J, Escousse A, Truchot RC. 1988. Comparative induction of rat liver bilirubin UDP-glucuronosyltransferase by ciprofibrate and other hypolipidemic agents belonging to the fibrate series: influence of the thyroid status. Colloque INSERM 173:311-315.

86. Gustafsson JA, Mode A, Nordstedt G, Hoekfelt T, Sonnenchein C, Eneroth P, Skett P. 1980. The hypothalamo-pituitary-liver axis: a new hormonal system in control of hepatic steroid and drug metabolism. In: Litwack G, editor. Biochem Actions Horm 7:47-89.

Chapter 7

IMPACT OF GENE VARIABILITY ON DRUG METABOLISM

7.1 INTRODUCTION

Dispensing of medicines is already strongly influenced by considerations of genetic factors that play a role in drug response. With the extremely rapid accumulation of knowledge of the genetic make-up of the human species and the recent technological advances accompanying it, this tendency is set to increase significantly in the future and therefore a basic knowledge of the principles of pharmacogenetics is essential to the health professional. This chapter sets out to provide a basic introduction to these disciplines, beginning with principles and nomenclature. Several specialised sub-disciplines (e.g. toxicogenomics, proteomics) are also outlined. A discussion of species-dependent biotransformations and their genetic control, illustrated with recent examples, follows. The discipline of pharmaco-informatics is briefly described and discussed, and finally the implications of genetics in the future dispensing of drugs are outlined.

7.2 BASIC PRINCIPLES OF PHARMACOGENETICS

Pharmacogenetics, still considered a relatively new field of clinical investigation, is the study of genetically determined variations in drug response; practically, it reflects the linkage between an individual's genotype and that individual's ability to metabolise a foreign compound [1-6]. The term was first proposed in 1959 [7].

Large inter-individual differences that may occur in the disposition of many drugs (or other xenobiotics) are controlled, at least in part, by genetic factors. In this context it is important to mention that while environmental factors including smoking, alcohol consumption and drug use, diet, occupational exposure to chemicals, and disease can vary during the course of drug therapy, genetic factors are constant throughout life. Most commonly, genetic variations reflect themselves in different rates and extents of drug elimination from the body. This explains, in the first place, the possibly marked differences in dosage requirements for many patients, resulting in the need to individualise doses and, in general, the therapeutic treatment. On the other hand, it is assumed that differences in metabolism of various therapeutic compounds as a consequence of genetic polymorphism, can lead to severe toxicity or even therapeutic failure, by altering the relation between dose and blood concentration of the pharmacologically active drug. This is determined by the absence, insufficiency or alteration of metabolising enzymatic systems, due to the genetic aberrations. Under these circumstances, it is evident that understanding the mechanisms of genetic variation in drug effects could be the key to applying pharmacogenetic principles to improve the therapeutic strategies, by ensuring greater efficacy and decreased risk of adverse reactions or toxicity. Thus, genetic variations will be important firstly for those genes encoding drug-response proteins that are expressed in a monogenic fashion. If a single locus determines the expression of a drug-response gene, then it is assumed that the genetic variation has the potential to contribute to inter-individual variation in drug response.

At the same time it should be noted that these inter-individual variations help to explain also the inter-individual differences and susceptibilities observed in disease states such as cancer, hypercholesterolaemia, alcoholism, and toxicity to environmental pollutants or industrial chemicals. Therefore, we mention here a relatively recent addition to the discipline of pharmacogenetics, generally known as 'ecogenetics', which deals with the dynamic interactions between an individual's genotype and environmental agents, including industrial chemicals, pollutants, plant and food components, pesticides, and other chemicals.

In this context, we should mention also the related field of toxicogenetics, dealing with an individual's predisposition to different toxic effects of drugs, including carcinogenesis and teratogenesis, for example.

Population (interethnic) differences in response to drugs give rise to the terms ethnopharmacology or pharmacoanthropology, that represent another area of relatively recent interest, having obvious implications for drug therapy especially in multiracial societies.

In keeping with very recent developments, we should introduce also the following topics [8]:

- toxicogenomics: Here, investigation based on the use of whole-genome and specialty microarrays yields information concerning the response to xenobiotics at the genomics level (mainly gene expression). In these studies, toxicity is classified on the basis of gene transcriptional patterns. The intention is to extrapolate the toxicities of new chemicals by comparing their patterns with databases of responses and well-documented toxicological endpoints. Toxicogenomics profiling is however limited by the fact that unless the structure of the new compound bears a strong resemblance to those existing in the database, it will not yield reliable predictions of toxicity. This is a definite drawback since the essence of modern drug design is the incorporation of novel structural features into potential drug molecules.

- proteomics: This may be defined as 'high-throughput separation, display and identification of proteins and their interactions' [8]. Analytically, the methodology used to achieve this includes 2D-PAGE electrophoresis separation, mass spectrometry and NMR spectroscopic detection. The method finds use both in the study of new targets for toxicants as well as in predictive profiling. An essential idea that underlies the application of proteomics to toxicogenomics is that specific groups of xenobiotics should induce specific patterns of protein expression. As proteomics permits the assaying of body fluids with rapid detection of biomarkers, there is some advantage over mRNA expression.

- metabonomics: This is the study of metabolic profiles at the organism (i.e. large-scale) level, where human metabolism is considered the basis of cellular organization and responsible for responses to stimuli through control of cellular signalling. Consequently, a measure as drastic as administration of a drug will determine the expression of metabolic enzymes qualitatively and quantitatively, these modifications being interrelated to the organism's responses to gene mutation, drug intervention and disease state. A distinct advantage of metabonomics technology is its treatment of changes in metabolite concentration [8].

- chrono-pharmacogenetics: This is based on a fairly recent concept, namely that changes in the expression level of genes vary with the time of day [9]. It is well known that biochemical, physiological and behavioural patterns vary with the time of day, such variation being at the basis of biological organization. In a recent study [9] expression levels in the liver of 3906 genes in Fischer 344 rats were determined as a function of time of day. While the maximum estimated changes observed for most genes were less than 1.5-fold, statistical tests revealed that 67 genes displayed significant alterations in expression as a function of time of day. Interestingly, these turned out to be genes playing important roles in key cellular pathways including drug metabolism and other major processes.

From all of the above considerations, we may conclude that even inter-individual variability in drug metabolism can be determined by several factors. However, the one still considered to be most important is the existence of genetic polymorphism in the genes encoding the metabolising enzymes. It should be borne in mind that protein structure, three-dimensional configuration and concentration – may also alter the action of drugs in various qualitative and quantitative ways. It should be stressed that successful predictions of effects of genetic variations and their consequent pharmacological and clinical implications are due to changes in the encoded amino acid sequences, with consequent modifications in the three-dimensional structure of the newly synthesised protein, and its subsequently modified function and properties [10]. It is already known from previous chapters that part of the fate of a drug entering the body involves (besides interaction with enzymes, a requirement for biotransformation), interactions with proteins and lipids as well. After passing through membranes (lipoprotein structures) by various mechanisms (see Chapter 1), they react with plasma and/or tissue proteins and interact with their specific receptors. Therefore, it is obvious that genetic mutations that alter even in a punctiform manner, the quality or quantity of these proteins, or characteristics of membranes or receptors, will result in disturbances of the pharmacokinetics of a drug or drug-cell interactions.

Consequences of pharmacogenetic variations in drug metabolising enzymes may include:

a. alteration in the kinetics and duration of action of certain drugs; these phenomena are due either to inherited deficiencies in metabolising enzymes, or to the reverse, an over-expression of them. In the case of deficiencies in metabolising enzymes, the altered kinetics result in retarded inactivation, increased blood concentrations and decreased clearance, leading to overdose with possible adverse reactions or even toxicity; on the other hand, with over-expression of metabolising enzymes, the consequence will be a decreased drug blood concentration, and subsequently, therapeutic inefficacy (sometimes resulting in the need for administering megadoses);
b. drug-drug interactions and
c. idiosyncratic adverse drug reactions (see Chapter 8).

Therefore, the objectives of research in pharmacogenetics are multiple and involve the following:

1. identification of genetically controlled variations in an individual's ability to metabolise a foreign compound (drug or other xenobiotic);
2. study of the molecular mechanisms causing these variations;
3. evaluation of clinical relevance and
4. development of simple methods to identify those individuals who may be susceptible to variable and abnormal responses to drugs administered in normal doses.

Some important, general principles to mention in support of the detailed aspects for the discussion that follows are:

(a) if in some patients the desired, expected effect is not obtained (or even worse, adverse reactions or toxicity appears) with standard, safe doses of a drug, then the most likely cause may be a genetic variability or an inherited metabolic effect;

(b) therefore, any unexpected or unusual (qualitative, but in particular quantitative) response of an individual to a drug should be a warning signal to investigate the genetic source of such variation.

Before detailing some modern procedures generally used in pharmacogenetics, it would probably be useful to define some of the terms commonly used in this area:

- *allele*: one of two or more alternative forms of a gene at the same site in a chromosome that determines alternative characteristics in inheritance;
- *autosome*: one of 22 pairs of chromosomes not connected with the determination of the sex of the individual;
- *autosomal dominant*: a trait that is expressed in the heterozygous (see below) state;
- *autosomal recessive*: a trait expressed only in the homozygous state;
- *gene*: a DNA segment (in a chromosome) that carries the encoded genetic information necessary for protein synthesis;
- *genotype*: a gene combination at one specific locus or any specified combination of loci;
- *heterozygous*: having different alleles at the genetic locus determining a given character;
- *homozygous*: having identical alleles at the genetic locus determining a given character;
- *isoenzymes*: electrophoretically distinct forms of an enzyme displaying the same catalytic role;
- *phenotype*: the visible expression of a gene;
- *polymorphism*: the coexistence of individuals with distinct qualities as normal members of a population [5].

Genetic polymorphisms with functional effects on drug metabolism are usually detected on the basis of discontinuous variation in phenotype, where phenotype represents either levels of enzymes or rate of metabolism. Pharmacogenetic polymorphisms in genes encoding xenobiotic-metabolising enzymes may have a variety of effects, depending on both the reaction catalysed and the type of substrate. That is the reason why, nowadays, in pharmacogenetic studies, one applies genotyping of polymorphic alleles encoding drug-metabolising enzymes to the identification of an individual's drug metabolism phenotype. This knowledge, when applied to drug selection or dosing, can avoid adverse reactions or therapeutic failure.

Phenotyping is accomplished by administration of a drug test, followed by measurement of the metabolic ratio. The main condition is that the metabolism of the drug should be solely dependent on the function of a specific drug-metabolising enzyme. Furthermore, defining the individual's phenotype, relative to a reference substrate, will allow the drug metabolism phenotype for other substrates of that enzyme to be predicted. Hence arises the clinical importance in predicting adverse or inadequate response to certain therapeutic agents. It is emphasised that in pharmacokinetic studies, phenotyping has the advantage over genotyping, in revealing drug-drug interactions or defects in the overall process of drug metabolism.

Genotyping involves identification of defined genetic mutations that will give rise to the specific drug metabolism phenotype. Included in these mutations, we may mention: the genetic alterations that lead to over-expression, the absence of an active protein product (also known as a "null allele"), or the production of a mutant protein with diminished catalytic activity.

As already mentioned, but important to recall here, is the fact that genetic polymorphism (such as genetic mutation or gene deletion) is a permanent cause of variation in drug metabolism phenotypes, while others are considered transient causes (enzyme inhibition or induction). With drugs, the consequences of a polymorphism may be either toxic plasma concentrations or lack of pharmacological response. Toxic plasma concentrations, associated with accumulation of specific drug substances, are autosomal recessive traits and characterise the so-called 'poor-metabolisers' (PM). In contrast, lack of response, characteristic for 'extensive' or, 'ultra-extensive metabolisers' (EM, UEM), is a consequence of increased drug metabolism, resulting in too rapid a rate of elimination. The UEM is an autosomal dominant trait arising from gene amplification [11].

For certain classes of therapeutic agents as well as environmental carcinogens, there is strong evidence that genetic polymorphism of drug-metabolising enzymes plays a significant role in adverse effects of therapeutic agents or incidence of exposure–linked cancer [12-14].

7.2.1 Species-dependent biotransformations and their genetic control

We have, in Chapters 2 and 3, already classified enzymes involved in drug metabolism either as phase I (non-synthetic) or phase II (synthetic or conjugative). The corresponding two reaction types often complement one another in function, in the sense that through catalysis of oxygenation, oxidation, reduction, and hydrolysis reactions, phase I enzymes generate functional groups that subsequently serve as a site for different conjugation reactions, catalysed by phase II enzymes. As a result, we find it opportune to classify the polymorphisms as well.

However, before proceeding to a fairly detailed presentation of these polymorphisms, a very important aspect that needs highlighting in this context concerns the pharmacogenomics in the newborn [15]. It is well established that deficiency in hepatic and renal drug metabolism and disposition are characteristics of the human newborn. Superimposed on genetic polymorphisms that determine drug metabolism and transport, the immaturity of drug-handling ability in the newborn could result in significant interpatient variability, both as regards dosage requirements and responses to medications. Hence, the role of pharmacogenomics in this situation is to individualise drug therapy for the newborn to minimise adverse effects and optimise drug efficacy.

A. Phase I polymorphisms

As already presented, the major route of phase I metabolism is oxidation by cytochrome P450 mixed-function monooxygenases (see Chapter 2). Owing to the diversity of this heme thiolate protein, quite a number of forms have been characterised in humans, with reference to their specificity and unique regulation. A relatively recent article reviews the impact of the cytochrome P450 enzyme system genetic polymorphism upon drug biotransformation and (most probably) incidence of drug-drug interactions [16].

It is well recognized that the pharmacokinetics of many drugs often vary considerably among individuals, precisely because of variations in the expression of different cytochrome P450 (CYP) enzymes. In this subsection we shall focus on some of the various polymorphic CYP enzymes, with emphasis on clinical implications and testing strategies.

Subfamily CYP2 [17]

CYP2D6 is an isoenzyme of particular importance because it metabolises a wide range of commonly prescribed drugs including antiarrhythmics, β-adrenergic blockers, antidepressants and antipsychotics. It is also, by far, the best characterised P450 enzyme demonstrating polymorphic expression in humans. The best-known polymorphism is the debrisoquine/sparteine polymorphism, which involves mutations in the CYP2D6 gene. This was first recognized following adverse reactions in sub-populations of patients receiving the antihypertensive debrisoquine or the oxytocic sparteine. Until recently, more than 50 mutations and 70 alleles have been described for this isoform, many of these resulting in an inactive protein.

This isoform comprises 2 to 6% of the total hepatic cytochrome P450 content, but is responsible for the biotransformation of many important drugs

[2]. As mentioned above, the earliest evidence of polymorphic expression was identified during clinical trials on the antihypertensive drug debrisoquine. Since then, several additional drugs have been identified for use in phenotyping studies, including dextromethorphan, and more recently, propafenone [5] and antidepressants [18].

There are interethnic differences in the prevalence of the phenotype of debrisoquine hyroxylase. The clinical significance of this drug metabolism polymorphism owes to the fact that about 5-10% of Europeans and 1% of Asians lack CYP2D6 activity, and these individuals are classified as 'poor metabolisers' (PMs). It is assumed that in the case of the Caucasian population, the most common mutated allele generating the PM phenotype is CYP2D6B, which in fact is almost absent in Orientals. The prevalence of 'extensive' (EM) and 'ultra extensive' metaboliser (UEM) phenotype in Caucasians is also relatively high (about 7%), and is the result of a partially deficient allele CYP2D causing the exchange of a proline (position 34) with a serine [19].

As for the interracial differences in CYP2D6 genotypes, two situations have been revealed:

- mutations giving decreased activity and,
- mutations giving increased activity.

In the case of mutations giving decreased activity, studies showed that a specific fragment of 11.5 kDa contains a deletion of the entire CYP2D6 gene, while another fragment (of 44 kDa) contains an inserted pseudogene [5,19]. The allele-specific polymerase chain reaction technique could distinguish a 'splicing mutation', present in most of the CYP2D6 genes of the 44 kDa fragments of Caucasians, this being in fact the reason why they are non-functional. This mutation, known also as the B mutation, accounts for about 75% of the mutant CYP2D6 alleles. Unlike the Caucasian population, among the Chinese people, the B mutation has not been detected and is in fact reflected in the low frequency (<1%) of PMs in this population [5,19].

In contrast, the gene deletion allele has been found to be similar in Caucasians, Chinese and Black people [20,21] indicating, in fact, that this gene deletion occurred before the evolutionary separation of the three races.

Mutations giving increased activity appeared in individuals having 10-12 extra copies of the CYP2D6 gene, this resulting in an ultra-rapid metabolism of substrates of CYP2D6 [22]. Extra genes seem to be present, with more or less the same frequency, within all three major races. As mentioned earlier, in this case what obtains is an increased rate of biotransformation, resulting in rates of elimination that are too rapid, and therefore possibly yielding no therapeutic effect from normal therapeutic doses. An interesting recent study dealt with different allele and genotype frequencies, including CYP2D6 in a random Italian population [23]. Here it was found that volunteers could be divided into four CYP2D6 genotype groups comprising 53.5% with no mutated alleles (homozygous EMs), 35.0% with one mutated allele (heterozygous EMs), 3.4% with two mutated alleles (PMs) and 8.3% with extra copies of a functional gene (UMs). Frequencies of CYP2D6 detrimental alleles in these subjects were similar to those of other Caucasian populations. In contrast, the prevalence of CYP3D6 gene duplication among Italians was very high, confirming the tendency for the higher frequency of CYP2D6 UMs in the Mediterranean area relative to Northern Europe.

We may thus conclude that a pronounced variation in the activity of CYP2D6 may be seen both within and between the three major races. This variation, basically caused by mutations in CYP2D6 locus, may manifest differently, as follows:

a) no encoded enzyme,
b) unstable enzyme, or,
c) enzyme with increased activity (gene duplication, triplication, or amplification).

Modified phenotypes may have important consequences both as regards the therapeutic efficiency, and/or exposure to various xenobiotic toxicities [24]. That is why genotyping could play a major role in preventing adverse reactions.

A large number of drugs (the average estimate is 25-30%) have been shown to be metabolised by CYP2D6, all of them being lipophilic bases, and the binding between drug and enzyme being of an ion-pair type. Some selected examples will be given in subsection C.

Recent studies revealed another important aspect that is very significant (given the large variety of drugs metabolised by this isoform), namely that its activity may be inhibited by concurrent administration of various chemicals (drugs or other xenobiotics). The consequences should be as expected *viz.* an increase in the metabolism of the co-administered drug [25]. As an example, we refer to effects of co-administration of drugs on the pharmacokinetics of metoprolol [25]. Celecoxib significantly increased the AUC of metoprolol and the extent of this interaction was more pronounced in individuals having two fully functional alleles relative to those with a single fully functional allele. Rofecoxib, on the other hand, had no significant effect on the pharmacokinetics of metoprolol. Thus, celecoxib evidently inhibits the metabolism of the CYP2D6 substrate metoprolol in this situation, whereas rofecoxib does not. Clinically, relevant interaction may occur between celecoxib and CYP2D6 substrates, particularly those with a narrow therapeutic index.

Another relevant example, involving another isoform, but based on the same principle, is the following which involves thioTEPA (*N*,*N*′,

N''-triethylenethiophosphoramide), an agent commonly administered in high-dose chemotherapy including cyclophosphamide. Following previous studies which concluded that thioTEPA partially inhibits cytochrome P4502B6 (CYP2B6)-catalysed 4-hydroxylation of cyclophosphamide, a study probing the detailed mechanism of this CYP2B6 inhibition was undertaken [26]. Potent inhibition of CYP2B6 activity was confirmed with bupropion as substrate. The inhibition of the isoform CYP2B6 by thioTEPA was established as being time- and concentration-dependent. Furthermore, the loss of CYP2B6 enzymatic activity was shown to be NADPH-dependent and could not be restored. One conclusion of the study was that the pharmacokinetic consequences of irreversible inactivation are more complex than those of reversible inactivation, since the metabolism of the drug itself can be affected; drug interactions will be determined not only by dose, but also by the duration and frequency of application.

A final, but very important aspect involving variation in CYP2D6 genotype, is its impact on non-response or even appearance of adverse reactions during treatment with various drugs. A well-documented example involves antidepressants [18]. Adverse effects or inadequate clinical response often accompany treatment with antidepressants, several of which are substrates for cytochrome P450 (CYP) 2D6. Depending on the polymorphism of the CYP3D6 gene, enzyme activity for individuals can span the range from PMs to UMs. In the study referred to, CYP2D6 genotyping was undertaken using a panel of polymerase chain reaction techniques. The study identified both poor and intermediate metaboliser alleles, as well as allelic duplications of the CYP2D6 isoform. Patients displaying adverse effects had two inactive alleles (PMs). For 19% of the non-responders, amplification of fully functional alleles was established. For psychiatric patients treated with CYP2D6-dependent antidepressants, the conclusion was that the CYP2D6 genotype is associated with adverse effects and non-response.

CYP2C19

The next best-characterised CYP-related drug metabolism polymorphism in humans is associated with the metabolism of the *(S)*-enantiomer of the anticonvulsivant mephenytoin [27]. As in the case of CYP2D6, specific genetic mutations lead to a PM phenotype, with respect to several common therapeutic drugs. The phenotype is inherited in an autosomal recessive manner [28] and, in contrast to the previously described polymorphism, no 'ultra extensive metaboliser' phenotype has been reported for this polymorphic enzyme.

As in the case of CYP2D6 polymorphisms, significant interethnic differences are characteristic for the PM type; approximately 3% of Caucasians and some black populations (e.g. Zimbabwean Shona) are poor metabolisers, while in the Oriental population the estimate is around 20% [29,30].

The *(S)*-mephenytoin hydroxylase reaction is catalysed by CYP2C19, and two mutant alleles associated with the defect have been identified [31]. The principal genetic defect in PMs of mephenytoin is a punctiform exchange of a guanine residue with an adenine residue in exon 5, resulting in an aberrantly spliced CYP2C19 mRNA. The direct consequence is that translation of this mRNA will lead to the production of a truncated, and consequently inactive protein. This is considered a null allele and is designated m_1 (or CYP2C19*2, after other authors [5]). Further evaluation of PM subjects revealed a second mutant allele, designated CYP2C19 m_2 (CYP2C19*3, after [5]), resulting from a G_{636} to A mutation, consequently leading to a premature stop codon. This mutation has been proven to be unique to Japanese individuals [31]. All Japanese PMs whose phenotype could not be explained by the m_1 mutation, were found to be either homozygous or heterozygous (m_1m_2) for the mutant allele [2]. Nevertheless, we must mention the existence of an EM phenotype, which comprises both the homozygous dominant and heterozygous recessive genotypes.

A noteworthy aspect is that individuals of the PM phenotype, due to decreased metabolism of specific drugs such as mephenytoin, are predisposed to CNS adverse effects [3].

Other drugs known to be CYP2C19 substrates include omeprazole [32], propranolol [33] and diazepam [34]. While substrates for CYP2D6 are all lipophilic bases, substrates for CYP2C19 could be bases (propranolol), acids (mephenytoin) or even neutral drugs (diazepam).

The clinical consequence of the CYP2C19 polymorphism has not been fully described. Yet, in about 20% of persons of certain ethnic origins that lack the isoenzyme, the consequences could be of considerable clinical importance.

Also important to stress is that one of the CYP2C19 substrates, omeprazole, is also a CYP1A2 inducer. Consequently, high serum levels of omeprazole (such as might appear in persons deficient in CYP2C19) may result in increased CYP1A2 activity [35]. CYP2C19 also appears to be the major enzyme that activates the antimalarial chloroguanide (proguanil) [36] by cyclization; therefore, in deficient individuals, this compound may be ineffective.

The CYP2D6 and CYP2C19 polymorphisms have been studied less extensively in Black than in Caucasian and Oriental populations. Nevertheless, such interracial differences should be considered during drug development. If the metabolising enzymes of a novel drug have been thoroughly investigated in, for example, a European country, the disposition might then be predicted for an Asian population (and further confirmed in a small phenotyped or genotyped population).

An important aspect to consider, as revealed by recent clinical observations, is that drug-induced hepatitis may be related to the consumption of Atrium – a combination preparation of phenobarbital, febarbamate and difebarbamate – in the PM phenotype of mephenytoin hydroxylase [37]. A decrease in the oral clearance of diazepam was described in Caucasian PMs after a single dose [34].

About 14 years ago, another important aspect was revealed: CYP2C19 polymorphism can be induced by different drugs, for example by rifampicin treatment [38]. More recently, it has been shown that CYP2C19 polymorphism is subject to interactions not only with co-substrates but also with a number of drugs that can inhibit its activity both *in vitro* and *in vivo* [39].

CYP2C9

This is an important CYP450 isoform, involved in the biotransformation of quite a range of therapeutically important drugs, including tolbutamide, *(S)*-warfarin, as well as a range of non-steroidal anti-inflammatory drugs, including diclofenac and ibuprofen [40].

The frequency of the various CYP2C9 allelic variants also varies among ethnic groups, as follows: in whites, an average of 0.06-0.10%; lower frequencies among African Americans, averaging 0.005-0.01%, and in the Chinese population, about 0.02% [41].

All of them are PMs, with the consequences already stated. Selected examples appear in subsection C.

CYP2E1

This is an ethanol-inducible isoenzyme, responsible for the metabolism and bioactivation of many procarcinogens [42] and certain drugs, including ethanol and acetaminophen [43,44]. Actually, it metabolises mainly low molecular weight compounds, such as acetone, ethanol, benzene and nitrosamines.

CYP2E1 is encoded by a single gene in humans, located on chromosome 10 [45]. Two alleles of this gene, C and c_2, have been identified in humans. For each location on the gene where polymorphic mutations have been observed, there is a designation for the wt allele as well as for the mutant allele. For example, the common wt allele with respect to the C mutant allele designating a simple point mutation located in intron 6 of CYP2E1, is designated D. Interestingly, the absence of this allele (C) has been associated with lung cancer in a Japanese control-study [46]. Mutation c_2, more rare, may potentially result in increased expression of functional protein, consequently leading to increased metabolism of CYP2E1 substrates.

A marker of CYP2E1 activity *in vivo* is provided by the skeletal muscle relaxant, chlorzoxazone [47]. However, the non-bimodal distribution of oral and fractional clearance values suggested that a single CYP2E1 allele is predominant in the population studied. A great limitation of this cohort study was that no individuals homozygous for the c_2 variant were identified in Caucasian subjects [48]. Apparently, the lack of c_2 alleles identified in this study is due to the interracial differences in the prevalence of the c_2 allele, first described in the Japanese [49].

Subfamily A

CYP3A

In humans, this family comprises the 3A3, 3A4, and 3A5 isoenzymes – in adults, and the 3A7 isoenzyme in foetal liver.

The most abundant isoenzyme in the adult is 3A4, accounting for 20-40% of the total hepatic CYP in humans, this being also the one with the widest range of drug substrates. The latter include benzodiazepines, erythromycin, cyclosporine and dihydropyridines [50].

Although levels of CYP3A4 activity vary considerably among individuals, no genetic basis for this polymorphic expression has been defined to date.

However, the closely related gene, CYP3A5 has been proven to show a polymorphism in its expression, detectable in only 10-20% of adult livers [51], but with the molecular basis still unclear. It shows a similar, but not identical, substrate specificity for CYP3A4.

In addition to the potential for genetic variability in expression or activity, CYP3A activity is also known to be induced on exposure to barbiturates and glucocorticoids and to be inhibited by macrolide antibiotics such as erythromycin [2]. Interestingly, extrahepatic expression of CYP3A can influence phenotyping approaches, depending on the route of test drug administration [52].

The cytochromes P450 CYP1A1 and 1A2, have also been suggested as showing polymorphism, with the molecular basis however not being identified. CYP1A1 is less important for drug metabolism, but of considerable importance in the activation of certain procarcinogens, such as benzo[α]pyrene. In contrast, the closely related CYP1A2 is of greater importance in drug metabolism, being involved in the biotransformation of important drug substrates like theophylline, imipramine, clozapine, phenacetin and acetaminophen [53]. The level of induction of CYP1A2 by aromatic hydrocarbons is less than for CYP1A1; still, some of the variation seen in CYP1A2 levels in non-smokers might reflect polymorphism in induction owing to passive smoking, diet, or even environmental factors.

Recent studies reveal the important role of some CYTP450 isoforms in the metabolism of certain drugs, as well as an incidence of drug-drug interactions. Based on previous studies indicating that CYP1A2 is the principal isoform responsible for lidocaine metabolism, a study was preformed to assess the effect of a cytochrome P450 (CYP) 1A2 inhibitor, fluvoxamine, on the pharmacokinetics of intravenous lidocaine and its pharmacologically active metabolites MEGX and GX [53]. A second aim of the investigation described was to establish whether fluvoxamine-lidocaine interaction was dependent on liver function. A randomised, double-blind, two-phase, crossover design was employed in the study, details of which appear in reference 53. The authors concluded that liver function did modify the effects of fluvoxamine co-administration, with lidocaine clearance reduced by 60% on average in patients with mild liver dysfunction, but practically unaffected in cases of severe liver dysfunction. The kinetics of formation of the metabolites MEGX and GX were affected in an analogous manner i.e. severely impaired in cases of healthy patients and those with mild cirrhosis, but there was practically no change for subjects with severe liver cirrhosis. Conclusions drawn from this study were (a) that CYP1A2 is the enzyme that is chiefly responsible for the metabolism of lidocaine in patients with normal liver function, and (b) that there is a reduction in fluvoxamine-lidocaine interaction as liver function gets worse. The latter effect was attributed to the likely decrease in the hepatic level of CYP1A2 accompanying this condition.

In an analogous study, the interaction between ciprofloxacin and pentoxifylline was examined, as was the possible role of CYP1A2 in this interaction. Furafylline was employed as a selective CYP1A2 inhibitor here [55].

Other phase I polymorphisms are either relatively common, but not of great importance in drug metabolism, or else rare, but important in the biotransformation of a limited range of drugs.

Examples include polymorphisms detected in some esterases (paraoxonase and cholinesterase) [56-58], epoxide hydrolases [59], and dehydrogenases [53].

Examples and details are presented in subsection C.

A very recent example describes the characterisation of a new CYTP450 isoform, 4F11, and its role in the metabolism of some endogenous compounds and drugs [60]. Its catalytic properties with respect to endogenous eicosanoids were examined. CYP4F11 was found to have a considerably different profile from that of CYP4F3A and was a better catalyst for many drugs including benzphetamine, ethylmorphine, chlorpromazine, imipramine and erythromycin, the latter being the most efficient substrate. Modelling of the structural homology led to the conclusion that for CYP4F11, a more open access channel exists than in CYP453A, this being a possible reason for its capacity to act on large substrate molecules such as erythromycin.

B. Phase II polymorphisms

For most commonly prescribed therapeutics, the major phase II-metabolising enzymes are, in general, the UDP-glucuronosyltransferases and the sulphotransferases. Although there is some evidence for the existence of polymorphisms in certain isoforms of both enzyme families, the molecular and genetic basis are not still very well understood. In contrast, two most common polymorphisms in genes encoding some phase II enzymes are well known for N-acetyltransferase 2 (NAT2) and glutathione S-transferase M1 (GSTM1).

UDP-glucuronosyltransferases

The importance of pharmacogenetic variation in the UDP-glucuronosyltransferases is still not very clear. However, few cases of inter-subject variation in activity in the general population have been reported. Both in Caucasian and Oriental populations, 5% of subjects show very low levels of glucuronide excretion [33]. A particular inborn error of metabolism will be presented in subsection C.

A very recent study investigated the possible involvement of two UGT isoforms, 1A9 and 1A8, in metabolism of particular drugs and the possible appearance of drug-drug interactions [61]. Inhibitory properties of a novel gastroprokinetic agent, Z-338, were examined and compared with those of cisapride to assess its potential for drug-drug interactions. In *in vitro* studies using human liver microsomes, no significant inhibition of terfenadine metabolism or of any of the isoforms of cytochrome P450 (CYP1A1/2, 2A6, 2B6, 2C9, 2C19, 2D6, 2E1 and 3A4) by Z-338 was evident. It was established that Z-338 was primarily metabolised to its glucuronide by UGT1A9 and UGT1A8, while showing no significant inhibition of P-glycoprotein activity. Cisapride, however, 'strongly' inhibited CYP3A4 and 'markedly' inhibited CYP2C9. It was further concluded that drug-drug interactions are unlikely to arise upon co-administration of the agent Z-338 with CYP substrates at clinically effective doses.

Sulphotransferases

These are phase II enzymes conjugating both endogenous and exogenous compounds, thus playing an important role in the biotransformation of a range of compounds. Five isoforms have been identified, but the molecular basis and the pharmacological effects of this variation are still unclear [63]. Nonetheless, the human sulphotransferase family is a complex one, this statement being supported by at least two facts:

- there are two separate genes (STP1 and STP2) encoding proteins, that show 96% homology and appear to be both phenol sulphotransferases [64];
- there is evidence for the existence of allelic variants of each of the phenol sulphotransferases and for the occurrence of two alternative promoters in STP1 [65].

N-Acetyltransferases

Acetylation reactions of different chemical groups are catalysed in humans by two N-acetyltransferases, designated as NAT1 and NAT2. Polymorphism has been detected in both of them, with a more significant impact on NAT2.

A variation in the ability of certain patients to metabolise different drugs, including isoniazid, sulphamethoxazole, hydralazine, caffeine, nitrazepam, sulphamethazine, procainamide and dapsone, is well known; in addition, the acetylation polymorphism probably is the best-known classic example of a genetic defect in drug metabolism. On the basis of the ability to acetylate these drugs, individuals are classified into two phenotypes, namely 'slow' and 'rapid' (or 'extensive') metabolisers. Family studies established that this ability is determined by two alleles at a single autosomal gene locus.

Slow acetylators have a deficiency of hepatic acetyltransferase and are homozygous for a recessive allele [66]. They maintain higher concentrations of un-acetylated drugs for longer periods in body fluids, thus resulting in a greater incidence of adverse drug reactions, due to accumulation of the administered drug (or its phase I metabolites). The precise percentage of slow acetylators in the population varies with ethnic origin: among most European and North American populations, the prevalence of slow acetylators is between 40-70%, whereas, among certain Asian populations, it is only about 10-30% [67]. Thus far, four variant alleles with low activity have been identified; it is assumed that these variations are due to amino acid substitutions, and their frequency varies also between ethnic groups, with NAT2*7A (in which the $_{857}$G is exchanged with A) most common among Japanese, and NAT2*14A (where the $_{191}$G is exchanged with A), common in individuals of African origin, but not in other ethnic groups [53].

Rapid acetylators presumably have a cytosolic N-acetyltransferase; a second cytosolic N-acetyltransferase, NAT1, has been proven to present selectivity for the metabolism of other types of compounds such as *p*-aminosalicylic acid and *p*-aminobenzoic acid, and to be strictly independent of the NAT2 polymorphism.

The acetylator polymorphism is important from the standpoints of both clinical responses to drugs and disease susceptibility, affecting both the efficacy and the occurrence of adverse effects for a number of drugs.

An interesting fact to note is the observation that the phenotypic expression of NAT2 may be influenced by AIDS [5]; patients afflicted with AIDS have been demonstrated to be slow acetylators, which may offer a good explanation for the high incidence of adverse drug reactions to sulphonamides among these patients.

Selected examples will be given in subsection C.

Glutathione S-transferases

These are phase II enzymes that catalyse the glutathione conjugation of both endogenous and exogenous compounds, generally having a detoxifying action (see Chapter 3).

Several polymorphisms have been identified. The most significant and well-characterised GST polymorphisms have been reported for the class μ-enzyme GSTM1 in the class Φ-enzyme GSTT1; there are also reports of polymorphisms in GSTM3, GSTP1 and GSTA2 [6].

The GSTM1 and GSTT1 polymorphisms are of more importance in toxicology than in drug metabolism, with the GSTM1 having a possible role in the metabolism of nitrogen mustard [68,69].

Methyltransferases

These enzymes catalyse methylation of both endogenous molecules, such as neurotransmitters, and of xenobiotics, using S-adenosylmethionine as a methyl donor group (see Chapter 3).

Methylation may occur at different heteroatoms (S, N, O), and it is assumed that at least four separate enzymes carry out these reactions. However, for relevance to drug therapy, polymorphism has been clearly established only for thiopurine S-methyltransferase. The interindividual differences are significant: approximately 0.3% of Europeans have undetectable activity, while about 11% present intermediate levels [70]. Examples appear in a subsection below.

C. Consequences of monogenic variability – selected examples

Monogenic variability takes place on a single specific gene and is due either to deletions or point mutations resulting in splicing defects.

In the case of CYP2D6, an isoenzyme of particular importance because it metabolises a wide range of commonly prescribed drugs, the most common deficient alleles, giving over 98% of the PMs, are represented by CYP2D6*3, *4, *5 and *6. In the *4 allele for example, a guanine is replaced by an adenine at the 3'-ending of intron 3. In the 5* allele, there is a complete deletion of the gene. PMs show, as already mentioned, higher plasma levels of several drugs, which consequently puts them at increased risk of adverse reactions. However, this also depends very much on the particular drug in question and the overall contribution of CYP2D6 to its metabolism. Mutant alleles of the CYP2D6 gene related to the PM phenotype have been studied in numerous laboratories over the last 20 years.

These studies identified the primary mutations (some of them already mentioned) that cause either null alleles or decreased-function alleles, resulting either in total loss of activity, or partial decrease in enzyme function. Knowing that for some of the numerous substrates of this cytochrome isoform, polymorphic oxidation may have important therapeutic consequences, it is obvious that for these specific drugs knowledge of the phenotype could be of utmost help in individualising the dose range required for optimal therapy. For example, pronounced differences in plasma half-life and metabolic clearance have been reported between EM and PM individuals in the biotransformation of flecainide [71]. The implications are that in the case of PMs the plasma steady-state concentrations are achieved only after 4 days of therapy, whereas in the case of EMs, the required time is halved. Moreover, PMs with impaired renal function will be at greater risk of developing flecainide toxicity because of the decreased renal clearance, resulting in potentially dangerous accumulation of the drug.

In the case of CYP2C19, two new allelic variants that contribute to the PM phenotype in Caucasians have been isolated: CYP2C19*2 and CYP2C19*3. The more common, accounting for about 80% of mutant alleles both in Europeans and Orientals is the first mentioned, namely, the CYP2C19*2. Both inactivating mutations are single-base-pair substitutions, with aberrant splice site created in CYP2C19*2 and a premature stop codon in CYP2C19*3. The most important consequence (already mentioned above) is the predisposition of the PM phenotype individuals to CNS adverse effects after administration of even a single 100 mg dose of mephenytoin.

CYP2C9 is another important drug-metabolising enzyme, with a large number of widely used drug substrates. Several single-base-pair substitutions that obviously result in amino acid changes account for the CYP2C9 polymorphism. The most common are $Arg_{144}Cys$ (CYP2C9*2) and $Ile_{359}Leu$ (CYP2C9*3) [72]. A particular example is that of tolbutamide biotransformation in homozygous for the recessive Leu allele (CYP2C9*3), who are PMs [73].

However, perhaps some of the most important and clinically relevant examples involving the existence of polymorphism are reflected by the N-acetylation polymorphism. For example, PMs of isoniazid are more likely to accumulate the drug to toxic concentrations and so are at risk of developing peripheral neuropathy [74]. In contrast, EMs might have to be given unusually high doses to attain efficacy. An important issue in this context is the increased occurrence of severe phenytoin toxicity in PMs of isoniazid, when both drugs are given simultaneously. The assumed mechanism is a non-competitive inhibition of the *p*-hydroxylation of phenytoin displayed by isoniazid [75].

A relatively recent example involves paraoxonase (PON1), a Ca^{2+}-dependent glycoprotein that is associated with high-density lipoprotein

(HDL). Two genetic polymorphisms, determined by punctual amino acid substitutions at positions 55 and 192, have been reported. The major determinant of the PON1 activity polymorphism is assumed to be the position 192 [76]. One of the most important consequences of PON1 polymorphisms is that they are important in determining the capacity of HDL to protect low-density lipoproteins against oxidative modifications, which may explain the relation between the PON1 alleles and coronary heart disease. In the same context, by protecting lipoproteins against oxidative modifications (most probably by hydrolysing phospholipid hydroperoxides), PON1 may also be a determinant of resistance to the development of atherosclerosis. Finally, PON1 polymorphs, by hydrolysing organophosphate insecticides may be responsible for determining the selective toxicity of these compounds in mammals.

7.3 PHARMACO-INFORMATICS

An active biological compound introduced into the body will generate a sequence of events. According to the informational causality laws (principles), it has been proven that both the desired therapeutic effects and the adverse reactions are of informational nature. It was definitively revealed that a specific drug may be toxic not necessarily due to the dose, rhythm of administration, variations in the metabolism profile determined especially by the genetic polymorphism of the enzymatic systems involved, but precisely because of the information transmitted, especially in correlation with the receptor substrate and the whole body.

It is assumed that the impact of the pharmacological information depends not only on the quality and quantity of the signal, but especially on the significance conferred by a specific type or subtype of receptor system.

Thus, from the essential characteristics of the receptors involved, we can mention:

- selectivity (meaning the strict preference for a specific molecular type of ligand);
- saturability (given by the identical number of sites in the receptor molecule which can attach the ligand);
- the cellular location (the receptor being generally located on the cells that will generate the biological response).

Consequently, a rational conception of a new therapeutic entity should primarily take into account the three-dimensional structural details of the receptor. Nowadays, there are several more or less routine procedures, such as X-ray diffraction, NMR spectroscopy and MS that can elucidate such structure. In a subsequent stage, computational techniques are used for the theoretical evaluation of possible interactions between these receptors and ligands capable of eliciting useful responses.

The most important conclusion from the above brief considerations is that a medicinal substance represents nothing outside the body. The drug in question may become a pharmacological signal only if its chemical structure allows its integration into the network of the receptor substrates of the body [77].

7.4 IMPLICATIONS FOR THIRD MILLENNIUM MEDICINE

The main conclusion that arises from all the above considerations is that both phenotyping and genotyping techniques can nowadays be successfully applied by the clinical laboratory for linking human genetics to therapeutic treatment.

In the last few years, through the methods of molecular genetics, and more recently with the discovery of the human genome map, many clinical observations concerning the therapeutic response to a particular treatment, the incidence of adverse reactions, or the toxicity of different drugs or their metabolites, can now be understood at the molecular level. This could be very helpful in more effective prescribing, particularly for compounds with narrow therapeutic index, so that the ratio of therapeutic effect/toxic risk may be increased to the benefit of the patient. Genotyping in particular could represent a good alternative to predicting the appearance of some undesirable secondary effects (or even of some pathologies following a treatment), allowing in this way the selection of the most efficient drug for the specified profile, in other words, the individualisation of the treatment. In fact, the announced perspective for the third millennium is precisely such 'personalised medicine', possibly through creation of an 'ID genetic card' for every patient, completed even in the first years of life; such a contingency would assist medical personnel to find the formula for indicating a therapeutic treatment as close as possible to the ideal i.e. one which is both efficient and devoid of adverse effects.

Finally, we should stress that both pharmacogenetics and pharmacogenomics have become rapidly emerging fields with implications not only for efficient and safe drug therapy, but also for drug discovery and development and for the assessment of the risk for developing certain diseases.

It should be borne in mind too that in the future, physicians should be more aware of these inherited variations of drug responsiveness, because as already highlighted, they are a constant factor throughout a patient's life.

Under the circumstances, new diagnostic procedures should be developed and appropriate dosage adjustments carefully made.

In the next (penultimate) chapter of this book, two special topics with significant clinical implications are considered, namely drug-drug interactions and adverse reactions. The discussion of those subjects will bring to a close the major thrust of this book on drug metabolism, leaving a final chapter of particular interest to the medicinal chemist.

References

1. Ritter JM. 1999. Pharmacogenetics. In: Ritter JM, Lewis LD, Mant TGK, editors. A Textbook of Clinical Pharmacology, 4th ed. London: Arnold (publisher)-member of the Hodder Headline Group, pp 108-118.

2. Linder MW, Prough RA, Valdes R Jr. 1997. Pharmacogenetics: a laboratory tool for optimizing therapeutic efficiency. Clin Chem 43:254-266.

3. Meyer UA. 2000. Pharmacogenetics. In: Carruthers SG, Hoffman BB, Melmon KL, Nierenberg DW, editors. Melmon and Morrelli's Clinical Pharmacology: Basic Priciples in Therapeutics, 4thed. New York: McGraw-Hill Medical Publishing Division, pp 1179-1205.

4. Bechtel P, Bechtel Y. 2000. Pharmacogenetics of biotransformations. In: Carruthers SG, Hoffman BB, Melmon KL, Nierenberg DW, editors. Clinical Pharmacology. Basic Principles in Therapeutics, 4th ed. New-York: McGraw-Hill Medical Publishing Division, pp 25-36.

5. Tischio JP. 2000. Pharmacogenetics. In: Gennaro AR editor. Remington's: The Science and Practice of Pharmacy, 20th ed. Philadelphia: College of Pharmacy and Science, pp 1169-1173.

6. Dally AK.1999. Pharmacogenetics, In: Woolf TF, editor. Handbook of Drug Metabolism. London: Marcel Dekker Inc., pp 175-189.

7. Vogel F. 1959. Moderne Probleme der Humangenetik. Ergebn Inn Med Kinderheilkd 12:52.

8. Rostami-Hodjegan, A, Ticker G. 2004. '*In silico*' simulations to assess the '*in vivo*' consequences of '*in vitro*' metabolic drug-drug interactions. Drug Discov Today: Technologies 1:441-448.

9. Desai VG, Moland CL, Branham WS, Delongchamp RR, Fang H, Duffy PH, Peterson CA, Beggs ML, Fuscoe JC. 2004. Changes in expression level of genes as a function of time of day in the liver of the rats. Mutation Res 549:115-129.

10. Thompson MA, Weinshilboum RM, El Yazal J, Wood TC, Pang Y-P. 2001. Rabbit indolethylamine N-methyltransferase three-dimensional structure prediction: a model approach to bridge sequence to function in pharmacogenomic studies. J Mol Model 7:324-333.

11. Gonzales FJ, Idle JR. 1994. Pharmacogenetic phenotyping and genotyping. Present status and future potential. Clin Pharmacokinet 26:59-67.

12. Daly AK, Cholerton S, Armstrong M, Idle JR. 1994. Genotyping for polymorphisms in xenobiotic metabolism as a predictor of disease susceptibility. Environ Health Perspect 102:55-69.

13. Poulsen HE, Loft S. 1994. The impact of genetic polymorphisms in risk assessment of drugs. Arch Toxicol 16:211-222.

14. Smith G, Stanley LA, Sim E, Strange RC, Woolf CR. 1995. Metabolic polymorphisms and cancer susceptibility. Cancer Surv 25:27-34.

15. Kapur G, Mattoo T, Aranda JV. 2004. Pharmacogenomics and renal drug disposition in newborn. Seminars in Perinatology 28:132-140.

16. Moerike K, Schwab M. 2003. The cytochrome-P450 enzyme system and its role in drug metabolism. Interaktionen und Wirkmechanismen Ausgewaehlter Psychopharmaka (2nded), pp 4-18.

17. Golstein JA, de Morais SMF. 1994. Biochemistry and molecular biology of the human CYP2C subfamily. Pharmacogenetics 4:285-292.

18. Rau T, Wohlleben G, Wuttke H, Thuerauf N, Lunkenheimer J, Lanczik M, Eschenhagen T. 2004. *Cyp2d6* genotype: impact on adverse effects and nonresponse during treatment with antidepressants – a pilot study. Clin Pharmacol Ther 75:386-393.

19. Bertilsson L. 1995. Geographical/Interracial Differences in Polymorphic Drug Oxidation. Current State of Knowledge of Cytochromes P450 (CYP)2D6 and 2C19. Clin Pharmacokinet 29:192-209.

20. Johansson I, Yue QY, Dahl ML, Heim M, Saewe J, Betilsson L, Meyer UA, Sjoeqviat F, Ingelman-Sundberg M. 1991. Genetic analysis of the interethnic difference between Chinese and Caucasians in the polymorphic metabolism of debrisoquine and codeine. Eur J Clin Pharmacol 40:553-556.

21. Evans WE, Relling MV, Rahman A, McLeod HL, Scott EP, Lin JS. 1993. Genetic basis for a lower prevalence of deficient CYP2D6 oxidative drug metabolism phenotypes in black Americans. J Clin Invest 91:2150-2154.

22. Dahl ML, Johansson I, Bertilsson L, Ingelman-Sundberg M, Sjoqvist F. 1995. Ultra rapid hydroxylation of debrisoquine in a Swedish population. Analysis of the molecular genetic basis. J Pharmacol Exp Ther 274:516-520.

23. Scordo MG, Caputi AP, D'Arrigo C, Fava G, Spina E. 2004. Allele and genotype frequencies of *CYP2C9*, *CYP2C19* and *CYP2D6* in an Italian population. Pharmacol Res 50:195-200.

24. Ingelman-Sundberg M. 2002. Polymorphism of cytochrome P450 and xenobiotic toxicity. Toxicology 181-182:447-452.

25. Werner U, Werner D, Rau T, Fromm MF, Hinz B, Brune K. 2003. Celecoxib inhibits metabolism of cytochrome P450 2D6 substrate metoprolol in humans. Clin Pharmacol Ther 74:134-137.

26. Richter T, Schwab M, Eichelbaum M, Zanger UM. 2005. Inhibition of human CYP2B6 by *N,N',N''*-triethylenethiophosphoramide is irreversible and mechanism-based. Biochem Pharmacol 69:517-524.

27. Kupfer A, Presig R. 1984. Pharmacogenetics of mephenytoin: a new drug hydroxylation polymorphism in man. Eur J Clin Pharmacol 26:753-759.

28. Ward SA, Goto F, Nakamura K, Jacqz E, Wilkinson GR, Branch RA. 1987. S-mephenytoin 4-hydroxylase is inherited as an autosomal recessive trait in Japanese families. Clin Pharmacol Ther 42:96-99.

29. Masimirembwa C, Bertilsson L, Johnansson I, Hasler JA, Ingelman-Sundberg M. 1995. Phenotyping and genotyping of S-mephenytoin hydroxylase (cytochrome P450 2C19) in a Shona population of Zimbabwe. Clin Pharmacol Ther 57: 656-661.

30. Nakamura K, Goto F, Ray WA, McAllister CB, Jacqz E, Wilkinson GR, Branch RA.1985. Interethnic differences in genetic polymorphism of debrisoquine and mephenytoin hydroxylation between Japanese and Caucasian populations. Clin Pharmacol Ther 38:402-408.

31. De Morais SMF, Wilkinson GR, Blaidsdell J, Nakamura K, Meyer UA, Goldstein JA. 1994. The major genetic defect responsible for the polymorphism of (S)-mephenytoin metabolism in humans. J Biol Chem 269:1541-1549.

32. Anersson T, Regardh CG, Dahl-Puustinen ML, Bertilsson L. 1990. Slow omeprazole metabolisers are also poor S-mephenytoin hydroxylators. Ther Drug Monit 12:48-62.

33. Ward SA, Walle T, Walle UK, Wilkinson GR, Branch RA. 1989. Propranolol's metabolism is determined both by mephenytoin and debrisoquin hydroxylase activities. Clin Pharmacol Ther 45:72-79.

34. Bertilsson L, Henthorn TK, Sanz E, Tybring G, Sawe J, Villen T. 1989. Importance of genetic factors in the regulation of diazepam metabolism: relationship to S-mephenytoin, but not debrisoquin, hydroxylation phenotype. Clin Pharmacol Ther 45:348-355.

35. Rost K, Brosicke H, Brockmoller J, Scheffler M, Helge H, Roots I. 1992. Increase of cytochrome P450 1A2 activity by omeprazole: Evidence by the ^{13}C[N-3-methyl]-caffeine breath test in poor metabolisers of S-mephenytoin", Clin Pharmacol Ther 52:170-181.

36. Ward SA, Helsby NA, Skjelbo E, Brosen K, Gram LF, Breckenbrige Am. 1991. The activation of the biguanide antimalarial proguanil co-segregates with the mephenytoin oxidation phenotype – a panel study. Br J Clin Pharmacol 31:689-692.

37. Horsmans Y, Lannes, Pessayre D, Larrey D. 1994. Possible association between poor metabolism of mephenytoin and hepatotoxicity caused by Atrium, a fixed combination preparation containing phenobarbital, febarbamate and difebarbamate. J Hepatol 21:1075-1079.

38. Zhou HL, Anthony LB, Wood AJJ, Wilkinson GR. 1990. Induction of polymorphic 4'-hydroxylation of S-mephenytoin by rifampicin. Br J Clin Pharmacol 30:471-475.

39. Flockhart DA. 1995. Drug interactions and the cytochrome P450 system: the role of cytochrome P450 2C19. Clin Pharmacokinet 29:45-51.

40. Daly AK, Cholerton S, Gregory W, Idle JR. 1993 Metabolic polymorphisms. Pharmacol Ther 57:129-141.

41. Wang S-L, Huang J-D, Lai M-d, Tsai J-J. 1995. Detection of CYP2C9 polymorphism based on the polymerase chain reaction in Chinese. Pharmacogenetics 5:37-45.

42. Guengerich FT, Kim D-H, Iwasaki M. 1991. Role of cytochrome P450 2E1 in the oxidation of many low molecular weight cancer suspects. Chem Res Toxicol, 4:168-179.

43. Raucy JL, Lasker JM, Lieber KS, Black M. 1989. Acetaminophen activation by human liver cytochromes 450 2E1. Arch Biochem Biophys 271:270-283.

44. Pattern CJ, Thomas PE, Guy RL, Lee M, Gonzales FJ, Guengerich FP, Yang CS. 1993. Cytochrome P450 enzymes involved in acetaminophen activation by rat and human liver microsomes and their kinetics. Chem Res Toxicol 6:511-518.

45. Umeno M, McBride W, Yang CS, Gelboin HV, Gonzales FJ. 1988. Human ethanol-inducible P450IIE1: complete gene sequence, promoter characterization, chromosome mapping, and cDNA-directed expression. Biochemistry 27:9006-9013.

46. Uematsu F, Kikuchi H, Motomiya M, Abe T, Sagami I, Ohmachi T, Wakui A, Kanamaru R, Watanabe M. 1991. Association between restriction fragment length polymorphism of the human cytochrome P450IIe1 gene and susceptibility to lung cancer. Jpn J Cancer Res 82:254-256.

47. Peter R, Bocker R, Beaune PH, Iwasaki M, Guengerich FP. 1990. Hydroxylation of chlorzoxazone as a specific probe for human liver cytochrome P450IIE1. Chem Res Toxicol 3:566-573.

48. Kim RB, O'Shea, Wilkinson GR. 1995. Interindividual variability of chlorzoxazone 6-hydroxylation in men and women and its relationship to *CYP2E1* genetic polymorphisms. Clin Pharmacol Ther 57:645-655.

49. Watanabe J, Hayashi S-I, Nakachi K, Imai K, Suda Y, Sekine T. 1990. *Pst*I and *Rsa*I RFLPs in complete linkage disequilibrium at the CYP2E gene. Nucleic Acids Res 18:7194.

50. Watkins PB. 1994. Non-invasive tests of CYP3A enzymes. Pharmacogenetics 4:171-183.

51. Aoyama T, Yamano S, Waxman DJ, Lapenson DP, Meyer UA, Fischer V, Tyndale R, Inaba T. 1995. Cytochrome P450hPCN3, a novel cytochrome P450 IIA gene product that is differentially expressed in adult human liver. J Biol Chem 264:10388-10395.

52. McKinnon RA, Burgess WM, Hall P de la M, Roberts-Thomson SJ, Gonzalez FJ, McManus ME. 1995. Characterisation of CYP3A gene subfamily expression in human gastrointestinal tissues. Gut 36:259-267.

53. Daly AK.1999. Pharmacogenetics. In: Woolf TF, editor. Handbook of Drug Metabolism. New York, Marcel Dekker Inc., pp175-202.

54. Orlando R, Piccoli P, De Martin S, Padrini R, Floreani M, Palatini P. 2004. Cytochrome P450 1A2 is a major determinant of lidocaine metabolism *in vivo*: effects on liver function. Clin Pharmacol Ther 75:80-88.

55. Peterson TC, Peterson MR, Wornell PA, Blanchard MG, Gonzalez FJ. 2004. Role of CYP1A2 and CYP2E1 in the pentoxifylline ciprofloxacin drug interaction. Biochem Pharmacol 68:395-402.

56. Humbert R, Adler DA, Disteche DM, Hasset C, Omiecinski CJ,.Furlong CF. 1993. The molecular basis of the human serum paraoxonase activity polymorphism. Nature Genet 3:73-77.

57. Lockridge O. 1990. Genetic variants of human serum cholinesterase influence metabolism of the muscle relaxant succinylcholine. Pharmacol Ther 47:35-39.

58. Mackness B, Durrington PN, Mackness MI. 1998. Human Serum Paraoxonase. Gen Pharmac 31:329-336.

59. Oesch F, Timms CW, Waker CH, Guenthner TM, Sparrow A, Watabe T.1994. Existence of multiple forms of microsomal epoxide hydrolases with radically different substrate specificities. Carcinogenesis 5:7-11.

60. Kalsotra A, Turman CM, Kikuta Y, Strobel HW. 2004. Expression and characterisation of human cytochrome *P*450 4F11: Putative role in the metabolism of therapeutic drugs and eicosanoids. Toxicol Appl Pharmacol 199:295-304.

61. Patel M, Tang BK, Kalow W. 1993. Variability of acetaminophen metabolism in Caucasians and Orientals. Pharmacogenetics 2:38-43.

62. Furuta S, Kamada E, Omata T, Sugimoto T, Kawabata Y, Yonezawa K, Wu XC, Kutimoto T. 2004. Drug-drug interactions of Z-338, a novel gastroprokinetic agent, with terfenadine, comparison with cisapride, and involvement of UGT 1A9 and 1A8 in the human metabolism of Z-338. Eur J Pharmacol 497:223-231.

63. Weinshilboum R. 1990. Sulphotransferase pharmacogenetics. Pharmacol Ther 45:93-99.

64. Her C, Raftogianis R, Weinshilboum RM. 1996. Human phenol sulphotransferase STP2 gene: Molecular cloning, structural characterisation, and chromosomal location. Genomics 33:409-416.

65. Bernier F, Soucy P, Luu-The V. 1996. Human phenol sulphotransferase gene contains two alternative promoters: Structure and expression of gene. DNA Cell Biol 15:367-376.

66. Evans DAP. 1960. Genetic control of isoniazid metabolism in man. Br Med J 2:485-491.

67. Liu HJ, Han C-Y, Lin KB, Bruce K, Hardy S. 1994. Ethnic distribution of slow acetylator mutations in the polymorphic *N*-acetyltransferase (NAT2) gene. Pharmacogenetics 4:125-134.

68. Smith MT, Evans CG, Doane-Setzer P, Castro VM, Tahir MK, Mannervik B. 1989. Denitrosation of 1,3-bis(2-chlorethyl)-1-nitrosourea (BCNU) by class m glutathione transferases and its role in cellular resistance in rat brain tumor cells. Cancer Res 49:2621-2627.

69. Evans CG, Bodell WJ, Ross D, Doane P, Smith MT. 1986. Role of glutathione and related enzymes in brain tumor resistance to BCNU and nitrogen mustard. Proc Am Assoc Cancer Res 37:267-274.

70. Weinshilboum R, Sladek SL. 1990. Mercaptopurine pharmacogenetics: Monogenic inheritance of erythrocyte thiopurine methyltransferase activity. Am J Hum Genet 32:651-658.

71. Funck-Brentano C, Becquemont L, Kroemer HK, Buhl K, Knebel NG, Eichelbaum M, Jaillon P. 1994. Variable disposition kinetics and electrocardiographic effects of flecainide during repeated dosing in humans: contribution of genetic factors, dose-dependent clearance, and interaction with amiodarone. Clin Pharmacol Ther 55:256-269.

72. Romkes M, Faletto MB, Blaisdell JA, Raucy JL, Goldstein JA. 1991. Cloning and expression of complementary DNAs for multiple members of the human cytochrome P450IIC subfamily. Biochem 30:3247-3253.

73. Scott J, Poffenbarger PL. 1979. Pharmacogenetics of tolbutamide metabolism in humans. Diabetes 28:41-46.

74. Devadatta S, Gangadharam PR, Andrews RH, Fox W, Ramakrishnan CV, Selkon JB, Velu S. 1960. Peripheral neuritis due to isoniazid. Bull WHO 23:587-598.

75. Kutt H, Brennan R, Dehejia H, Verebely K. 1970. Diphenylhydantoin intoxication: a complication of isoniazid therapy. Am Rev Respir Dis 101:377-384.

76. Mackness B, Durrington PN, Mackness MI. 1998. Human serum paraoxonase. Gen Pharmacol 31:329-336.

77. Hiroyuki O. 2002. Pharmaco informatics. Farumashia 38:120-124.

Chapter 8

DRUG INTERACTIONS AND ADVERSE REACTIONS

8.1 INTRODUCTION

Two principal aspects of drug metabolism are addressed in this chapter, namely drug-drug interactions and adverse reactions. Since drug-drug interactions can occur at various stages following drug administration, these are systematically subdivided into interactions associated with the pharmacodynamic phase, pharmacokinetic interactions, and interactions occurring during the biotransformation phase. Known interactions between drugs and food, alcohol and tobacco smoke are treated separately. A special feature of the present chapter is an extensive tabulation of drug-drug interactions which serves as a useful reference to those occurring most frequently, together with their biological consequences. In the treatment of adverse reactions that follows, these are first defined and an attempt to classify them according to various criteria is presented. A significant emphasis is given to allergic reactions and associated toxicity in the extensive discussion that follows. The latter is supported by a wide range of examples. Finally, a brief outline of some of the modern approaches to predicting drug metabolism is presented.

8.2 DRUG-DRUG INTERACTIONS

8.2.1 Definitions, concepts, general aspects

Today, with the increasing complexity of therapeutic agents available, and widespread polypharmacy (a particular problem especially in the elderly, who receive more medications than younger individuals), the potential for drug interactions is enormous. Drugs can interact to alter the absorption, distribution, metabolism or excretion of a drug, or interact in a synergistic or antagonistic fashion altering their pharmacodynamics. Generally, the outcome of an interaction can be harmful, beneficial or clinically insignificant. Although clinically often unrecognized, many of the drug interactions are responsible for increased morbidity.

Drug interactions are of utmost importance in clinical practice, since they account for 6-30% of all adverse reactions (ADRs). In some cases, drug interactions can be useful, and it is already a relatively current practice for prescribers to use known interactions to enhance efficacy in the treatment of several conditions such as epilepsy, hypertension or cancer [1]. An example illustrating beneficial effects rather than ADRs, involves the co-administration of carbidopa (an extracerebral dopadecarboxylase inhibitor), together with levodopa to prevent its peripheral degradation to dopamine [2]. On the other hand, association of theophylline with ciprofloxacin, for instance, causes a two- to threefold increase in theophylline serum level, resulting in theophylline toxicity [3].

A drug interaction is a measurable modification in magnitude or duration of the pharmacological response of one drug, due to the presence of another drug that is pre- or co-administered. Many drug interactions involve an effect of one drug on the action or disposition of another, with no recognizable reciprocal effects [4]. Usually, this modification of the action of one drug by another is a result of one or more of four principal mechanisms: a) pharmaceutical, b) pharmacodynamic, c) pharmacokinetic, and d) metabolic [5].

It should be stressed that usually the term 'drug interactions' refers to drug-drug interactions, although it can be taken to include interactions between drugs and food constituents, alcohol, or environmental factors. In addition, the term may include even interferences by drugs in clinical laboratory tests, with important consequences for diagnoses [6]. Drugs may also interact with diseases, potentially worsening their symptoms [7].

A definition with important implications was given a decade ago by Thomas [8], according to which a drug interaction is "considered to occur when the effects of giving two or more drugs are qualitatively and quantitatively different from the simple sum of the observed effects when the same doses of the same drugs are given separately". The implications mentioned above may involve different aspects: either increased or decreased activity of two drugs given concurrently (in a purely quantitative manner), qualitative change in the effect of a drug, antagonism of the effects of one drug by another, resulting in annulment of beneficial effects of therapy, and potentiation of an unwanted effect. Especially for this last possible effect, as has often been observed, patients are in many cases exposed to unnecessary risk, by pre- or co-administration of therapeutic agents that are assumed to interact adversely. Fortunately, many interactions are predictable, and avoidance of unwanted effects or therapeutic ineffectiveness is thus possible [9,10]. Therefore, a priority of the clinical pharmacist today, to increase the likelihood of identifying or preventing an adverse drug reaction, primarily involves knowing or predicting those situations in which a potential drug interaction is likely to have clinically significant consequences, recognizing these clinical settings, and understanding the mechanisms by which they occur. In this case it is important either to recommend steps that may be taken to avoid them (e.g. altering sequence of administration and time interval between administration of two drugs), or preferably alternative treatments.

8.2.2 Interactions associated with the pharmacodynamic phase

Pharmacodynamic interactions, assumed to be the most common drug interactions in clinical practice, are those for which the effects of one drug are altered by the presence of another drug at its site of action [3]. At the same time, some of the most clinically important ADRs also result from pharmacodynamic interactions [6].

Most of these interactions have a simple mechanism, consisting either of summation or opposition of the effects, and therefore being either synergistic or antagonistic. Most of these interactions are intuitively evident; thus, it is not surprising for instance, that two drugs with sedative properties (e.g. alcohol and benzodiazepines) can potentiate each other's sedative action. Other synergistic interactions include those between a diverse range of drugs, such as tetracyclines, clofibrate, and estrogens, with warfarin, leading to increased anticoagulation [6]. When two drugs concomitantly administered share similar adverse effects, their association can produce additive side effects. For example, hydrocortisone and hydrochlorothiazide together can produce additive side effects of hyperglycaemia or hypokalemia [6,8]. Another example is the increased risk of bleeding in anticoagulant patients taking salicylates [3].

In some instances, important interactions may occur between drugs acting at a common receptor. When used deliberately, many of these interactions can generally be useful. For example, we may mention the use of naloxone to reverse opiate intoxication. Less directly, by the local increase in acetylcholine (caused by cholinesterase inhibition), muscular relaxation by tubocurarine can be reversed [4].

The antagonistic interactions can be partial, when the global antagonistic effect is smaller than that of the sum of the individuals, or total, when the global effect is null [12]. A well-known example is that of the antiparkinsonian levodopa, whose action can be antagonised by certain dopamine-blocking drugs, such as haloperidol and chlorpropamide [11]. Usually, such interactions are due to direct effects at the receptors (the same or different), but more often can also occur by indirect mechanisms, due either to an interplay of receptor effects or combined interferences with biochemical or physiological mechanisms [2-6]. If generally, both drugs compete directly for the same receptor, it has to be stressed that often such interactions involve a more complex interference with physiological mechanisms [11].

Often, through a pharmacodynamic mechanism, the risk of certain toxic effects can be potentiated. A general example is that of diuretic-induced hypokalemia and hypomagnesemia that may act to increase the risk of dysrhythmias caused by digoxin [5,8,11].

Besides the additive (or synergistic) and/or antagonistic interactions, in the category of pharmacodynamic interactions, could be also included interactions due to changes in drug transport mechanisms [2,11]. For example, the tricyclic antidepressants are able to potentiate the action of epinephrine and norepinephrine through their blocking the neuronal re-uptake of amines. In contrast, the antihypertensive effects of certain adrenergic neurone blocking drugs (e.g. debrisoquine, bethanidine) are prevented or even reversed by tricyclic antidepressants, most probably by the same mechanism as above. Other drugs, such as aminoglycosides and especially, local anaesthetics, may exert weak inhibiting effects on neuromuscular transmission. In patients with normal neuromuscular transmission, the effect is more or less evident, unlike in those who received e.g. neuromuscular blocking drugs, or patients with *myasthenia gravis*. In such circumstances, these drugs may produce apnoea or even neuromuscular paralysis.

Also noteworthy are the indirect pharmacodynamic interactions, several of them with potential clinical significance [2,6]. Well-known examples involve co-administration of aspirin and NSAIDs with anticoagulants, such as warfarin. Because these drugs are known to cause gastrointestinal lesions, including ulcerations, it is obvious that such concomitant administration may provide a focus for bleeding. In co-administration with anticoagulants, salicylates may also lead to enhanced tendency to bleeding, through inhibition of platelet aggregation. Another interesting pharmacodynamic interaction through an indirect mechanism, worth mentioning in this context, is that between propranolol and glycogen: propranolol reduces the breakdown of glycogen (the major energy- storage polysaccharide in mammals), subsequently delaying the elevation in blood glucose levels after hypoglycaemia.

Finally, it is necessary to include here the situation when pharmacodynamic interactions involve unknown and multiple mechanisms, for instance by involving different sites of action, or by inhibiting the P-glycoprotein efflux transporter [11].

To illustrate how pharmacodynamic drug interactions may arise, some examples in more detail follow.

• NSAIDs and corticosteroids: both are known to cause gastrointestinal irritation, subsequently leading to bleeding and ulceration. Under these circumstances, it is easily predicted that in association, the incidence and risk of bleeding will rise significantly (presumably due to a simple additive effect). Therefore, during concomitant use, close monitoring of patients is strongly recommended. Ultimately, concomitant treatment for gastrointestinal damage should be advisable [11,12].

• NSAIDs, loop diuretics and antihypertensive agents: reducing renal sodium excretion, NSAIDs increase renal prostaglandins that accompany administration of certain diuretics, such as furosemide. Through the same mechanism, the antihypertensive efficacy of ACE inhibitors and β-adreno receptor blockers may be reduced as well [2,5].

• neuromuscular blockers and/or aminoglycoside antibiotics and anaesthetics:

• the aminoglycoside antibiotics are known to display, as an additional pharmacological action, that of potent neuromuscular blockers; therefore, their concomitant administration with neuromuscular blockers may lead to prolonged, or even fatal respiratory depression. It is assumed that these effects are additive to the conventional neuromuscular blockers that act on the post-synaptic membrane. Consequently, concurrent use must be avoided. In the case of anaesthetics that may cause prolonged neuromuscular blockade, it is recommended that the postoperative period be closely monitored [3,8,11,12].

• phenothiazines and antihypertensives: certain phenothiazines, such as promazine and chlorpromazine, have been shown to cause postural hypotension. Under these circumstances, if the patient is also taking an antihypertensive drug, the reaction may be exaggerated. Such cases have been reported following co-administration of a phenothiazine with various antihypertensive agents, including captopril, nadolol, clonidine and nifedipine [3,12].

• corticosteroids and digitalis glycosides: administered systemically, corticosteroids (particularly cortisone, deoxycortone and hydrocortisone, occurring naturally as well) have been proven to increase potassium loss, concomitant with sodium and water retention. Subsequently, oedema and hypertension result, which can lead to cardiac failure in some individuals. Under these circumstances, if these drugs are co-administered with digitalis glycosides, it is advisable to monitor the patient well [11].

8.2.3 Pharmacokinetic interactions: incidence and prediction

Pharmacokinetic drug interactions can occur during any of the processes assumed to represent the fate of a drug in the body and contributing to the drug's pharmacokinetic profile. The positive aspect is that for a new drug candidate, thorough preliminary studies can be undertaken before it appears on the market. This would confirm either the presence or absence of possible pharmacokinetic interactions that such a drug could cause.

Such interactions may affect: a) absorption of orally administered drugs, through different mechanisms such as: chemical interactions (chelation and complexation), alteration of gastrointestinal motility, changes in gastrointestinal pH, perturbation of gastrointestinal flora; b) distribution; c) drug metabolism (epecially through enzyme induction or inhibition effects, discussed separately in subchapter 8.2.4); and d) excretion [2,3,5,11].

a) Absorption

Generally, as already outlined in Chapter 1, the process of absorption from the gastrointestinal tract in the case of orally administered drugs is variable and complex. Consequently, drug interactions of this type are difficult to predict. However, the significant advances made in this area, especially during the last decade, permit (through mathematical approaches to prediction based on competitive enzyme inhibition) an early assessment of potential drug-drug interactions in patients that are taking concurrent medications [3,9,10]. Most of these interactions refer to the rate of absorption, although, in some instances the extent of absorption may be affected as well. Changes in the rate of absorption – in most cases, delay of the process, can be of real clinical significance when referring either to drugs having a short half-life, or when achievement of rapid and high plasma levels may be critical (as may be the case with analgesics or hypnotics). Usually, this phenomenon is expected to appear if inappropriate combinations are administered without sufficient separation in time; therefore, most of these interactions can be avoided by simply allowing a two or three hour interval between the administration of the interacting drugs [3,6,12].

The mechanisms of generating such interactions are various:

- certain drugs given orally can sometimes react directly within the gastrointestinal tract, leading either to chelates or complexes, forms which are not readily absorbed. Examples include:
  - interaction of tetracyclines or fluoroquinolone antibiotics with metal ions (e.g. aluminium and magnesium in antacids, or iron salts), resulting in reduced drug absorption due to formation of a chelate

complex within the gut; this chelation with divalent or trivalent ions, leading to insoluble complexes, may result in severely reduced plasma levels of the administered drugs and thus, therapeutic inefficacy.

- interaction of digoxin, warfarin or thyroxine with cholestyramine and related anion exchange resins, with the same consequence of reduced absorption due to binding/complexation in the gut (in fact, the adsorption of the former onto cholestyramine).

Nevertheless, such effects may sometimes be used to therapeutic advantage:

- activated charcoal, which acts as an adsorbent agent within the gut (although it can affect the absorption of certain drugs), may be used with good efficacy in the management of poisoning;
- cholestyramine and related anion exchange resins, binding cholesterol metabolites and bile acids, prevent their re-absorption in the intestinal lumen, thus lowering plasma levels of total cholesterol.

• altering the rate of gastric emptying is assumed to generally alter the rate of drug absorption as well; drugs that retard gastric emptying may delay or attenuate the rate of absorption of other co-administered drugs. For example, drugs with anticholinergic effects (anticholinergic agents, antihistamines, and phenothiazines), tricyclic antidepressants, and opioids, that decrease the rate of gastric emptying will consequently increase the necessary time to achieve the therapeutic plasma levels of drugs administered concurrently. In some instances, bioavailability of the affected agent may be reduced as well [2].

On the other hand, attention is drawn to drugs that increase the rate of gastric emptying, resulting in an accelerated absorption of certain co-administered drugs. For example, metoclopramide has been shown to accelerate the absorption of diazepam, propranolol, paracetamol, and conversely, to reduce that of digoxin. Other drugs that enhance gastric emptying include cisapride and domperidone, and as a consequence of their effects, may cause earlier and higher peak concentrations, which could be dangerous especially in the case of index drugs. The rate of gastric emptying is especially important when a rapid onset of effect of the drug is desired: rapid relief pain or onset of sedation, and in instances where parenteral administration is not feasible. Among the factors that slow gastric emptying, apart from the concurrently administered drugs already referred to above, we should also mention food, heavy exercise, and autonomic neuropathy [3].

• changes in bacterial flora, generally caused by broad-spectrum antibiotics, may affect the absorption of any drugs subject to metabolism by bacterial enzymes. As a clinically relevant example (although the mechanism has not been fully elucidated), we should mention the reduction in oestrogen levels resulting from diminished bacterial flora that results in an increased risk of contraceptive failure [2].

• changes in gastrointestinal pH. As already discussed in Chapter 1, the gastrointestinal mucosa having an essentially lipid-based structure, drugs will usually pass through them by simple diffusion, if they are in a lipid-soluble form. However, it is known that drugs vary in their lipid solubility, and many of them may act as weak acids or bases; in the latter case, a proportion of the dose exists in dissolved (ionised) form, some still remaining unionised, in a dynamic equilibrium. Therefore, in such circumstances, changes in gastric pH can affect the solubility and absorption of ionisable drugs, shifting the balance of this equilibrium very significantly. Drugs such H_2 antagonists, proton pump inhibitors and antacids, by increasing gastric pH, will markedly reduce the bioavailability of certain drugs such as ketoconazole, for example, which requires an acidic medium for adequate absorption.

Some more detailed examples follow:

• digoxin and metoclopramide: on concurrent administration, the serum levels of digoxin have been shown to be reduced by about a third [3]. Apparently, metoclopramide increases the mobility of the gut to such an extent, that both full dissolution and absorption of digoxin remain incomplete by the time it is eliminated in the faeces. Under these circumstances, two options exist: either to increase the digoxin dose, which would not be advisable since digoxin is an index drug, or to administer them with a sufficient time interval between doses.

On the other hand, propantheline seems to exert quite the opposite effect, increasing digoxin plasma levels, through reduction in gut motility. In either case, the patient could be placed outside the desired range for plasma levels, either for obtaining the expected therapeutic effect, or instead being subject to increased risk of toxic effects [12]. Since digoxin has a narrow therapeutic index, its levels require very close monitoring.

• ketoconazole and antacids, H_2 blockers and proton pump inhibitors: for adequate absorption, ketoconazole, being a poorly soluble base needs to be converted into a soluble salt; usually, this is mediated by the acid in the stomach, resulting in the corresponding hydrochloride salt. In this situation, it is obvious that co-administration of antacids (which raise the pH in the stomach), or H_2 blockers, agents that reduce gastric acid secretion, will cause a reduction in both the dissolution and absorption of ketoconazole. Clinical observations confirmed dramatic reductions in ketoconazole plasma levels upon its co-administration with ranitidine or cimetidine [2]. For managing this interaction, two methods may be suggested: either to administer ketoconazole when the stomach contents are most acidic, or to ensure a suitable temporal separation between ketoconazole and H_2 blockers or antacids. In both situations, to ensure ketoconazole efficacy, it is advisable to monitor the effects of treatment.

• regarding fluoroquinolone antibiotics and divalent/trivalent metallic ions: on concurrent administration with antacids containing calcium, magnesium or aluminium, clinical observations indicated reduced absorption of these antibiotics, reflected in their reduced plasma levels. The most probable mechanism involves the interaction of certain functional groups on the antibiotics with the metallic di- or trivalent ions, forming insoluble chelates within the gut, that are not absorbed to any great extent, and in addition, appear to be relatively inactive as antibacterials.

Iti is appropriate to mention here that a relatively new product, iron-ovotransferrin, through its ability to combine directly with the transferrin receptors of intestinal cells, will consequently release little ionic iron into the gut. This would presumably reduce the incidence of combination with quinolones, as was confirmed for iron-ovotransferrin on co-administration with ciprofloxacin [3].

b) Distribution

The major distributional process that may contribute to drug interactions is binding to plasma proteins [13]. Following absorption, and after passing through the liver, a drug reaches the systemic circulation and is distributed throughout the body, including its site of action. This phase of distribution depends on several factors, including the ionic composition, lipid-solubility, and protein-binding characteristics of the drug. Protein binding may refer to either plasma albumin binding, or, outside the bloodstream, to tissue proteins, and directly influences the pharmacokinetics of a drug. It is well known that only free drug can exert a pharmacologic effect. Drugs that are generally highly bound to plasma proteins are also potentially subject to displacement from their specific carrier proteins by a concurrently administered drug that might display a higher affinity for the same protein. Such a displacement interaction, involving reduction in the extent of plasma protein binding of one drug by the presence of a co-administered one, consequently results in an increased unbound fraction of the displaced drug [2,3,8,13,14]. The unbound (i.e. free in solution) molecules are pharmacologically active, while the bound ones form a circulating, but pharmacologically inactive reservoir. Since the two forms exist in a dynamic equilibrium, biotransformation and excretion of free, active molecules, results in their immediate replacement by molecules from the inactive reservoir.

There are several examples of clinically important interactions that are attributed entirely to protein-binding displacement, the most frequently cited example probably being that between warfarin and NSAIDs. The anticoagulant effect of warfarin is potentiated in co-administration with different NSAIDs, most probably because of displacement of the former from its protein-binding sites [2]. Another example is the marked diuresis observed in patients with nephrotic syndrome when they were given clofibrate [3].

c) Interactions due to altered biotransformations (see following subchapter)

d) Excretion (elimination interactions)

The renal excretion of drugs (or their metabolites) may be affected by a co-administered drug in various ways [1-3,6,11]. A change in glomerular filtration rate, tubular secretion or urinary pH can alter the elimination of some drugs.

Selected examples:

- some acidic drugs lower the urinary pH, whereas some antacids (or bases) cause an increase in the pH of urine. Therefore, the excretion of other ionisable compounds that display appreciable renal clearance is expected to be influenced in some way. Aciduria, for example, will increase the renal clearance of certain basic drugs, such as amphetamine, antihistamines and tricyclic antidepressants. Conversely, for acidic drugs, including salicylic acid, phenobarbital, and nitrofurantoin, the renal clearance will increase with increasing urine pH.
- many drugs share a common transport mechanism in the proximal tubules, and consequently can reduce one another's excretion by competition. In practice, the clearance of drugs actively secreted into the tubular lumen can be significantly inhibited by other drugs. Examples include the reduction in renal clearances of penicillins and indomethacin by co-administration of probenecid, and of methothrexate by salicylates and NSAIDs. However, it is important to stress that in certain situations this type of interaction can be used to advantage: for example, by decreasing the clearance of penicillin, probenecid actually prolongs its duration of action. On the other hand, the opposite consequence may be reported as well: methotrexate toxicity can be caused by inhibition of its tubular secretion, by some of the drugs mentioned above.
- of course, diuretics are certainly expected to exert such effects. These compounds reduce sodium absorption, a phenomenon leading indirectly to increased proximal tubular re-absorption of monovalent cations. In certain instances, this increased re-absorption can cause accumulation and potentially fatal toxicity (e.g. in patients treated with lithium salts) [5].
- digoxin excretion can be reduced by several drugs including amiodarone, quinidine, spironolactone and verapamil, and this will increase its toxicity.

8.2.4 Interaction during the biotransformation phase

Because of significant inter-patient variation, the biotransformation of one drug can be dramatically affected by other pre- or co-administered drugs. Actually, it is assumed that most clinically important drug-drug interactions result from perturbations of drug metabolism, involving either induction or inhibition of metabolising enzymes. When two drugs are involved with the same range of enzymes, this can lead to changes in the extent of metabolism of either or both, either increasing or decreasing, with consequent changes in plasma levels.

a) Enzyme inhibition
Decreasing enzyme activity, which is an extremely common mechanism underlying the interaction of two drugs, often results in high drug plasma concentrations, exaggerated and prolonged responses, and subsequently, an increased risk of toxicity [1-3,6,10-12,14]. The direct consequences of inhibitory interactions may be more severe than those from induction, which often lead to only diminished efficacy. Clinically significant interactions of this type generally involve the most common enzyme system, namely the hepatic microsomal mixed function oxidases, the most representative being the cytochrome P450 isozymes.

Several different mechanisms mediate inhibition-based interactions. Among these, probably the most common and significant is substrate competitive inhibition. Other recognized mechanisms include interference with drug transport, alteration of the conformation (or expression) of the P450 enzyme, as well as interfering either with the energy or cofactor supply [6]. Sometimes, competition can even result in irreversible inactivation, a mechanism that leads to the most enduring effects [3].

Drugs that are able to inhibit the MMFOs, by competitive binding to cytochrome P450, usually form a stable complex with it, which obviously will prevent access of other agents to the P450 enzyme system [2,11,12,14]. Drugs commonly involved in such types of interactions (due to enzyme inhibition) include amiodarone, azapropazone, chloramphenicol, cimetidine, ciprofloxacin, diltiazem, disulfiram, enoxacyn, erythromycin, ethanol, fluconazole, fluoxetine, fluvoxamine, isoniazid, itraconazole, ketoconazole, metronidazole, miconazole, nefazodone, omeprazole, oral contraceptives, paroxetine, phenylbutazone, propoxyphene, quinidine, sulphinpyrazone, sulphonamides, valproate and verapamil.

The clinical significance of this type of interaction depends on various factors; these refer either to the drugs involved (e.g. dosage, alteration in pharmacokinetic properties of the affected drug), or to patient characteristics, such as disease state. Interactions of this type are again most likely to affect drugs with a narrow thepeutic range.

Representative examples:

- the association of cimetidine or ciprofloxacin – both enzyme inhibitors, with theophylline, which could result in a doubling in plasma concentration of the latter [2,3,8,9,12].
- a severe interaction occurred following co-administration of the enzyme inhibitors erythromycin, ketoconazole and terfenadine, as first described by Honig et al. [15]. Further studies demonstrated that terfenadine is converted by a specific P450 enzyme, namely CYP3A4, to an active metabolite. On the other hand, ketoconazole being a potent inhibitor of CYP3A4 isoform, on concurrent administration with terfenadine will dramatically reduce the latter's metabolism, resulting in increased concentration of the parent drug; in this situation a quinidine-like action may result, leading to ventricular arrhythmias and prolongation of QT interval [2,12].
- an interesting example that involves a stereoselective inhibition is the association warfarin/enoxacyn [12]. Warfarin exists in two enantiomeric forms, (R)-warfarin and (S)-warfarin, the (S)-enantiomer being more active then the (R)-enantiomer. In humans, the (S)-enantiomer is almost totally eliminated as the (S)-7-hydroxylated-metabolite, while the (R)-enantiomer is predominantly biotransformed to the (R)-6-hydroxylated metabolite. Co-administration of enoxacyn inhibits metabolism of the less potent (R)-enantiomer, causing a reduction in its clearance. On the other hand, the co-administration of phenylbutazone inhibits the metabolism of the more potent (S)-warfarin predominantly, resulting in a greater proportion of it in plasma, and subsequently, in increased anticoagulant effects (the anticoagulant potency of the (S)-enantiomer being five times greater than that of the (R)-enantiomer) [2]. In the same context, though with less significant clinical consequences, we should mention the association warfarin/cimetidine. This is also a stereoselective inhibition involving the (R)-enantiomer. This form is, however, less active then the other enantiomer, so it is assumed that interaction will produce only a weak effect upon the anticoagulant effect of warfarin [12].
- indinavir and ketoconazole: *in vitro* studies on rat hepatic microsomes indicated that ketoconazole inhibits the biotransformation of indinavir by a competitive mechanism, with a K_i value of about 2.5 μM. As a result, on pre-administration of ketoconazole, both the bioavailability and AUC value of indinavir increased significantly [12].
- fluoxetine and imipramine: both being co-substrates for the same P450 isoform, CYP2D6, on co-administration of fluoxetine, the plasma

concentration of imipramine increases several fold, due to the same competitive inhibitory mechanism, as above [12].

- terfenadine and erythromycin: similar mechanism of action as above; terfenadine is metabolised by participation of another P450 isoform, namely, CYP3A4. On concurrent administration of erythromycin, the plasmatic levels of terfenadine increase, both drugs being co-substrates for the same enzyme isoform.

b) Enzyme induction

The phenomenon of induction of cytochrome P450, as a mediator of metabolic drug interactions, has also been recognized for sometime[1-3, 6, 8,11,14].

Enzyme induction may occur by a number of different mechanisms, but generally results in increased amounts of enzyme, and thus, in increased rate of biotransformation reaction [16]. In general, two major consequences arise with induction-based interactions: either increased metabolic clearance, leading to reduced therapeutic efficacy, or the opposite, namely metabolic activation, yielding a toxic metabolite, resulting in increased toxicity. As an example we quote the increasing risk of acetaminophen-induced hepatotoxicity on co-administration of isoniazid, due to an increase in the formation of the toxic metabolite of the former [17]. It is useful to note that the phenomenon of enzyme induction primarily affects phase I metabolism, although there is evidence that some phase II reactions may also be affected [2]. The effects of enzyme induction vary considerably between individuals, depending on various factors such as age, concurrent drug treatment, genetic factors and disease state. Enzyme induction is generally dose-dependent and represents the process of temporary adaptative increase of a specific enzyme concentration. The process is essentially attributable either to the increase in the rate at which the enzyme is synthesised, or to a decrease in its degradation rate. The enzyme inducers encountered most commonly in clinical practice include barbiturates, carbamazepine, griseofulvin, phenytoin and rifampicin.

Some of the best recognized examples and most widely studied drug interactions of this type include:

- warfarin and phenobarbital: an interaction that is well-documented and often cited. Phenobarbital is known as a potent inducer of many P450 isoforms, including those involved in warfarin's biotransformation. As a consequence of enzyme induction, the plasma levels of warfarin will decrease; in order to maintain the therapeutic effect, a substantial increase in the therapeutic dose will be needed. Under these circumstances, close monitoring of the patient is strongly recommended [12].

- antiepileptic drugs, frequently administered in combination: some combinations involve true interactions by reciprocal effects [18], and some of the consequences are therefore quite complex. Close monitoring of plasma levels of co-administered drugs, should however enable the consequences of these interactions to be recognized and, if not avoidable, at least minimised.
- nelfinavir and rifampin: both are used in HIV-patients. Nelfinavir, a non-nucleoside reverse transcriptase inhibitor, is partially metabolised by the P450 isoenzyme CYP3A. The antitubercular drug rifampin is a very potent inducer of this isoform, consequently increasing nelfinavir's biotransformation, which results in a greater clearance from the body. The AUC is dramatically reduced (by about 80%) and avoidance of this combination is therefore strongly recommended [3]. An alternative could be the co-administration of rifabutin, also an enzyme inducer, but far less potent than rifampin. The AUC of nelfinavir in this case is reduced by about 30% only. Other inducers of the isoform CYP3A, such as carbamazepine, phenobarbital, and phenytoin are expected to produce similar reduction phenomena, and such combinations are best avoided.

8.2.5 Other selected, miscellaneous recent examples

The constant interest in possibly undesirable effects that might arise from drug-drug interactions is reflected in the numerous studies and clinical observations that aim to reveal, predict and minimise such effects. Under the circumstances, attention has focused on new possible co-administrations, the potential interactions, and consequences of therapeutic or toxicological significance [19]. Examples of these follow:

- a recent article reviewed pharmacokinetic herb-drug interactions, taking in account that in recent years, the number of such interactions has increased [20]. Assuming that most herbal medicines have a broad therapeutic range, in order to identify, and predict such interactions in practice, systematic *in vitro* screenings as well as more clinical studies have been proposed.
- some interactions between statins and various drugs or even foods have also been reviewed recently [21]. Statins, medicines currently used for the treatment of hyperlipidemias, competitively inhibit HMG CoA reductase – an enzyme found in the liver; statins also display affinities for various P450 isoenzymes (CYP3A4, 2C8, 2C9). Because of this, they might be expected to be involved in metabolism-type drug interactions, and more. Recent studies confirm that all statins are absorbed orally, so the impact of food present in the stomach could be extremely important in achieving the

desired therapeutic effect. Usually, an individualization of the treatment is recommended to avoid interactions and generally to improve this form of treatment of hyperlipidemia. A limited number of clinical observations showed that the anticoagulant effects of warfarin can be increased in some patients on concurrent administration of lovastatin. The asumed mechanism is one of enzyme inhibition, resulting in increased anticoagulant effects of warfarin, with subsequent bleeding and increased prothrombin times reported. Itraconazole, a potent inhibitor of the CYP3A4 isoform, acts more predictably. Some of the statins commercially available and in therapeutic use, such as lovastatin and simvastatin, are metabolised by CYP3A4; on co-administration of itraconazole, the serum levels of the former are dramatically increased due to itraconazole's enzyme inhibiting action. The most common recommendation in the case of co-administration is dosage reduction if either is given concurrently with itraconazole [3].

- an interesting study concerned possible pharmacokinetic and pharmacodynamic interactions of drugs for internal diseases, such as analgesics, antiallergics antibiotics, anticoagulants, anticonvulsants, antihypertensives, β-blockers, gastroenterologic drugs, nonsteroidal antirheumatics and a series of new antidepressants, in an attempt to evaluate their clinical relevance [22]. Being co-substrates of the same P450 isoform, several other drugs were also shown to give potential interactions with antidepressants [23].
- many studies in recent years have been motivated by the need to individualize therapeutic schemes and avoid interactions and dangerous adverse effects as well. These studies had their origin in inter-individual differences in the activities of metabolising enzymes, as a consequence of pharmacogenetic factors, which in fact have been proven to play an important role in the response of individuals receiving the same, specific treatment (same dose, intervals of administration, and so on). A particular study focused on the large inter-individual variability in the human biotransformation of risperidone, a drug mainly metabolised to the corresponding 9-hydroxylated metabolite by specific P450 isoforms, and in particular CYP2D6. Because a large number of drugs have been described to be biotransformed by the same isoform, evaluation of the possible drug interactions on the enzyme appeared as an important issue, as did the consequent clinical significance of this phenomenon [24].
- the use of immunosuppresants prescribed to prevent rejection of transplanted organs or tissues, as well as in the treatment of autoimmune disorders, is on the increase. Therefore, a sound knowledge of the pharmacokinetics of these drugs is helpful in avoiding different drug-drug

interactions that might occur on co-administration of other drugs, such as tacrolimus, sirolimus, monoclonal antibodies and glucocorticoids [25].

- a mechanistic approach to drug interactions involving antiepileptics, frequently administered in combination and many of them involving interactions by reciprocal effects, has been revisited recently [26]. The study focused on the most common antiepileptic drug interactions, which are pharmacokinetic in nature. Interactions involving various antiepileptic drugs are expected to appear either by enzyme induction or inhibition, or displacement of protein binding. Such interactions are discussed in detail.
- an interesting relatively recent study, revealed interactions between NSAIDs and angiotensin converting enzyme inhibitors (ACEI) on concurrent administration [27]; the latter are indicated in the treatment of hypertension, myocardial infarction and congestive heart failure.
- oral contraceptives have been shown to be involved in many drug-drug interactions, that consequently will reduce their efficacy. An extensive study focused on oral contraceptive interactions with various drugs: e.g. anticonvulsants, antibiotics, adsorbents, analgesics and corticosteroids [28].
- the neuromuscular blocker succinylcholine hydrochloride (used as adjunct in surgery) is biotransformed not typically in the liver, but in the serum, by the circulating enzyme pseudocholinesterase. On co-administration of cyclophosphamide, the latter irreversibly inhibits this enzyme, reducing the biotransformation of succinylcholine. As a result, respiratory insufficiency and prolonged apnoea may appear [11].
- significant clinical interactions have been shown to appear on co-administration of lithium salts with a large range of medicines, including antidepressants, neuroleptics, anticonvulsants, antibiotics, muscle relaxants, chemotherapeutics and hormones [19]. The clinical observations have been reviewed and the conclusions summarised in a relatively recent account [29].
- some methotrexate interactions; being excreted mainly unchanged in the urine, methotrexate is a likely candidate for displaying excretion-type interactions. Some of the drugs that have been seen to interact with methotrexate include anion exchange resins, NSAIDs, penicillins, uricosurics and urinary alkalinisers and the interactions presumably involve excretion mechanisms. For some of these, detailed discussions are offered [11]: the NSAIDs are known to inhibit prostaglandin (PGE_2) synthesis, which will result in a fall in renal perfusion. As a consequence, a rise in methotrexate serum levels is observed, consequently leading to increased toxicity. It has also been suggested that protein-binding displacement may play a part. In attempting to avoid these interactions, if the concomitant use

of NSAIDs is thought appropriate, it is strongly recommended that treatment be monitored closely, and folinic acid rescue therapy should be available.

Another interesting example with severe clinical consequences is afforded by the co-administration of methotrexate and penicillins. These, acting like weak acids, will compete with methotrexate in the kidney tubules for excretion. Since penicillins have been proven to cause marked reductions in the clearance of methotrexate from the body, severe toxicity and even death have occurred as a result of such interactions. To avoid or minimise these unwanted effects, the same recommendations as above are suggested. When excreted in bile, methotrexate is then re-absorbed through the enterohepatic cycle, in the gut. Marked falls in methotrexate plasma levels have been reported in patients given concomitantly cholestyramine orally. The assumed mechanism involves binding of methotrexate to the cholestyramine in the gut, thereby preventing re-absorption. Concurrent use should be monitored and dosage adjustments made as necessary [11].

• a recent study has focused on methadone interactions [30]. Methadone is biotransformed almost exclusively by the liver, the main biotransformation of both methadone enantiomers being N-demethylation. In methadone metabolism, more P450 isoforms are involved *viz.* CYP3A4 and CYP1A2 (being inducible isoforms) and CYP2D6 (not inducible, but subject to genetic polymorphism). Often, the main metabolic substrates of the same CYP are administered concurrently; consequently, the drug that has a higher affinity for that CYT isoform partly prevents the biotransformation of the other drugs. Since most drugs are substrates for the CYP isoforms involved in the metabolism of methadone, interactions are expected to take place readily. Drugs that could be co-administered during methadone maintenance treatment and that are assumed to produce drug-drug interactions of the kinetic type include anticonvulsants, antidepressants, antifungals, benzodiazepines and macrolide antibiotics. Some of these drugs are inhibitors, inducers or substrates of CYP3A4 or CYP2D6. Specific examples show that generally the effects on methadone are either increased or decreased plasma levels, usually moderate in severity, delayed or rapid in onset, and involving different mechanisms on the CYP isoforms implicated (most of them are inducers or co-substrates, competing with methadone). Particular mention is made of the interaction with the antidepressant fluvoxamine, which inhibits both CYP3A4 and 2D6; consequently, although administered in therapeutic doses, plasma concentrations of methadone will correspond to those that are inhibitory *in vitro*.

Another noteworthy point in this context is that maintenance treatment with methadone remains the best choice in HIV-positive heroin addicts; therefore, the most frequent interactions that can take place and that are of utmost clinical significance are those between methadone and antiretroviral drugs. The antiretrovirals are usually metabolic inducers of the liver CYP3A4 isoform, implying an increase in enzymatic activity, with consequent decrease in the amount of methadone available. The antiretrovirals, including abacavir, amprenavir, didancosine, efavirenz, indinavir, nelfinavir, nevirapine, ritonavir, stavudine and zidovudine, generally determine, as already mentioned, a decrease (minor or moderate) in methadone plasma concentrations, a few with rapid onset (didanosine, stavudine), and the majority with delayed effect (even up to 8-10 days, e.g. efavirenz). Interesting co-administrations involve association of methadone with different combinations of antiretroviral drugs, such as: ritonavir + lopinavir, ritonavir + nelfinavir, ritonavir + nelfinavir + nevirapine, nevirapine + efavirenz etc. The effects on methadone are, just as in the previous case, minor or moderate decreases in its plasma levels.

The selected examples presented above illustrate that clinically important interactions may occur when methadone is taken concomitantly with other drugs. Because these pharmacokinetic interactions are generally extremely variable among patients, it is recommended that in the course of long-term treatments, the daily dose should be personalised.

- the interaction between terfenadine and ketoconazole, and the corresponding clinical consequences [15] were outlined above. More recently, another interesting interaction has been communicated [31]. The inhibitory properties of a novel gastroprokinetic agent (Z-338) were investigated and compared with those of cisapride, to evaluate its potential for drug-drug interactions. While there was no notable inhibition of terfenadine metabolism or of any of the P450 isoforms involved in biotransformation, the study showed that, on the other hand, cisapride markedly inhibited both of the main CYP isoforms involved in metabolisation, namely CYP3A4 and CYP2C9. From the prediction method used (based on K_i and PK parameters), it was concluded that this novel gastroprokinetic agent is considered unlikely to cause significant drug-drug interactions when co-administered with CYP substrates at clinically effective doses.

- recent studies demonstrated the stimulative action of acetaminophen on the peroxidative metabolism of anthracyclines by a common effect of enzyme induction [32]. Frequently administered concurrently with various anthracyclines, such as daunorubicin and doxorubicin, acetaminophen has been proven to stimulate their oxidation by lacto- and myeloperoxidase systems strongly, resulting in irreversibly altered products. The phenomenon has considerable clinical significance because the biological properties of transformed anthracyclines are quite different from those of the corresponding parent drugs. It is possible that this enhanced acetaminophen induced degradation might interfere with the therapeutic effects of these drugs (*viz.* anticancer and/or cardiotoxic).

- omeprazole and clarithromycin is an association commonly used in the treatment of *Helicobacter pylori* associated gastroduodenal ulcer. It has been demonstrated that a pharmacokinetic interaction occurs between these two co-administered drugs, with consequences on omeprazole's biotransformation: the combination resulted in a significantly reduced value (almost one-half) of the 5-hydroxylated metabolite and increased levels of unchanged omeprazole, with mean value of AUC increased about twofold [33]. The clearance and volume of distribution of omeprazole were dramatically reduced on co-administration with clarithromycin. The conclusion of the study was that the concurrent administration of clarithromycin and the proton pump inhibitor omeprazole, resulting as it does in markedly increased levels of omeprazole, can consequently improve the therapeutic response to this drug. Interestingly, no significant changes in the pharmacokinetics of pantoprazole and corresponding metabolites were observed.
- by an inhibitor mechanism, fluvoxamine was demonstrated to modify the pharmacokinetics of lidocaine and its two pharmacologically active metabolites [34]. The study also revealed (confirming *in vitro* studies) that the main isozyme involved in the biotransformation of lidocaine is a P450 isoform, CYP1A2. Inhibiting the isoform responsible for the metabolism of lidocaine, concurrently administered fluvoxamine resulted in significant decreases in lidocaine clearance, depending also on the state of health of the liver of the subjects. The main conclusion of the study was that the extent of fluvoxamine-lidocaine interaction decreases in patients with liver dysfunction, most likely because of the concomitant decrease in the hepatic level of CYP1A2.
- possible interaction between ciprofloxacin and pentoxifylline was recently investigated [35]. In the murine hepatic microsomes, previously incubated with ciprofloxacin, the metabolism of pentoxifylline was found to decrease significantly, suggesting a possible inhibitory effect of the former.
- an interesting study revealed the induced biotransformation of zolmitriptan in rats, as well as the interaction between six drugs and this highly selective 5-HT receptor agonist used in acute oral treatment for migraine [36]. Studies were carried out in rat hepatic microsomes treated with different inducers. Earlier clinical observations revealed that potential drug interactions can take place on co-administration with diazepam, propranolol, and moclobemide; thus this study continued investigations with other drugs that might possibly interact with zolmitriptan, namely

fluvoxamine, cimetidine and diphenytriazol. The *in vitro* model approach employed is increasingly used in drug development to enable early predictions of possible clinically significant drug interactions during co-medication. Fluvoxamine showed a potent inhibitory effect on CYP1A2, the main P450 isoform involved in the biotransformation (by N-demethylation) of zolmitriptan, resulting in increased plasma levels and reduced clearance. Diphenytriazol appeared to display the same effect. Propranolol, metabolised by the same CYP1A2, competes for the active site of the enzyme when administered with zolmitriptan, displaying a competitive inhibitory effect with the same consequences (increased mean C_{max} and AUC and prolonged mean $t_{1/2}$ for the former). Moclobemide, a MAO-A inhibitor, decreased the clearance of zolmitriptan, subsequently elevating plasma concentration indirectly.

- a profound interaction between tacrolimus and a combination of lopinavir and ritonavir in three liver-transplanted patients has recently been described [37]. The clinical observations based on tacrolimus blood concentrations and half-life revealed that the combination of antiretroviral agents led to a much greater increase in tacrolimus blood concentrations than did the use of a single protease inhibitor, such as nelfinavir for example. From the clinical observations, it was concluded that, depending on liver function, when therapy with the combination of antiretroviral agents is initiated, a dose of 1 mg/wk or less of tacrolimus may be sufficient to maintain adequate blood tacrolimus concentrations, and patients may not need a further dose for 3 to 5 weeks.
- an interesting example of one drug inhibiting the metabolism of a specific CYP2D6 substrate is represented by the effect of celecoxib on the pharmacokinetics of metoprolol [38]. Because celecoxib inhibits the metabolism of metoprolol, it is expected to increase the area under the plasma time-concentration curve of metoprolol, which in fact is almost doubled. In contrast, a comparative study revealed that this effect is not observed with rofecoxib. The interactions that may occur between celecoxib and different CYP2D6 substrates can be of important clinical relevance, especially with drugs having a narrow therapeutic index.

8.2.6 Other frequent and relevant interactions

In the following subsection we present in tabulated format, a selection of the most frequent and important drug-drug interactions and the consequent biological effects (Table 8.1):

Tab.8.1 Selected examples of frequent drug-drug interactions and consequent biological effects

Drug	Drug(s) of interaction	Interaction consequences
Acetaminophen (Paracetamol)	alcohol	severe hepatotoxicity with therapeutic doses of acetaminophen in chronic alcoholism (proposed mechanism: increased formation of hepatotoxic acetaminophen metabolites and glutathione depletion) [39]
	anticoagulants, oral	increased anticoagulant effect (mechanism not established) [40]
	barbiturates	acetaminophen hepatic toxicity (mechanism not established) [41]
	benzodiazepines	possible diazepam toxicity (mechanism not established) [42]
	cholestyramine	decreased acetaminophen effect (decreased absorption) [43]
	isoniazid	acetaminophen toxicity (increase in toxic metabolites) [44]
	probenecid	possible acetaminophen toxicity (decreased metabolism and renal excretion) [45]
	zidovudine	granulocytopenia (mechanism not established) [46]
Acyclovir	narcotics: meperidine	possible meperidine toxicity (decreased renal excretion) [47]
	probenecid	possible acyclovir toxicity (decreased renal excretion) [48]

	zidovudine	lethargy (unknown mechanism) corticosteroids [49]
Alkylating agents	azathioprine	liver necrosis (mechanism not established) [50]
	corticosteroids	decreased effect (increased metabolism) [51]
	cyclosporins	nephrotoxicity (mechanism not established) [52]
Aminoglycoside antibiotics	antifungals	nephrotoxicity (synergism) [53]
	cephalosporins	nephrotoxicity (mechanism not established) [54]
	cisplatin	nephrotoxicity (mechanism not established) [55]
	cyclosporins	renal toxicity (possibly additive or synergism) [56]
	digoxin	decreased digoxin effect (possible decreased absorption) [57]
	furosemide	ototoxicity and nephrotoxicity (additive) [58]
	polymyxins	nephrotoxicity; increased neuromuscular blockade (possibly additive) [59]
	vancomycin	possible nephrotoxicity and ototoxicity (possibly additive) [60]
Antacids	antihistamine, H_2-blockers	decreased cimetidine, ranitidine and nizatidine effect (decreased absorption) [61]
	benzodiazepines	decreased oral clorazepate effect (decreased absorption) [62]
	β-adrenergic blockers	decreased oral effect (decreased absorption) [63]

	cephalosporins	possible decreased effect (decreased absorption) [64]
	corticosteroids	decreased oral corticosteroid effect (decreased absorption) [65]
	digoxin	decreased digoxin effect (possible decreased absorption) [66]
	hypoglycaemics	possible hypoglycaemia (increased absorption and accelerated insulin response) [67]
	quinidine	possible quinidine toxicity (decreased renal excretion) [68]
	quinine	possible quinine toxicity (decreased renal excretion) [69]
	tetracyclines	decreased oral tetracycline effect (decreased absorption) [70]
	vitamin C	possible aluminium toxicity (possibly increased absorption) [71]
	vitamin D	possible bone toxicity with aluminium compounds (increased deposition of aluminium in bone, possibly due to increased absorption) [72]
Anticoagulants, oral	Antifungals (griseofulvin)	decreased anticoagulant effect (mechanism not established) [73]
	antihistamine, H_2-blockers	increased anticoagulant effect (decreased metabolism) [74]
	barbiturates	decreased anticoagulant effect (increased metabolism) [75]
	carbamazepine	decreased anticoagulant effect (increased metabolism) [76]

	chloral hydrate	increased anticoagulant effect (displacement from binding) [77]
	cholestyramine	decreased anticoagulant effect (binding of drug in intestine) [78]
	disulfiram	increased anticoagulant effect (decreased metabolism) [79]
	fluoroquinolones	increased anticoagulant effect (mechanism not established) [80]
	glutethimide	decreased anticoagulant effect (increased metabolism) [81]
	macrolide antibiotics	increased anticoagulant effect (possibly decreased metabolism) [82]
	nalidixic acid	increased anticoagulant effect (displacement from binding) [83]
	NSAIDs	increased bleeding risk (inhibition of platelets, other mechanisms) [84]
	phenytoin	phenytoin toxicity (decreased metabolism) [85]
	propafenone	increased warfarin effect (probably decreased metabolism) [86]
	spironolactone	decreased anticoagulant effect (hemoconcentration) [87]
	sulphonamides	increased anticoagulant effect (decreased metabolism and displacement from binding sites) [88]
	tetracyclines	increased anticoagulant effect (mechanism not established) [89]

	thyroid hormones	increased anticoagulant effect (increased clotting factor catabolism) [90]
	valproate	increased anticoagulant effect (probably displacement from binding sites) [91]
	vitamin A	increased anticoagulant effect with large doses (mechanism not established) [92]
	vitamin C	decreased anticoagulant effect (mechanism not established) [93]
	vitamin E	increased anticoagulant effect with large doses (mechanism not established) [94]
Barbiturates	chloramphenicol	possible barbiturate toxicity (decreased metabolism) [95]
	contraceptives, oral	decreased contraceptive effect (increased metabolism) [96]
	corticosteroids	decreased corticosteroid effect (increased metabolism) [97]
	narcotics: meperidine	increased CNS depression with meperidine (increased meperidine metabolites) [98]
	quinine	possible phenobarbital toxicity (probably decreased metabolism) [99]
	valproate	phenobarbital toxicity (decreased metabolism) [100]
Chloramphenicol	hypoglycaemics	increased hypoglycaemic effect (mechanism not established) [101]
	phenytoin	phenytoin toxicity (decreased metabolism) [102]
	metronidazole	dystonic reactions (mechanism not established) [103]

	pheno-thiazines	possible chlorpromazine toxicity (decreased metabolism) [104]
Digitoxin	diltiazem	possible digitoxin toxicity (mechanism not established) [105]
	ethacrynic acid	digitoxin toxicity (potassium and magnesium depletion) [106]
	furosemide	digitoxin toxicity (potassium and magnesium depletion) [107]
	neuromuscular blocking agents	increased incidence of arrhythmias (mechanism not established) [108]
	thiazide diuretics	digitoxin toxicity (potassium and magnesium depletion) [109]
Estrogens	phenytoin	decreased estrogen effect (increased metabolism) [110]
	vitamin C	increased serum concentration and possible toxicity of estrogens with 1 gram/day of vitamin C (decreased metabolism) [111]
Fluoroquinolones	antacids	decreased fluoroquinolone effect (decreased absorption) [112]
	iron	decreased fluoroquinolone effect (decreased absorption) [113]
	penicillins	possible ciprofloxacin toxicity with azlocillin (decreased metabolism) [114]
	theophyllines	theophylline toxicity (decreased metabolism) [115]

	zinc	decreased ciprofloxacin or norfloxacin effect (decreased absorption) [116]
Haloperidol	barbiturates	decreased haloperidol effect (increased metabolism) [117]
	isoniazid	possible haloperidol toxicity (probably decreased metabolism) [118]
	lithium	encephalopathy, lethargy, fever, confusion, extrapyramidal symptoms [119]
	methyldopa	dementia (mechanism not established) [120]
	phenothiazines	agranulocitosis (mechanism not established) [121]
	tacrine	possible parkinsonian symptoms (possibly additive) [122]
Insulin	angiotensin-converting enzyme inhibitors	increased hypoglycaemic effect (probable increased insulin effect) [123]
	naltrexone	possible increase in insulin requirements (mechanism not established) [124]
	NSAIDs	possible increased hypoglycemic effect with large doses of salicylates (mechanism not established) [125]
Methotrexate	azathioprine	azathioprine toxicity (fever, rash, muscle pain) (mechanism not established) [126]
	cisplatin	methotrexate toxicity (decreased renal clearance) [127]
	cyclosporin	toxicity of both drugs (decreased elimination) [128]

	NSAIDS	methotrexate toxicity (decreased renal excretion) [129]
	penicillins	possible methotrexate toxicity (decreased excretion) [130]
	sulponamides	possible methotrexate toxicity (displacement from binding sites and decreased renal excretion) [131]
	tetracyclines	possible methotrexate toxicity (displacement from binding) [132]
	trimethoprim	pancytopenia (probably additive inhibition of folate metabolism) [133]
Nifedipine	NSAIDs	possible decreased antihypertensive effect with indomethacin (mechanism not established) [134]
	phenytoin	phenytoin toxicity (mechanism not established) [135]
	rifampin	decreased antihypertensive effect of nifedipine (possibly increased metabolism) [136]
	selective serotonin reuptake inhibitors	nifedipine toxicity with fluoxetine (probably decreased metabolism) [137]
Omeprazole	digoxin	possible digoxin toxicity (increased absorption) [138]
	disulfiram	confusion, catatonic reaction, and disorientation (mechanism not established) [139]
	phenytoin	possible oral phenytoin toxicity (decreased metabolism) [140]

Penicillins	anti-coagulants, oral	decreased anticoagulant effect with nafcillin (increased metabolism) [141]
	cephalosporins	possible cefotaxime toxicity with mezlocillin in patients with renal impairment (decreased excretion) [142]
	lithium	hypernatremia with ticarcillin (decreased renal excretion) [143]
	neuromuscular blocking agents	recurrent neuromuscular blockade with IV piperacillin (mechanism not established) [144]
Rifampin	barbiturates	decreased barbiturate effect (increased metabolism) [145]
	corticosteroids	marked decrease in corticosteroid effect (increased metabolism) [146]
	haloperidol	decreased haloperidol effect (increased metabolism) [147]
	isoniazid	hepatotoxicity (possibly increased toxic metabolites) [148]
	trimethoprim-sulfamethoxazole	possible rifampin toxicity (possibly decreased metabolism) [149]
Sulphonamides	barbiturates	increased thiopental effect (decreased albumin binding) [150]
	cyclosporins	decreased cyclosporin effect with sulphadiazine (possibly increased metabolism) [151]
	hypoglycaemics	increased hypoglycaemic effect (mechanism not established) [152]

	phenytoin	possible phenytoin toxicity (decreased metabolism) [153]
Tetracyclines	digoxin	possible digoxin toxicity (decreased gut metabolism and increased absorption) [154]
	lithium	lithium toxicity (decreased renal excretion) [155]
	phenytoin	decreased doxycycline effect (increased metabolism) [156]
	theophyllines	possible theophylline toxicity (mechanism not established) [157]
	zinc	decreased tetracycline effect (decreased absorption) [158]
Trimethoprim	azathioprine	possible azathioprine toxicity with sulfasalazine (decreased metabolism) [159]
	cyclosporin	nephrotoxicity (synergism) [160]
	dapsone	dapsone toxicity, methemoglobinemia (probably decreased metabolism) [161]
	digoxin	possible digoxin toxicity (decreased renal excretion and possibly decreased metabolism) [162]
Verapamil	carbamazepine	carbamazepine toxicity (decreased metabolism) [163]
	clonidine	A-V Block (possible synergy) [164]
	digoxin	digoxin toxicity (probably decreased biliary excretion) [165]

Conclusions:

- From the above examples it can be observed that drug-drug interactions may occur due to different mechanisms and in most cases these are known at the molecular level; these variations may have as targets the absorption, protein binding or excretion processes, but most frequently the biotransformation process. For some interactions, the mechanisms are not yet established, but the interactions were revealed by clinical observations.
- At each level, the modifications are an increase or a decrease in the effect by different mechanisms, most of them described in the Table.
- In most cases, the interaction modifies the effect of only one of the co-administered drugs. Unfortunately, there are cases when such interactions may cause different pathologies as well, including for example nephrotoxicity, hepatotoxicity, ototoxicity, A-V block, methemoglobinemia, leucopenia, recurrent neuromuscular blockade, pancytopenia, confusion, catatonic reaction, and disorientation.
- In this context, besides limiting the polytherapy, it is strongly recommended that patients known to suffer from e.g. hepatic or renal impairments, or hypersensitivities, be monitored.
- The importance of knowing the mechanism of such interactions is obvious for predictions and limitations of the phenomena. As already mentioned in different chapters or subsections of the present work, the essential role is played by the enzymatic systems involved in drug biotransformation, the two primary mechanisms being enzyme induction and enzyme inhibition.
- Both *in vitro* and *in vivo* methods are available for evaluating the potential interactions between different drugs (or drugs and other entities) administered concurrently.

8.3 INTERACTIONS BETWEEN DRUGS AND OTHER ENTITIES

8.3.1 Drug-food interactions

The presence of food in the stomach is very important especially for the absorption process, causing irregularities in absorption or lowering the stomach pH. It is important as well to highlight the importance of the emptying rate of the stomach and the avoidance of certain nutrients during a specific treatment. Recent studies reviewed the pharmacokinetic drug-food interactions influencing drug plasma levels, as well as the bioavailability changes resulting from concomitant intake of drugs and meals [166]. Thus, following the normally administered dose of a drug, a decrease in the desired and expected effect should occur.

Conversely, in a very few instances, concomitant food intake can be beneficial. Liedholm and Melander [166], in 1986, concluded that "concomitant food intake can increase the bioavailability of propranolol by transient inhibition of its presystemic primary conjugation".

New knowledge relating to interactions between drugs and diet is steadily accumulating. In this context, pharmacokinetic interactions encompass not only drug-drug interactions, or those mediated by herbal medicines, but also interactions involving several foods and even beverages [20]. Recently, possible interactions between food and statins (medicines used for the treatment of hyperlipidemias) have been investigated [21]. This is of utmost importance since diet plays an essential role and exerts considerable influence on the prevention and/or treatment of these pathologies. As a consequence, therapy is commonly begun in combination with dietary advice. This is a very important aspect for several reasons: first, all statins are orally absorbed, so food intake is extremely important in achieving the appropriate therapeutic effect; avoidance of interactions between statins and foodstuff, and consequent alterations in the therapeutic benefits, is necessary; and last, but not least, statins are substrates for different CYTP450 isoforms, thus making them possible candidates for interactions with different components of foodstuff that are co-substrates for the same enzymes. A well-known example is that of grapefruit juice. This beverage, commonly consumed by the general population, is an inhibitor of the intestinal CYP3A4 isoform responsible for the first-pass biotransformation of many drugs. The possible interactions that might occur would lead to increased serum levels and/or decreased clearance, increasing the risk of overdosing. Some of its most notable effects concern the cyclosporins and some calcium anatagonists [167]. More recently, the risk of grapefruit juice interactions has been reviewed, with emphasis on aspects of pharmacokinetics and mechanisms of elimination, which could play a critical role when this beverage is consumed with certain drugs, especially those that are substrates for the CYP3A4 isoform [168]. Although these interactions may not necessarily alter the drug response in most instances, the recommendations are either an alternative medication (one that evidently does not interact with grapefruit juice), or avoidance of the combination to prevent toxicity [168].

Other drug-nutrient interactions were the subject of study in the case of patients receiving enteral nutrition (EN). A recent review [169] discusses problems in extrapolating available data to current practice and provides recommendations for managing interactions of this type.

Finally, it is interesting to note that sometimes the concurrent consumption of food and certain drugs may be beneficial, the drug-nutrient interactions (DNI) resulting either in an increase of drug effect or reduced toxicity. A recent review focuses on specific nutrients that enhance drug effects or reduce their toxicity [170].

8.3.2 Interactions with alcohol

The effects interactions between alcohol and different drugs have been the subject of numerous studies and clinical observations for many years. The first point to highlight is that the subsequent effects depend on whether the consumption of alcohol is chronic or acute.

For example, with chronic alcohol abuse, the effect of anticoagulants may be dramatically decreased, because of the increased rate of their biotransformation; in this situation, alcohol is an enzyme inducer. This phenomenon was first reported in the early 1970s [171]. On the other hand, in acute alcohol intoxication, increased anticoagulant effects of both oral anticoagulants and heparin have been reported. The assumed mechanism is that of decreased metabolism; under these circumstances alcohol acts like an enzyme inhibitor [172].

Important interactions involve barbiturates, benzodiazepines, beta-adrenergic blockers, chloral hydrate, cycloserine, cyclosporin, felodipine, hypoglycaemics, isoniazid, nifedipine, NSAIDs, phenothiazines, phenytoin, tetracyclines and verapamil. In some instances the interaction may involve only the effect of the drug e.g. decreased sedative effect of barbiturates with chronic alcohol abuse (due to increased barbiturate metabolism) [173], or increased CNS depression in case of concurrent consumption with benzodiazepines [174]. Sometimes, however the consequences are more serious, either increasing the toxicity of certain drugs (e.g. cyclosporin) [175] or even causing adverse reactions e.g. orthostatic hypotension (with felodipine) [176], increased incidence of hepatitis (with isoniazid) [177] or hepatotoxicity (with methotrexate) [178], bleeding (with aspirin) [179], and impaired motor coordination (with phenothiazines and phenylbutazone) [180].

A recent cohort study focused on patients with schizophrenia and related psychoses, who frequently use, abuse and become dependent on psychoactive substances. The most frequently abused substances were nicotine, alcohol and cannabis. Among the results of interest in this subsection is the reported situation that patients with psychotic disorder and current substance abuse (dual diagnosis, DD) are of enhanced risk for alcohol abuse [181].

Interactions of toxicological significance between alcohol and psychiatric drugs have been reviewed recently, with a focus on antidepressants and antipsychotics [182]. The study revealed that either acute or chronic consumption of alcohol, when combined with psychiatric drugs, may result in clinically significant toxicological interactions, including those leading to fatal poisoning. It is assumed that these toxicological effects characterise, in fact, overdosing due to decreased biotransformation, delayed by acute alcohol ingestion.

More updated information on this topic can be found in the Handbook of Drug Interactions [183].

8.3.3 Influence of tobacco smoke

Cigarette smoking remains highly prevalent in most countries. Tobacco smoke, deliberated inhaled, is considered a self-inflicted effector of drug metabolism. Inhalation of tobacco smoke, with its more than 3000 chemical components, may be considered a different way of ingesting pyrolisis products. It affects drug therapy by both pharmacokinetic and pharmaco dynamic mechanisms.

Pharmacokinetic drug interactions are assumed for example for theophylline, tacrine, insulin, imipramine, haloperidol, pentazocine, flecainide, estradiol, propranolol, diazepam, chlordiazepoxide, while pharmacodynamic interactions have been described for antihypertensive and anti-anginal agents, antilipidemics, oral contraceptives and histamine-2-receptor antagonists [184]. Commonly, pharmacokinetic interactions may call for larger doses of certain drugs due to an increase in plasma clearance, although other accepted mechanisms involve a decrease in absorption, an induction of main drug-metabolising enzyme systems, or a combination of all three factors. In contrast, pharmacodynamic interactions may increase the risk of adverse events in smokers with certain pathologies, such as cardiovascular or peptic ulcer disease [184].

However, the most common effect of tobacco smoke is assumed to be an increase in drug biotransformation through induction of specific enzyme activities. More, or less marked, effects on plasma levels of different therapeutics can be seen following tobacco smoking. Measurements of plasma levels of certain drugs, due to increased metabolism either by the intestinal mucosa or first-pass through the liver, confirm this.

Examples include phenacetin (much lower plasma levels in smokers compared with non-smokers, due to increased metabolism) [185], antipyrine (increase in drug clearance) [186], estrogens (possible decreased estrogen effect, increased metabolism) [187], tricyclic antidepressants (decreased antidepressant effect, increased metabolism) [188], propranolol (decreased propranolol effect, increased metabolism) [189], phenylbutazone (decreased phenylbutazone effect, increased metabolism) [190], mexiletine (decreased mexiletine effect, increased metabolism) [191].

More recent studies provided novel evidence that cigarette smoking accelerates chlorzoxazone and caffeine metabolism, by markedly enhancing oral clearance [192].

Other studies focused on the influence of tobacco smoke on specific isoforms of CYTP450. Owing to its inducer effect, tobacco smoke may increase the risk of cancer by enhancing the metabolic activation of carcinogens. From the numerous compounds present in tobacco smoke, the polycyclic aromatic hydrocarbons (PAHs) are believed to be responsible for the induction of CYP1A1, CYP1A2, and CYP2E1. Conversely, and still with no evidence in humans yet, other components such as carbon monoxide and cadmium displayed inhibitory effects on CYP enzymes [193]. The same studies revealed that due to nicotine, which is known to display a stimulant action, cigarette smoking may cause heart-rate and blood pressure lowering. Furthermore, nicotine, due to the cutaneous vasoconstriction induced, may slow the rate of insulin absorption after i.v. administration.

8.4 ADVERSE REACTIONS

8.4.1 Classification criteria

Adverse drug reactions are unwanted effects caused by normal therapeutic doses [194-197]. From the total reported adverse reactions, about 7% are severe, with an average of 0.32%, being fatal. Adverse reactions share some characteristics such as:

- they may be induced by the majority of drugs,
- they may appear immediately (the allergic reactions), or after a certain period of time (carcinogenity, mutagenity, teratogenity),
- they may appear more frequently in certain situations, such as self-medication, during co-administration of several drugs, in children (immaturity of the enzymatic systems), in the elderly (decrease in enzymatic activity), in particular physiological states (pregnancy), in pathological states pre-existing or co-existing with drug administration including renal, digestive, hepatic dysfunctions, cardiovascular diseases, dysfunctions in the routes of biotransformation, malnutrition, excessive consumption of alcohol, tobacco, coffee, greater individual sensitivity and reactivity (usually caused by genetic enzymatic deficiencies).

Several criteria have been proposed for a classification of the adverse reactions, but this remains a difficult task due to the complexity of the mechanisms involved, and the incidence and/or variable severity. In principle, the following criteria may be used:

- predictibality,
- some clinical and experimental characteristics,
- the producing mechanism, and
- the location criteria [194,196,197].

According to the first mentioned criterion, adverse reactions may be grouped into:

- predictable and,
- unpredictable adverse reactions.

The first group covers the so-called type A adverse reactions, which in fact constitute the great majority of adverse drug reactions and are usually a consequence of the drug's main pharmacological effect. They are dose-related and usually mild, although they may sometimes be severe, or even fatal. A term often applied to this type of adverse reaction is that of 'side-' or 'collateral effect'. Such a reaction may be either a consequence of incorrect dosage or of impaired drug elimination.

In contrast, the so-called type B adverse reactions, are not predictable from the drug's main pharmacological action, are not dose-related, and generally are severe, with a considerable incidence of mortality. These types of adverse reactions, also called 'idiosyncratic', occur rarely and usually have either a genetic or an immunological basis [194,197].

Based on the second criterion, the adverse reactions have similarly been classified into two types:

- experimentally reproducible,
- irreproducible adverse reactions.

In fact, they correspond to the first established groups, as follows:

- the first of these, being experimentally reproducible, are of course predictable; as predictable adverse reactions, according to the previous characterization, they are dose-related; the main consequence (with potential benefit for the patient) is the possibility of reducing the unwanted effects of such reactions by simply reducing the administered doses;
- as for the second type, corresponding obviously to the unpredictable ones, with no dose-relation and severe manifestations, the strong recommendation would be to stop the therapy.

As for the third criterion, which is more didactic, three classes are distinguished:

- adverse reactions of toxic type,
- 'idiosyncratic' type adverse reactions and
- adverse reactions of allergic type.

The first group refers to functional and morphological unwanted disturbances, which may appear in some patients under similar conditions of administration and usual doses. Determinant factors include the different individual reactivities, the drug-drug interactions, the pathological state of the organism, the state of the enzymatic systems involved in drug biotransformation and the small therapeutic index of certain drugs. The most severe adverse reactions of this type are assumed to be the mutagenic, teratogenic and carcinogenic effects.

The 'idiosyncratic' type adverse reactions are unusual reactions, qualitatively and quantitatively different from the common effects of a drug in the majority of a population, and most commonly are determined by genetically inherited enzymopathies. Many of them are strain-dependent. Included here are also the so-called type D reactions – delayed reactions, such as carcinogenesis induced by alkylating agents, or retinoid-associated teratogenesis. Other important examples are the blood dyscrasias, including thrombocytopenia, anaemia and agranulocytosis [197].

As for the adverse reactions of allergic type, we should stress that drugs may cause a variety of allergic responses, and moreover, a single drug can sometimes be responsible for more than one type of allergic response. It is assumed that this type of adverse reaction involves immune mechanisms, in the sense that most drugs, which are in general of low molecular weight, can however combine with substances of high molecular weight (usually proteins), forming an antigenic haptene conjugate. Most commonly, after the reaction Ag-Ac (antigen-antibody) takes place, serotonin, histamine as well as other chemical mediators are liberated, causing an allergic response. According to the immune mechanism involved, this type may be subdivided into the following subtypes:

- subtype I: anaphylactic reactions, due to the production of reaginic IgE antibodies. They commonly occur with foreign serum or penicillin, but may also occur with some local anaesthetics and streptomycin;
- subtype II: cytotoxic reactions, due to antibodies of class IgG and IgM, which (on contact with antibodies on the cell surface) are able to fix complement, causing cell lysis;
- subtype III: immune complex arthus reactions; these soluble, circulating complexes can fix in the small vessels and basal membranes, activating the complement, and subsequently determining various inflammatory phenomena.
- subtype IV: delayed hypersensitivity reactions, due to the drug forming an antigenic conjugate with dermal proteins and sensitised T-cells reacting to drug, causing a rash [194,197].

Other proposed categories include:

- continuous reactions due to long-term drug use (e.g. analgesic neuropathy) and
- end-of-use reactions, such as withdrawal syndromes following discontinuation of a treatment (with e.g. benzodiazepines, tricyclic antidepressants or β-adrenoreceptor antagonists) [194,197].

Summarising the factors involved in adverse drug reactions, we should classify them also according to so-called 'patient' factors, 'prescriber' factors and 'drug' factors.

The patient factors may be intrinsic (age, sex, genetic abnormalities, presence of organ dysfunction etc.) or extrinsic (environment, malnutrition, xenobiotics).

The prescriber factors refer generally to incorrect dosage or drug combination, duration of therapy etc., while the drug factors refer mostly to drug-drug interactions.

Although it is probably not possible to avoid allergic drug reactions altogether, the following measures can decrease their incidence:

- the drug history is essential whenever treatment is anticipated;
- drugs given orally are less likely to cause severe allergic reactions than those given by injection;
- prophylactic skin testing should become more routinely practised because it could probably reduce the risk of anaphylaxis or other less severe reactions [194,196,197].

As indicated in the first three chapters, and illustrated by numerous examples in Chapters 2 and 3, the processes of drug metabolism result in biotransformation of the drug to metabolites that differ chemically from the parent drug, consequently displaying altered affinities for the drug receptor. This change in the structure of the drug may be beneficial or detrimental.

For example, when 'inactive' drugs, such as prodrugs (inert species whose pharmacological effect depends entirely on metabolism) are biotrans formed yielding active metabolites, the process is obviously beneficial and is called pharmacological activation.

However, in general, biotransformation of a drug prepares it for excretion and in this case, the process of metabolism results in pharmacological deactivation. When certain toxins or potentially toxic drugs are involved, such metabolism leading to 'detoxication' is obviously of invaluable benefit.

On the other hand, when a drug (or other xenobiotic) is transformed into a toxic metabolite, the reaction is called 'toxicological activation' or, 'toxication', and this is obviously detrimental to health. Such a metabolite may act or react in a number of ways to elicit a variety of toxic effects at different levels, as will be evident from examples cited in the next section.

It is essential to stress that the occurrence of a toxication reaction at the molecular level does not necessarily imply toxicity at the levels of organs and organisms. On the other hand, when metabolic toxication reactions occur, they are always accompanied by competitive and/or sequential reactions of detoxication that compete with the formation of the toxic metabolite. This may lead to its inactivation. The existence of essential survival mechanisms should also be borne in mind. These act to repair molecular lesions by removing them immunologically and/or by regenerating lesioned areas.

From the above discussion, it is evident that the process of drug metabolism may either decrease or increase toxicity of a given drug compound depending on the biological potencies of the drug and its metabolites; for this reason we focus in the present subchapter on a more detailed examination of the toxicological aspects of drug metabolism.

Classification criteria refer to the adverse reactions and toxicological consequences, the most severe of them including hepatotoxicity and nephrotoxicity, pulmonary toxicity, carcinogenesis and teratogenesis.

8.4.2 Selected examples

Oxidation of some secondary hydroxylamines may yield nitroxide and other reactive metabolites, possibly accounting for the hepatotoxicity of these chemicals. If the metabolic intermediates of such compounds undergo N-oxygenation they may possibly form complexes with the cytochrome P450 enzymes, inhibiting them reversibly.

A well-known, medically relevant example is given by norbenzphetamine, which undergoes a two-step N-oxidation (Figure 8.1). In the first step, it is hydroxylated to the corresponding hydroxylamine and in the second, the product is oxidised to a nitrone intermediate [198]. The reaction is catalysed mainly by a FAD-containing monooxygenase. The nitrone intermediate is susceptible to further oxidation to the corresponding nitroso derivative under CYTP450 catalysis. It is assumed that this nitroso derivative is the metabolic intermediate (MI) responsible for formation of a complex with the CYTP450, resulting in a (usually) reversible inhibition of the enzyme [199]. The binding of such metabolic intermediates to CYTP450 involves the presence of the enzyme in reduced form; under such conditions, the nitrogen atom will interact with the iron cation [200] (Figure 8.2):

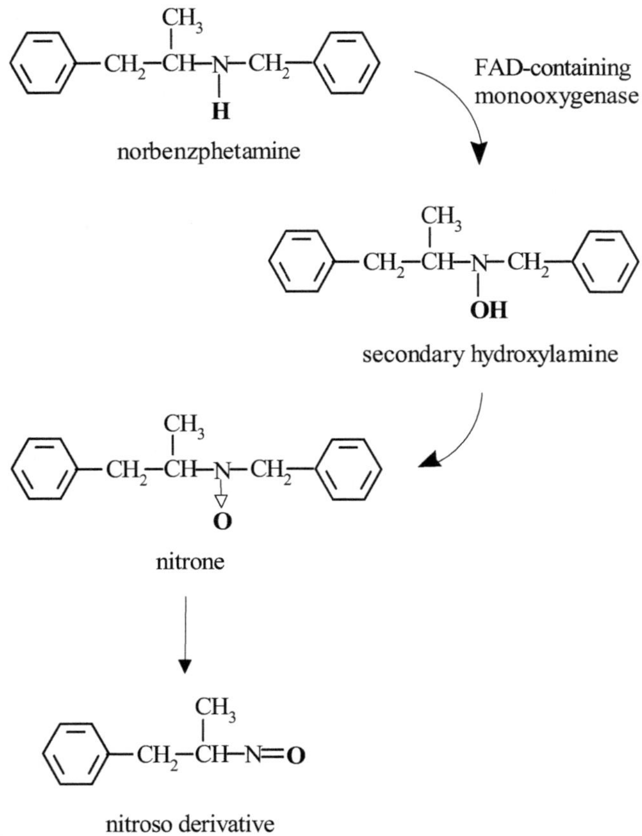

Fig.8.1 Oxidation of norbenzphetamine yielding a nitrone and a nitroso species

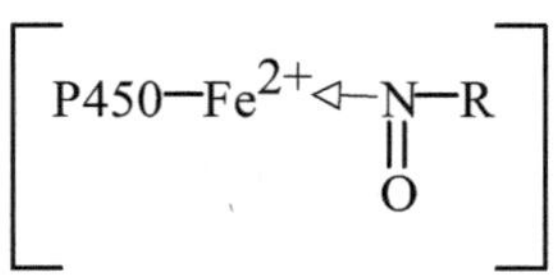

Fig.8.2 Formation of the complex with the CYTP450 enzyme

Many more studies have focused on the biotransformation of primary arylamines, given their toxicological significance. An interesting and representative example in this context is procainamide, studied in the early 1980s (Figure 8.3). In human liver microsomes, procainamide was shown to be metabolised to a hydroxylamine intermediate, which further undergoes non-enzymatic oxidation to the corresponding nitroso-compound; this is assumed to covalently react with glutathione (and thiol groups in proteins), forming sulphinamide adducts [200]. Two other possibilities exist for the intermediate nitroso derivatives: they can either bind to a hydroxylamine, yielding an azoxy derivative, or even to the parent primary amine forming an azo compound (Figure 8.3). As in the previous case of secondary hydroxylamines, the nitroso metabolites of primary arylamines can also form complexes with reduced cytochrome P450 [201].

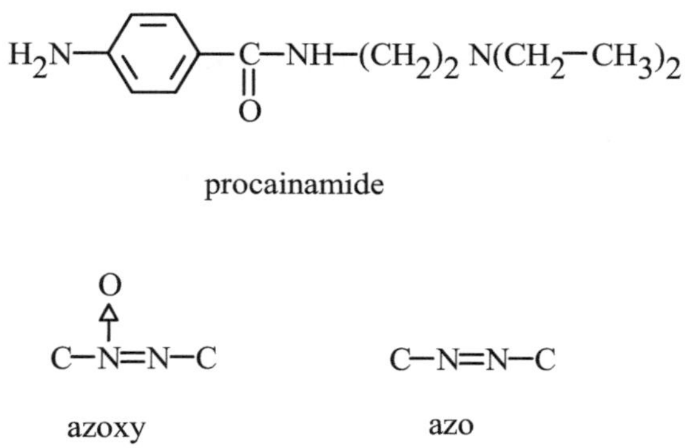

Fig.8.3 Structure of procainamide and other intermediary metabolic groups responsible for the toxicity of the drug

The toxicological significance of primary arylamines, and polycyclic arylamines in particular, is very considerable due to the carcinogenic and mutagenic potential of their intermediates, involving highly reactive species, namely, nitrenium ions (aryl-N^+-H) [202].

There is evidence that such nitrenium ions may be responsible for the covalent binding to DNA of certain drugs [203]. These highly reactive ions are known to exist in two states, namely singlet and triplet [204]. This has turned out to be of fundamental importance, since the nitrenium ions of non-toxic amines exist preferentially in the triplet state whereas singlet states have been attributed to nitrenium ions of mutagenic/carcinogenic amines. Therefore, it was concluded that for the initiation of either carcinogenic or mutagenic process, the nitrenium ions must exist in the singlet state.

An important conclusion drawn from the above examples is that hydroxylamine formation may generally be considered as a route of toxication. Among the compounds known to be N-hydroxylated (rather than forming other intermediates, such as e.g. N-oxides) much interest is focused on carcinogens occurring as amino acid pyrolysates, in cooked or charred foods. A representative example is the mutagenic and carcinogenic compound IQ (2-amino-3-methylimidazo[4,5-*f*]quinoline) [205] (Figure 8.4):

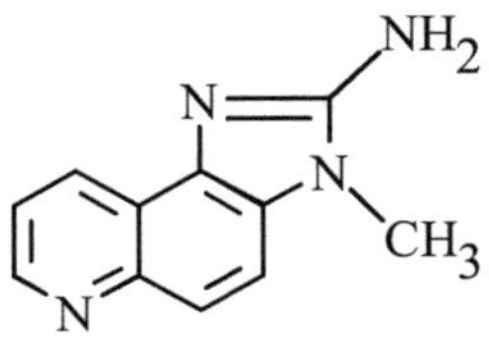

Fig.8.4 Structure of the carcinogenic and mutagenic compound IQ

Another interesting example involves the N-hydroxylation of the endogenous purine base, adenine; it is assumed that the 6-N-hydroxylated derivative is genotoxic and carcinogenic [206] (Figure 8.5):

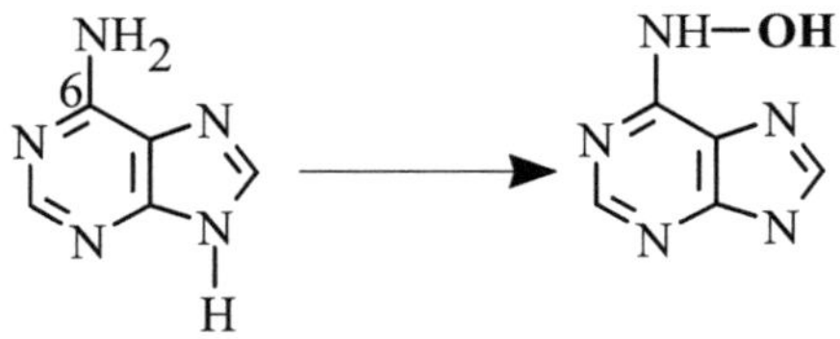

Fig.8.5 6-N-hydroxylation of adenine

It is again assumed that highly reactive intermediate nitrenium ions are implicated in mutagenic and carcinogenic effects associated with such heterocyclic hydroxylamines.

1,2-disubstituted hydrazines can also be N-oxygenated, yielding first the corresponding azo intermediates, which may either rearrange to hydrazones, or be further oxygenated to azoxy derivatives. The product hydrazones are reversibly hydrolysable, forming primary amines and aldehydes [207].

For some alkyl azo- and azoxy-derivatives, toxicity results from further activation by α-carbon hydroxylation, occurring after the initial hydrogen abstraction [208]. In contrast, certain aromatic azo-compounds with a *para*-amino group are potentially carcinogenic due to the activation of the amino group, while the azo-group has been shown to undergo reduction

[209].

A compound of particular significance and medicinal interest (first studied some twenty years ago) is the anticancer drug cyclophosphamide. Here, toxication is determined by the N-C oxidative ring cleavage and takes place mainly in hypoxic tumor cells. According to Borch [210] the first step is a preferential oxidation at the 4-position, yielding the 4-hydroxycyclophosphamide. This carbinolamine intermediate is in equilibrium with aldophosphamide, its open-ring tautomer. Subsequent dehydrogenation of these intermediates deactivates the drug and yields the corresponding urinary metabolites 4-oxocyclophosphamide and carboxyphosphamide. The undehydrogenated aldophosphamide remains in a keto-enol equilibrium with aldophosphamide, another urinary metabolite [211]. Under the relatively anaerobic conditions within tumor cells, both the biologically active metabolite phosphoramide mustard and the toxic metabolite acrolein are generated from the aldophosphamide.

N-nitroso derivatives (nitrosamines) comprise a special group of xenobiotics, intensively studied and comprehensively reviewed, whose biotransformation can lead to highly reactive metabolites. This accounts for their potential hepatotoxicity and carcinogenicity. A much studied and potent mutagen and carcinogen, representative for the toxication of dialkyl- and alkylarylnitrosamines, is dimethylnitrosamine, a substrate of CYP2E1 [212].

The biotransformation is complex, toxication beginning with an N-dealkylation that produces a C-centred radical and an α-nitrosamino alcohol. The latter is a highly reactive intermediate, readily decomposing to the N-dealkylated species, formaldehyde and diazomethane. Following elimination of dinitrogen, the diazo intermediate decomposes to give a carbonium ion [213] (the methyl cation, in the given example), which may react as a strongly electrophilic species at different nucleophilic sites of biomolecules. If the biomolecule happens to be e.g. DNA, a 'molecular injury' is taking place, simultaneously initiating a sequence of events that possibly may lead to hepatotoxicity, carcinogenicity, or other toxic effects.

As regards the C-centred radical formed initially, it breaks down spontaneously to nitric oxide and N-methylformamidine. This latter intermediate hydrolyses to form methylamine and formaldehyde, while the nitric oxide is oxidised to nitrite [214].

Another important mechanism of toxication involves the cytochrome P450-catalyzed oxidation of sulphur-containing compounds, yielding as reactive, electrophilic species, the corresponding sulphenic acids.

An interesting example of metabolic toxication is that of the sulphur-containing steroidal drug, spironolactone (Figure 8.6):

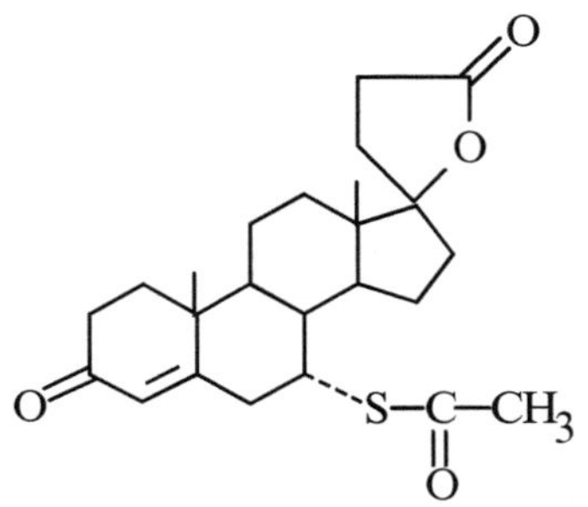

Fig.8.6 Structure of spironolactone

Following a sequence of metabolic reactions, this aldosterone antagonist will finally yield sulphenic acid. In the course of the biotransformation, cytochrome P450 is destroyed, the intermediates accounting for this toxication being assumed to be the thiyl radical and/or the sulphenic acid [215]. The proposed mechanism involves thiol oxidation to disulphides, sulphinic and sulphonic acids.

Another example of biological and toxicological interest is the oxidation of certain 4-alkylphenols to the corresponding quinone methides. These intermediates, seen in hepatic and pulmonary microsomes of some species, act like strongly alkylating agents which may undergo additions at the exocyclic methylene carbon, thereby binding covalently to macromolecular, soluble nucleophiles. This reactivity with nucleophiles was shown to correlate with hepatotoxicty [216].

Benzyl S-haloalkenyl sulphides having the general structure presented in Figure 8.7 are substrates of CYTP450-catalysed S-dealkylation that yields an unstable thiol [217]. The latter easily rearranges to mutagenic thioacylating intermediates, such as thioketenes and/or thioacyl chlorides [218].

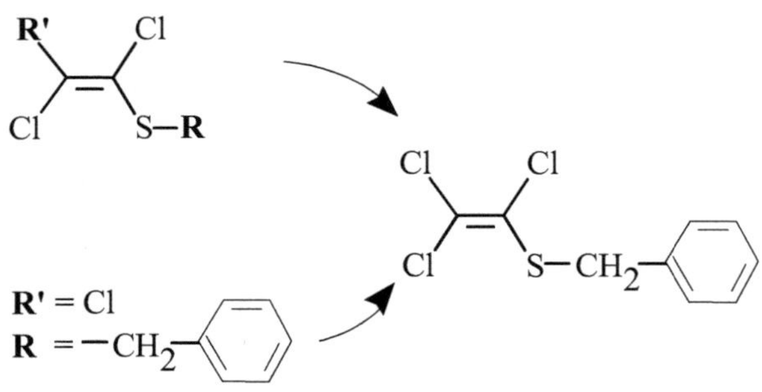

Fig.8.7 General structure of benzyl S-haloalkenyl sulphides

S-haloalkenyl-L-cysteine conjugates can be activated to the same unstable thiols by the action of cysteine-conjugate β-lyase, a pyridoxal phosphate-dependent enzyme, found mainly in the kidney. It cleaves L-cysteine conjugates to thiols, NH_3 and pyruvic acid [219], being of interest from the toxicological point of view in the context of kidney-selective delivery of thiol-containing drugs.

Such an example of renal activation by S-C cleavage, is given by S-(6-purinyl)-L-cysteine, to the corresponding 6-mercaptopurine (Figure 8.8):

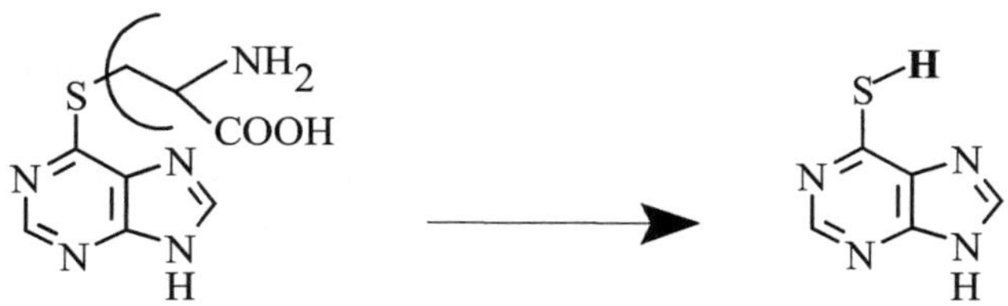

Fig.8.8 β-lyase catalysed S-C cleavage of S-(6-purinyl)-L-cysteine

Usually, the S-haloalkenyl-L-cysteine conjugates are formed from glutathione in the liver, but the high reactivity of β-lyase accounts for their nephrotoxicity by activating them to thiols in the kidney.

The sulphoxidation of thioamides is also of considerable interest due to the potential toxicity of some metabolites, in particular their hepatotoxicity and carcinogenicity. Studies have been made on different good substrates for the FAD-containing monooxygenase [220], such as those presented in Figure 8.9:

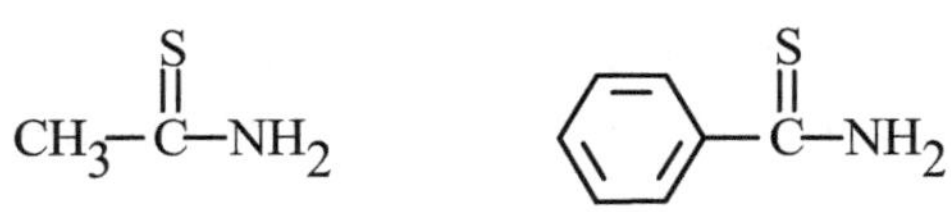

thioacetamide thiobenzamide

Fig.8.9 Substrates for FAD-containing monooxygenase catalysed sulphoxidation, yielding potentially toxic metabolites

The reason for such thioamides being good substrates of this enzyme is their resonance, which increases the nucleophilic character of the sulphur atom [221] (Figure 8.10):

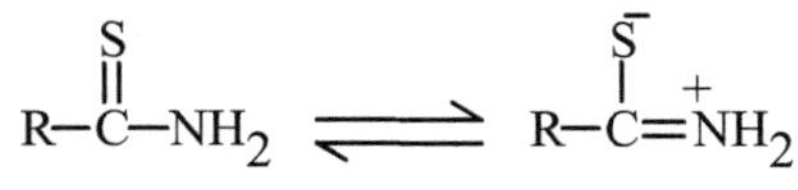

Fig.8.10 Resonance structures for thioamides, increasing the nucleophilic character of the sulphur atom

The monooxygenase-mediated oxygenation of thioamides yields different intermediates such as sulphines and sulphenes, and as end-products, acetamide, other polar compounds, and microsomal-bound material.

Very recently the mechanisms of covalent binding of reactive species and examples of bioactivation were updated [222].

The role and implications in pharmacological interactions of one of the most important drug-metabolising enzyme systems, CYTP450, were highlighted in a recent review [223].

Metabolic induction develops following repeated administration of a drug, with the synthesis of new enzyme and with the increase of its activity. The result is an increase in the metabolism of the drug involved in the interaction and a decrease in the quantity of drug available for pharmacological activity. In order for this to take place, one or two weeks are usually needed. On the other hand, enzymatic inhibition develops quickly since it takes a short time for the drug to bind to the enzyme. Inhibition of activity of the enzyme decreases the metabolism of the drug and therefore increases its pharmacological activity. Pharmacokinetic interaction can also occur when two or more drugs that are metabolic substrates of the same CYP are administered concurrently. In this case the drug that has the greatest affinity for that cytochrome can prevent in part the metabolism of the other drugs. Most drugs are substrates of only five isoenzymes (CYP3A4,1A2, 2C9,2C19,2D6); therefore, interactions can take place readily. The drugs that during absorption undergo a considerable first-pass effect or that have a low therapeutic index are the ones most often subject to significant interactions. Many interactions are not clinically apparent because plasma concentrations with therapeutic doses are lower than those used to cause the interaction *in vitro*. A very recent review refers to the role of the same enzymatic system in chemical toxicity and oxidative stress, based on studies with the CYP2E1 isoform [224].

As already stressed at the beginning of the chapter, drug allergies, known also as hypersensitivities, are reactions with a special nature. Clinical manifestations are very different and of various severities and include agranulocitosis, anaphylaxis, bronchospasm, dermatitis, fever, granulocytopenia, haemolytic anaemia, lupus erythematosus, nephritis and thrombocytopenia.

Commonly, the pathophysiology of such adverse reactions involves the presence of an organic molecule, generally larger than most drug molecules, recognized as non-self, and thus, inducing an immune response.

Sometimes however, even small, non-immunogenic organic molecules, covalently bound to an endogenous macromolecular carrier, may form a conjugate that will elicit an immune response. It is important to note that such drug-carrier conjugates may be formed if the drug or its decomposition products that might arise during manufacturing are chemically reactive, or if the drug is biotransformed into reactive intermediates. This is exemplified by carbamazepine.

As this anticonvulsant is known to be associated with frequent incidence of hypersensitivity, it would obviously be of interest to understand the molecular basis of such reactions. This question has been investigated [225], the authors postulating that reactive metabolites are responsible in many cases, including incidences of agranulocytosis and lupus. It is postulated that many drug hypersensitivity reactions, especially agranulocytosis and lupus, are due to reactive metabolites generated by the myeloperoxidase (MPO) (EC 1.11.1.7) system of neutrophils and monocytes. This led to a study of the metabolism and covalent binding of carbamazepine with MPO/H_2O_2/Cl- and neutrophils. Metabolism and covalent binding were observed in both systems and the same pathway appeared to be involved; however, the metabolism observed with the MPO system was approximately 500-fold greater than that observed with neutrophils. The metabolites identified were an intermediate aldehyde, 9-acridine carboxaldehyde, acridine, acridone, choloroacridone, and dichloroacridone. It was postulated that the first intermediate in the metabolism of carbamazepine is a carbonium ion formed by reaction of hypochlorous acid (HOCl) with the 10,11-double bond. Though there was no direct proof for the proposed carbonium ion, its presence was considered to be consistent with the likely mechanism for the observed ring contraction. Iminostilbene, a known metabolite of carbamazepine, was metabolised by a similar pathway leading to ring contraction; however, the rate was much faster and the first step possibly involves N-chlorination and a nitrenium ion intermediate. The data confirmed that carbamazepine is metabolised to reactive intermediates by activated leukocytes. Such metabolites could be responsible for some of the adverse reactions associated with carbamazepine, especially reactions such as agranulocytosis and lupus which involve leukocytes [226].

One of the best-defined models of hypersensitivity reactions is penicillin allergy. The hypersensitivity reactions, with an apparent prevalence of about 2%, may be divided into:

- immediate (anaphylaxis, asthma, urticaria);
- accelerated (urticaria, laryngeal oedema, asthma, local inflammatory reactions);
- late (commonly with urticaria, fever, haemolysis, granulocytopenia, eosinophilia and rarely, with acute renal insufficiency and thrombocytopenia) [194,197]. An interesting and useful approach to adverse reactions might be that of following the specific organ systems involved:

- probably the most common is dermatologic toxicity, and it may be mentioned that drug-induced cutaneous reactions may occur as solitary manifestations, or be part of a more severe systemic involvement. Most frequently associated with allergic skin reactions are the penicillins, sulphonamides, and blood products [197].
- many drugs as well as other xenobiotics (industrial chemicals and solvents) are associated with impaired auditory or vestibular function. Beginning with streptomycin in the 1940s, ototoxicity has become a major clinical problem. Indeed, in the last twenty years it has been suggested that more than 130 drugs and chemicals are associated with this type of adverse reaction [227]. Major classes include aminoglycosides, antimalarials, anti-inflammatory drugs, diuretics and some topical agents [228-231].
- other organs responsive to influences from both topical and systemic medications are the eyes. Ocular toxicity involves blurred vision, disturbances of colour vision, degeneration of the retina and other untoward effects on the cornea, sclera, or optic nerve [232]. It is important to note that the untoward responses are sometimes genetically determined [233].
- many drugs have been shown to cause renal dysfunction, either through reactive intermediates, or because of drug-drug interactions. While this point was mentioned earlier, it is worth emphasising the importance of close monitoring of patients with impaired renal function, or in case of administration of certain drugs known to cause renal injury. Noteworthy in this context, because of their wide therapeutic utility, are the NSAIDs, which are often associated with fluid retention, hyperkalemia, deterioration of renal function, interstitial nephritis, papillary necrosis and even chronic renal failure (especially with prolonged use of high doses) [234]. Interstitial nephritis, for instance, can be produced by numerous therapeutic agents, most frequently by penicillins, cephalosporins, sulphonamides, rifampicin, cimetidine, allopurinol, diuretics and, as mentioned, NSAIDs. A lower incidence of nephrotoxicity seems to attend the use of aminoglycosides [235]; in this case the adverse reaction appears to be more closely related to the length of time rather than to concentrations. However, with other drugs, a solution for reducing nephrotoxicity was found. For example, alternative formulations with amphotericin B have been produced, in which the drug is

encapsulated into liposomes or other lipid carriers [236]. Special attention should be given to cyclosporins because of their narrow therapeutic index, marked variability in clearance, variable bioavailability and extensive drug interactions [237]. In view of these features, the need for therapeutic drug monitoring is obvious.

- hemopoietic toxicity is a very important type of adverse reaction, since the hemopoietic system is notably vulnerable to the toxic effects of drugs.

Adverse effects may involve platelet and coagulation defects, aplastic anaemia, thrombocytopenia and agranulocytosis. Unfortunately, there are a great number of drugs associated with (or presumed to be associated with) haematologic disorders; almost 20 years ago, they were summarised by Verstraete and Boogaerts [238]. They included in this category aspirin, carbenicillin, ticarcillin, cephalosporins, chloramphenicol, phenylbutazone, sulphonamides, heparin, dipyrone, mianserin, sulfasalazine, the group of penicillins, cimetidine and the thiouracil derivatives. As a severe adverse reaction we mention, with more detail, the thrombocytopenia, caused by inceased platelet destruction or by bone marrow suppression. Such hypersensitivity can be caused by a large number of drugs and commonly these patients presumably have drug-related antibodies of both IgG and IgM classes, known to be involved in the destruction of platelets. Clinical observations revealed that transient, mild thrombocytopenia occurs in about one-quarter of patients receiving heparin, probably due to heparin-induced platelet aggregation. The severe form follows the formation of heparin-dependent antibodies.

- hepatotoxicity can also be caused by numerous drugs in common use [239]. It is noteworthy that while many agents cause asymptomatic liver injury, chronic and acute hepatic injury may develop as well. Usually, drug-induced hepatotoxicity may be either predictable, causing hepatocellular necrosis, or idiosyncratic. In the first case, the injury is due to intrinsic toxicity of the drug or its metabolite(s) and the injury is dose-related and commonly reproducible in animals. As a well-known example we mention acetaminophen. Idiosyncratic reactions are generally unpredictable, do not relate to drug dose and usually occur because of hypersensitivity (with the usual clinical implications).

A very interesting and important aspect that should be mentioned in this context is that even dietary supplements can result in hepatotoxicity, developing with cirrhosis or even fulminant hepatic failure [240].

- pulmonary toxicity; commonly, adverse pulmonary reactions to drugs are considered likely when the cause of a respiratory illness is not clear. Clinical syndromes are heterogeneous, including hypersensitive lung disease, drug-induced lupus with potential for lung involvement, bronchiolitis obliterans, pneumotitis-fibrosis and noncardiogenic pulmonary oedema. Causative agents are numerous and include ampicillin, carbamazepine, hydralazine, imipramine, isoniazid, nitrofurantoin, penicillin, phenytoin, sulphadiazine, methotrexate, griseofulvin, oral contraceptives, phenylbutazone, procainamide, quinidine, sulphonamides, sedatives or opioid overdose. The most frequent, severe pulmonary toxicity reactions are associated with amiodarone (Amiodarone Trials Meta-Analysis Investigators 1997). Severe pulmonary diseases may also be caused by cytotoxic drugs, through different mechanisms. For example, bleomycin may cause adverse effects by generating reactive oxygen metabolites, while for mitomycin, appearance of adverse effects is associated with the alkylating properties of the drug [241]. Main classes of drugs that induce pulmonary parenchymal disease include cytotoxic antibiotics, nitrosoureas, alkylating agents, cyclophosphamide, chlorambucil, methotrexate, azathioprine, 6-mercaptopurine, cytosine-arabinoside, procarbazine, and vinca alkaloids [197].

Of course, particular attention should be paid to special categories of patients, such as pregnant and breast-feeding women, infants and children, and the elderly [194-197].

- Differences in drug effects in pregnancy are usually explained by altered pharmacokinetics [242]: increased volume of distribution, hepatic metabolism and renal excretion all tend to reduce drug concentration, while decreased plasma albumin levels increase the ratio of free drug in plasma. Under these circumstances, it is obvious that prescription of drugs to a pregnant woman warrants cautious consideration in order to strike a balance between possible adverse drug effects on the foetus and the risk of leaving maternal disease inadequately treated. Therefore, usual recommendations stipulate the following: minimise prescribing; use 'tried and tested' drugs whenever possible in preference to new agents; use the smallest effective dose; remember that the foetus is most sensitive in the first trimester.

It is well known that the most severe drug–induced consequence, especially in the first trimester (period of organogenesis), is teratogenity (foetal malformation). Commonly used drugs that have demonstrated teratogenity in humans include anticonvulsants, lithium, warfarin, phenytoin, sodium valproate, carbamazepine, sex hormones and retinoic acids. For some of them, the mechanism is known at the molecular level e.g. carbamazepine and phenytoin are metabolised to arene oxides; these are reactive, electrophilic compounds that may bind to foetal macromolecules, which consequently may be implicated in the production of malformations [243]. Moreover, arene oxides are known to be metabolised by epoxide hydrolase, and therefore a genetic defect in epoxide hydrolase activity may also be associated with phenytoin-induced teratogenity [244]. After the first trimester, the risk of anatomic defects decreases, the impact of drugs simultaneously moving from structural to physiological effects. From the numerous drugs that might possibly be required to be administered during pregnancy, antibiotics are the most common. Those that are considered safe include penicillin, ampicillin, amoxicillin, erythromycin and cephalosporins. On the other hand, certain antibiotics should be avoided; these include chloramphenicol, tetracyclines, aminoglycosides, sulphonamides, metronidazole and ciprofloxacin [245].

Another problem in this context is that drugs given to a mother who is breast-feeding her infant may pass into the breast milk and consequently into the baby. Most drugs enter breast milk by passive diffusion; therefore, small molecules are expected to cross more easily than large ones. Nonetheless, it should noted that there are several factors that influence the transfer of drugs from mother to infant in breast milk. Some of them affect the concentration of drug in the mother (drug dose, frequency, route, clearance rate, plasma protein binding); others affect the transfer across the breast (breast flow rate, metabolism of drug within the breast, molecular weight, degree of ionization, water/lipid solubility of the drug, relative binding affinity to plasma and milk protein). Finally, others affect drug concentration in the infant (frequency and duration of feeds, volume of milk consumed, ability of the infant to metabolise the drug – directly dependent on the development of drug-metabolising enzymatic systems involved). Among drugs absolutely contraindicated during breast-feeding because of their negative effects on the infant, should be mentioned ciprofloxacin, chloramphenicol, doxepine, cyclophosphamide, cytotoxic drugs, iodine-containing compounds, androgens, ergotamine and laxatives. The corresponding effects are arthropathy, bone marrow suppression, respiratory suppression, neutropenia, cytotoxicity, effect on thyroid, androgenisation of the infant, vomiting, convulsions and diarrhoea.

As final conclusions and recommendations, the following are noted:

- drug concentrations should be monitored monthly:
- women taking medication during pregnancy should have a detailed ultrasound scan at 20 weeks' gestation in order to identify any foetal abnormality [197];
- women receiving phenobarbitone should avoid breast-feeding.

- Infants and children. Pediatric patients represent a condition of unstable pharmacokinetics [246]. A knowledge of age-related changes in drug absorption, distribution, and clearance is essential to optimise drug efficacy and minimise or even avoid the risk of toxicity. Under these circumstances, special attention must be paid to the pharmacokinetic variable. For example, concerning absorption, diminished intestinal motility and delayed gastric emptying in neonates and infants will result in a longer period of time for a drug to reach appropriate therapeutic plasma concentration. Drug distribution and protein binding in neonates and children are also influenced by changes in body composition that accompany development. For instance, the extracellular water compartment of body weight is almost double in the neonate compared with the adult; this may have clinically important consequences, especially with water-soluble drugs that are distributed throughout the extracellular water compartment [247]. The other determinant of drug distribution is its protein binding. Since the free (unbound) drug concentration is responsible for drug effects, age-related changes in protein binding may exert important influences on drug efficacy and toxicity, especially in drugs with a narrow therapeutic index [248]. Under these conditions, drug biotransformation is strictly dependent on the development of the enzymatic systems involved. Usually, the decreased ability of neonates to metabolise drugs, due to the immaturity of their enzymatic systems, results in prolonged elimination half-lives. As a consequence, this can predispose neonates to adverse drug reactions, caused by relative overdosing.

Another aspect that merits emphasis is that many of the drugs prescribed for neonates or children can potentially inhibit or enhance the metabolism of other drugs. The clinical significance of these interactions is dictated by the magnitude of the increase or decrease in the clearance of the index drug. Also worth stressing are the possible consequences of co-administration of drugs, especially those with inhibitory effects on hepatic drug metabolism together with a hepatically metabolised drug having a narrow therapeutic range, resulting in increased serum concentrations, overdosing and even toxicity.

Finally, we refer to possible drug interactions due to altered renal function. Most of them are undesirable - for instance enhancing methotrexate toxicity by inhibition of its tubular secretion in co-administration with salicylates. However, some of these interactions can, on the other hand, be beneficial e.g. probenecid reduces renal penicillin excretion.

In conclusion, in pediatric patients drug concentrations are directly and strongly dependent on various factors such as drug dosage, the pharmacokinetic properties dictated by the liver and kidney functions, and genetic variability in drug metabolism.

- Drugs in the elderly. The elderly constitute a particularly heterogenous patient group, who are at increased risk of appearance of adverse reactions, for several reasons:

- elderly people take more drugs (at least three to four different drugs daily). The most commonly prescribed are diuretics, analgesics, tranquillisers, and antidepressants, hypnotics and digoxin. As already mentioned, all of these are associated with a high incidence of important adverse reactions:

- pharmacokinetics change with increasing age (and often, concomitant disease), leading generally to higher plasma concentrations of drugs, and consequently, increased susceptibility to side-effects;
- with advancing age, homeostatic mechanisms become less effective, so these individuals are less able to compensate for adverse effects;
- increasing age produces changes in the immune response (increased risk of allergic reactions); also, the central nervous system becomes more sensitive to the actions of sedative drugs.

Of the main reasons listed above that determine increased risk of drug toxicity in the elderly, the most important by far is considered to be the pharmacokinetics, potentially modified by ageing. By influencing drug disposition, these age-related changes might be expected to alter the response to drugs, which consequently may explain why older patients seem to be more susceptible to both the therapeutic and the toxic effects of many drugs.

As far as absorption is concerned, it is well known that the elderly exhibit several alterations in GI function that might result in impaired or delayed absorption of a drug. However, relatively recently it was demonstrated that very few drugs displayed delayed or reduced absorption after oral administration in the elderly [249]. In contrast, the active transport of calcium, iron, thiamine and vitamin B_{12} declines with age, due either to decreased intestinal blood flow rate (up to 50%), or to increased gastric motility. However, unless GI pathology is present, it appears that age *per se* does not affect drug absorption to a significant extent.

Body composition is one of the most important factors that may produce altered distribution of drugs in elderly patients, ageing being generally associated with loss of weight and lean body mass, increased ratio of fat to muscle, and decreased body water [250]. Therefore, hydrophilic drugs that are commonly distributed mainly in body water or lean body mass should have higher concentrations in blood in the elderly, especially when the dose is based either on the total weight or surface area. Conversely, highly lipophilic drugs tend to have larger volumes of distribution in older persons due to increased proportion of body fat. This may be partly responsible for the age-related increase in the volume of distribution of thiopental and some of the benzodiazepines [251].

As far as hepatic metabolism is concerned, a decrease in the rate of hepatic clearance of some but not all drugs was noted with advancing age. This is first determined by age-related decreases in liver size and blood flow [252]. On the other hand, it is well known that hepatic clearance of drugs is strongly dependent on the enzymatic activity, both microsomal and non-microsomal enzymes being involved in both phases of drug biotransformation. It was established that the activities of phase I pathways are often reduced in the elderly, whereas phase II pathways are generally unaffected [253]. Obviously, the reduced rate of clearance of certain drugs may lead to important clinical consequences, such as accumulation of drug (relative overdosing), leading to adverse reactions [254, 255, 256].

The most consistent effect of age on pharmacokinetics is the age-related reduction in renal excretion, with both glomerular and tubular functions being affected. As a consequence, it is generally assumed that drugs that are significantly excreted by the kidney will display diminished clearance of plasma from the elderly. Drugs with decreased renal excretion in old age, and consequently with potentially severe toxic effects, include amantadine, ampicillin, atenolol, captopril, chlorpropamide, cimetidine, digoxin, doxycycline, enalapril, furosemide, lithium carbonate, penicillin, procainamide, phenobarbital, ranitidine and tetracycline [257].

These alterations in pharmacokinetics in the elderly (especially impaired renal elimination and hepatic metabolism of drugs) may also contribute to exacerbated consequences of drug-drug interactions. A well-known example is the effect of dexamethasone on phenytoin metabolism, when these drugs are concurrently administered. Both are substrates for the same metabolising enzymes, resulting in increased serum phenytoin levels, and even toxicity [258]. Similarly, a drug-drug interaction causing a decrease in renal drug excretion (in addition to an already poor renal elimination capacity) could result in increased toxicity in a vulnerable elderly patient.

8.5 SUMMARY

From all the above considerations and examples, it emerges, as a first major conclusion, that a significantly important step in assessing the potential toxicity of a drug and its metabolites is the prediction of entry and fate of the compound in human body. With this in mind, three distinct approaches should be stressed. The first refers to the predictions that can possibly be made based on both *in vitro* and *in vivo* data concerning metabolic transformations of a particular drug. The second implies knowing and understanding the enzymatic systems involved in these biotransformations, while the third concerns the possibility of extrapolating data from *in vitro* or *in vivo* results on the one hand, and *interspecies* results, on the other.

There are predictions of *in vivo* drug-drug interactions based on *in vitro* data gathered from the literature [9]. They are based, in principle, on mathematical models using measurable, specific parameters, introduced for the calculation of the hepatic intrinsic *in vivo* clearance for a particular drug. Some of them are successful and some are not. Nowadays, *in vitro* experiments use human microsomes, hepatocytes, liver slices and isoform-specific microsomes from expressed systems. It is important to note that if studies are based on human microsomes, some accounting should be made for the interindividual variability in the expression of the target isoform [10]. As a successful prediction, we may mention the well-known tolbutamide-sulfaphenazole case, which concurrently administered may cause severe side effects, such as hypoglycaemic shock. An approximately five-fold increase in both AUC and $t_{1/2}$ of tolbutamide in co-administration with sulphaphenazole was reported [259]. Both drugs being co-substrates for the same CYTP450 isoform (CYP2D9), on concurrent administration of sulphaphenazole, the biotransformation of tolbutamide is inhibited. The main metabolisation pathway of tolbutamide *in vitro* is a CYP2D9-mediated hydroxylation. *In vivo*, the hydroxylated metabolite follows sequential biotransformations, resulting in a carboxylated metabolite. The K_i value for sulphaphenazole, a specific inhibitor of the isoenzyme involved in the main biotransformation pathway is extremely small (~0.1-0.2 μM). As a consequence, the inhibitor's affinity as co-substrate for the same enzyme will be very strong, as will its inhibitory action. The inhibition of this metabolic pathway results in a reduction of the total clearance of tolbutamide of about 80% (relative overdosing induced by co-administration of sulphaphenazole). If a substantial inhibition of an isoform-specific probe is observed, it is strongly recommended (as a matter of genuine practical interest) that the magnitude of the same effect for other substrates of that isoform be assessed. The above interaction is species-dependent, following more or less the same pattern in rats, but the opposite in rabbits (total clearance increased by 15-30%) [259].

A well-studied and previously mentioned interaction in this context is that between terfenadine and ketoconazole [260]. Terfenadine is extensively and rapidly metabolised by CYP3A4 isoenzymes. On the other hand, ketoconazole is known to be a potent CYP3A4 inhibitor; consequently, co-administration of ketoconazole results in a dramatic decrease in the biotransformation of terfenadine, with consequent increase in plasmatic levels.

Other examples based on the same mechanism (competitive inhibition of one drug as co-substrate for the same enzyme) include caffeine-ciprofloxacin [261] and cyclosporin-erythromycin [262].

An interesting mechanism, different from the 'classic' ones presented above, is that of mechanism-based inhibition, in which the inhibitor is biotransformed into a metabolite that covalently binds to the enzyme, resulting in its irreversible inactivation [263]. However, a special case, when the inhibition is not called 'mechanism-based inhibition', is that when the inhibitor is metabolically activated by one enzyme and inactivates another. Such an example, which determined 5-fluorouracil (5-FU) toxicity caused by high blood concentrations, is its interaction with sorivudine (an antiviral drug). Sorivudine is sequentially biotransformed into a metabolite that is rate-limiting in the metabolism of 5-FU. More attention to this type of interaction is needed because the inhibitory effect remains after the elimination of sorivudine from blood and tissues, possibly leading to serious side-effects [264].

In this context, it should also be mentioned that many drugs (other than sorivudine) are reported to be mechanism-based inhibitors e.g. macrolide antibiotics (erythromycin, troleandomycin – against CYP3A4) [265], orphenadrine (against CYP2B1) [266], and furafylline (against CYP1A2) [267].

Another important problem that warrants mention is that some enzymes act not only in the liver but in the gut as well, therefore playing an important role in the first-pass metabolism following oral administration. Such an important isoform is CYP3A4, an enzyme that metabolises many drugs, including cyclosporins [268]. As examples we can quote the decreased bioavailability of cyclosporin after co-administration of rifampicin – an inducer of CYP3A4, and its increased bioavailability by co-administration of ketoconazole, an inhibitor of the same isoform [269, 270].

A new and fashionable approach in drug discovery is predictive ADMET (Absorption, Distribution, Metabolism, Excretion, and Toxicity) [271]. Here, candidate drugs may be designed and their structures optimised by constructing computational models that associate structural variations with changes in response [272]. The process relies on large databases containing ADMET data for known structures. An offshoot of this approach is the possibility of predicting human ADMET properties from human *in vitro* and animal *in vivo* ADMET data. The success or otherwise of this approach is limited by the quality of the database and the level of sophistication of the modelling methods.

The status of *in-silico* prediction of drug metabolism and related toxicity has been reviewed [273]. Problems that are encountered include prediction of the nominal metabolic transformations for a given molecule and gaining an understanding of the nature of the enzymes that might be involved as well as possible alternative routes of transformation. In so-called 'rule-based' metabolism prediction studies, the aim is to predict both metabolic pathways as well as metabolites that might be generated [272]. Following formation of the first predicted metabolite, the possibility of its being metabolised to multiple products must be considered, as must the subsequent metabolism of those products. This could rapidly result in an unmanageable number of metabolites and therefore limitations on this number can be imposed, based on e.g. the probability of a particular metabolite being formed, the stability of the metabolite, and so on. Ideally, such predictions should be implemented at an early stage of drug discovery to eliminate 'junk' leads. Nonetheless, the problem remains a challenging one due to the complexity of the human body, including such factors as differences in enzymatic phenotype that could alter drug metabolism. Input from other areas is necessary to gain a comprehensive picture of the metabolic fate of a given compound.

QSAR modelling usually examines the interactions between small drug molecules and macromolecules such as their metabolising enzymes [273]. Recent developments in this area include QSAR models for substrates of the major drug-metabolising enzymes, including human cytochrome P450s. QSAR modelling of metabolic stages following that above is also an important area for investigation.

Concluding remarks

A considerably detailed treatment of drug-drug interactions and adverse reactions has been presented above, together with older and more recent examples of each. Much of this material relates to known, well-documented cases that are of general interest. However, a crucial aspect of drug-drug interactions and adverse reactions is the possibility of predicting their occurrence for new drug candidates. Some indication as to how this is being addressed by modern methods, including computational approaches, has been given above. The final chapter, dealing with certain aspects of drug design, draws on concepts presented in the previous chapters, the intention being to demonstrate how various aspects of drug metabolism are taken into consideration in deriving new drugs with predictable and controllable biotransformation.

References

1. Thomas A, Routledge PA. 2003. Drug interactions in clinical practice. Pharmacovigilance Bul 34:1-7.

2. Rollins DE. 2000. Adverse Drug Reactions. In: Genaro AR, editor. Remington: The Science and Practice of Pharmacy. Philadelphia: Lippincott Williams & Wilkins, pp 1165-1168.

3. Morris JS, Stockley IH. 2000. Fundamentals of drug interactions. In: Sirtori CR, Kuhlmann J, Tillement J-P, Vrhovac B, Reidenberg MM, editors. Clinical Pharmacology. London: McGraw-Hill International (UK) Ltd., pp 51-64.

4. Neis AS. 2001. Practical Principles. In: Hardman JG, Limbird LE, Gilman GA, editors. Goodman & Gilman's The Pharmacological Basis of Therapeutics, 10th ed. New York: McGraw-Hill (Medical Publishing Division), pp 45-66

5. Ritter JM. 1999. Drug interactions. In: Lewis LD, Mant T.KG, editors. A Textbook of clinical Pharmacology, 4th ed. Oxford University Press Inc., pp 97-107.

6. Hooper WD. 1999. Metabolic Drug Interactions. In: Woolf TF, editor. Handbook of Drug Metabolism. New-York: Marcel Dekker, Inc., pp 229-238.

7. Delafuente JC. 2003. Understanding and preventing drug interactions in elderly patients. Crit Rev Oncol Hematol 48:133-143.

8. Thomas J. 1994. Drug interactions. In: Thomas J, editor. Australian Prescriptions Product Guide, 23rd ed. Melbourne: Australian Pharmaceutical Publishing Co., p 62.

9. Ito K, Iwatsubo T, Kanamitsu S, Ueda K, Suzuki H, Sugiyama Y. 1998. Prediction of Pharmacokinetic Alterations Caused by Drug-Drug Interactions: Metabolic interaction in the liver. Pharmacol Rev 50:387-411.

10. Bertz RJ, Granneman GR. 1997. Use of *In Vitro* and *In Vivo* Data to Estimate the Likelihood of Metabolic Pharmacokinetic Interactions. Clin Pharmacokinet 32:210-258.

11. Wright JM. 2000. Drug interactions. In: Carruthers SG, Hoffman BB, Melmon KL, Nierenberg DW, editors. Clinical Pharmacology. Basic Principles in Therapeutics, 4th ed. New-York: McGraw-Hill Medical Publishing Division, pp 1257-1266.

12. Oniga O, Ionescu C. 2004. Interacţiuni medicamentoase. In: Reacţii adverse şi interacţiuni medicamentoase. Romania, Cluj-Napoca: "I.Hatieganu" Med. Univ. Press., pp 157-215.

13. Rolan PE. 1994. Plasma protein binding displacement interactions – why are they still regarded as clinically important? Br J Clin Pharmacol 37:125-128.

14. Griffin JP, Ferrero JD, Hughes CM, D'Arcy PF. 1996. Mechanisms of Drug Interactions. In: D'Arcy PF, McElnay JC, Welling PG, editors. Handbook of Experimental Pharmacology (122), pp 151-171.

15. Honig PK, Gillespie BK. 1995. Drug interactions between prescribed and over-the-counter medication. Drug Safety 13:296-303.

16. Okey AB. 1990. Enzyme induction in the cytochrome P-450 system. Pharmacotherapy 45:241-298.

17. Zand R, Nelson SD, Slaterry JT, Thummel KE, Kalhorn TF, Adams SP, Wright JM. 1993. Inhibition and oxidation of cytochrome P4502E1-catalyzed oxidation by isoniazid in humans. Clin Pharmacol Ther 54:142-149.

18. Prescott LF. 1987. Clinically important drug interactions. In: Speight TM, editor. Avery's Drug Treatment: Principles and Practice of Clinical Pharmacology and Therapeutics, 3rd ed. Auckland: ADIS Press, p 255.

19. Loghin F. 2002. In: Toxicologie generală. Romania, Cluj-Napoca: "I.Hatieganu" Med. Univ. Press, pp 139-173.

20. Butterweck V, Derendorf H, Gaus W, Nahrstedt A, Schulz, V, Unger M. 2004. Pharmacokinetic herb-drug interactions: Are preventive screenings necessary and appropriate? Planta Medica 70:784-791.

21. De Andres S, Lucena A, de Juana P. 2004. Interactions between foods and statins. Nutr Hosp 19:195-201.

22. Koenig F. 2003. Interactions of new antidepressants, phytopharmaceuticals, and mood stabilizers with drugs for internal diseases. Interaktionen und Wirkmechanismen Ausgewaehlter Psychopharmaka, 2nd ed., pp 136-169.

23. Hesse LM, Venkatakrishnan K, Court MH, von Moltke LL, Duan SX, Shader RI, Greenblatt DJ. 2000. CYP2B6 mediates the in vitro hydroxylation of bupropion: potential drug interactions with other antidepressants. Drug Metab Dispos 28:1176-1183.

24. Berecz R, Dorado P, De La Rubia A, Caceres MC, Degrell I, Lerena AL. 2004. The role of cytochrome P450 enzymes in the metabolism of risperidone and its clinical relevance for drug interactions. Curr Drug Targ 5:573-579.

25. Fireman M, DiMartini AF, Armstrong SC, Cozza KL. 2004. Med-Psych drug-drug interactions update: immunosuppressants. Psychosomatics 45:354-360.

26. Anderson GD. 2004. A mechanistic approach to antiepileptic drug interactions. Neurol Dis Ther 64:107-138.

27. Minchun J, Xiaofang G, Minde Y. 2003. Drug interaction between nonsteroid anti-inflammatory drugs and angiotensin converting enzyme inhibitors. Zhongguo Linchuang Yaolixue Zazhi 19:310-314.

28. Back DJ, Orme ML. 1994. Drug interactions [with oral contraceptives]. Pharmacol Contracept Steroids, pp 407-425.

29. Kaschka WP. 2003. Interactions of lithium salts with other drugs. Interaktionen und Wirkmechanismen Ausgewaehlter Psychopharmaka 2nd ed., pp 78-93.

30. Ferrari A, Coccia CPR, Bertolini A, Sternieri E. 2004. Methadone-metabolism, pharmacokinetics and interactions. Pharmacol Res 50:551-559.

31. Furuda S, Kamada E, Omata T, Sugimoto T, Kawabata Y, Yonezawa k, Cheryl Wu X, Kurimoto T. 2004. Drug–drug interactions of Z-338, a novel gastroprokinetic agent, with terfenadine, comparison with cisapride, and involvement of UGT1A9 and 1A8 in the human metabolism of Z-338. Eur J Pharmacol 497:223-231.

32. Krzysztof JR, Britigan LH, Rasmussen GT, Wagner BA, Burns CP, Britigan BE. 2004. Acetaminophen stimulates the peroxidative metabolism of anthracyclines. Arch Biochem Biophys 427:16-29.

33. Calabresi L, Pazzucconi F, Ferrara S, di Paolo A, Del Tacca M, Siroti C. 2004. Pharmacokinetic interactions between omeprazole/pantoprazole and clarithromycin in healthy volunteers. Pharmacol Res 49:493-499.

34. Orlando R, Piccoli P, De Martin S, Padrini R, Floreani M, Palatini P. 2004. Cytochrome P450 1A2 is a major determinant of lidocaine metabolism in vivo: effects of liver function. Clin Pharmacol Ther 75:80-88.

35. Peterson TC, Peterson MR, Wornell PA, Blanchard MG, Gonzales FJ. 2004. Role of CYP1A2 and CYP2E1 in the pentoxifylline ciprofloxacin drug interaction. Biochem Pharmacol 68:395-402.

36. Yu L-S, Yao T-W, Zeng S. 2003. In vitro metabolism of zolmitriptan in rat cytochromes induced with β-naphthoflavone and the interaction between six drugs and zolmitriptan. Chem-Biol Interact 146:263-272.

37. Jain AB, Venkataramanan R, Eghtesad B, Marcos A, Ragni M, Shapiro R, Rafail AB, Fung JJ. 2003. Effect of coadministered lopinavir and ritonavir (Kaletra) on tacrolimus blood concentration in liver transplantation patients. Liver Transplant 9:954-960.

38. Werner U, Werner D, Rau T, Fromm MF, Ninz B, Brune K. 2003. Celecoxib inhibits metabolism of cytochrome P450 2D6 substrate metoprolol in humans. Clin Pharmacol Ther 74:130-137.

39. Zimmerman HJ, Maddrey WC. 1995. Acetaminophen hepatotoxicity with regular intake of alcohol. Hepatology 22:767-773.

40. Kwan D, Bartle WR, Walker SE. 1999. The effects of acetaminophen on pharmacokinetics and pharmacodynamics of warfarin. J Clin Pharmacol 39:68-75.

41. Pirotte JH. 1984. Apparent potentiation by Phenobarbital of hepatotoxicity from small doses of acetaminophen. Ann Intern Med 101:403.

42. Mulley BA, Potter BI, Rye RM, Takeshita K. 1978. Interactions between diazepam and paracetamol. J Clin Pharmacol 3:25-31.

43. Dordoni B, Willson RA, Thompson RP, Williams R. 1973. Reduction of absorption of paracetamol by activated charcoal and cholestyramine: a possible therapeutic measure. Br Med J 3:86-87.

44. Crippin JS. 1993. Acetaminophen Hepatotoxicity: potentiation by isoniazid, Am J Gastroenterol 88:590-592.

45. Kamali F. 1993. The effect of probenecid on paracetamol metabolism and pharmacokinetics. Eur J Clin Pharmacol 45:551-553.

46. Steffe EM, King JH, Inciardi JF, Neil F, Goldstein E, Tonjes TS, Benet LZ. 1990. The effect of acetaminophen on zidovudine metabolism in HIV-infected patients. J Acquir Immune Defic Syndr 3:691-694.

47. Johnson R, Douglas J, Corey L. 1985. Adverse effects with acyclovir and meperidine. Ann Inter Med 103:962-963.

48. Laskin OL, De Miranda P, King DH, Page DA, Longstreth JA, Rocco L, Lietman PS. 1982. Effects of probenecid on the pharmacokinetics and elimination of acyclovir in humans. Antimicrob Agents Chemother 21:804-807.

49. Bach MC. 1987. Possible drug interaction during therapy with azidothymidine and acyclovir for AIDS. N Engl J Med 316:547.

50. Shaunak S, Munro JM, Weinbren K, Walport MJ, Cox TM. 1988. Cyclophosphamide-induced liver necrosis: a possible interaction with azathioprine. Q J Med 67:309-317.

51. Moore MJ. 1988. Rapid development of enhanced clearance after high-dose cyclophosphamide. Clin Pharmacol Ther 44:622-628.

52. Ghany AM, Tutschka PJ, McGhee RB Jr, Avalos BR, Cunningham I, Kapoor N, Copelan EA. 1991. Cyclosporin-associated seizures in bone marrow transplant recipients given busulphan and cyclophosphamide preparative therapy. Transplantation 52:310-315.

53. Churchill DN, Seely J. 1977. Nephrotoxicity associated with combined gentamicin-amphotericin B therapy. Nephron 19:176-181.

54. Trvedegaard E. 1976. Interaction between gentamicin and cephalothin as cause of acute renal failure. Lancet 2:581.

55. Engineer MS, Bodey GP, Newman RA, Ho DHW. 1987. Effects of cisplatin-induced nephrotoxicity on gentamicin pharmacokinetics in rats. Drug Metab Dispos 15:329-334.

56. Morales JM, Andres A, Prieto C, Diaz Rolon JA, Rodicio JL. 1988. Reversible acute renal toxicity by toxic synergic effect between gentamicin and cyclosporin. Clin Nephrol 29:272.

57. Lindenbaum J, Maulitz RM, Butler VP Jr. 1976. Inhibition of digoxin absorption by neomyicin. Gastroenterology 71:399-404.

58. Schwartz GH, David DS, Riggio RR, Stenzel KH, Rubin Al. Ototoxicity induced by furosemide. 1970. N Engl J Med 282:1413-1414.

59. 1997. In: Rizach MA, editor. The Medical Letter. Handbook of Adverse Drug Interactions. New York: The Medical Letter, Inc., pp 28.

60. Pauly DJ, Musa DM, Lestico MR, Lindstrom MJ, Hetsko CM. 1990. Risc of nephrotoxicity with combination vancomycin-aminoglycoside antibiotic therapy. Pharmacotherapy 10:378-382.

61. Steinberg WM, Lewis JH, Katz DM. 1982. Antacids inhibit absorption of cimetidine. N Engl J Med 307:400-404.

62. Shader RI, Georgotas A, Greenblatt DJ, Harmatz JS, Allen MD. 1978. Impaired absorption of desmethyldiazepam from clorazepate by magnesium aluminum hydroxyde. Clin Pharmacol Ther 24:308-315.

63. Laer S, Neumann J, Scholz H. 1997. Interaction between sotalol and an antacid preparation. Br J Clin Pharmacol 43:269-272.

64. Hughes GS, Heald DL, Barker KB, Patel RK, Spillers CR, Watts KC, Batts DH, Euler AR. 1989. The effects of gastric pH and food on the pharmacokinetics of a new oral cephalosporin, cefpodoxime proxetil. Clin Pharmacol Ther 46:674-685.

65. Uribe M, Casian C, Rojas S, Sierra JG, Go VL. 1981. Decreased bioavailability of prednisone due to antacids in patients with chronic active liver disease and in healthy volunteers. Gastroenterology 80:661-665.

66. Allen MD. 1981. Effect of magnesium-aluminum hydroxide and kaolin-pectin on absorption of digoxin from tablets and capsules. J Clin Pharmacol 21:26-30.

67. Kivisto KT, Neuvonen PJ. 1992. Effect of magnesium hydroxide on the absorption and efficacy of tolbutamide and chlorpropamide. Eur J Clin Pharmacol 42:675-679.

68. Zinn MB. 1970. Quinidine intoxication from alkali ingestion. Texas Med 66:64-66.

69. Haag HB, Larson PS, Scwartz JJ. 1943. The effect of urinary pH on the elimination of quinine in man. J Pharmacol Exp Ther 79:136.

70. Garty M, Hurwitz A. 1980. Effect of cimetidine and antacids on gastrointestinal absorption of tetracycline. Clin Pharmacol Ther 28:203-207.

71. Domingo JL, Gomez M, Llobert JM, Richart C. 1991. Effect of ascorbic acid on gastrointestinal aluminum absorption. Lancet 338:1467.

72. Colussi G, Rombola G, De Ferrari ME, Minola E, Minetti L. 1987. Vitamin D treatment: a hidden risk factor for aluminum bone toxicity? Nephron 47:78-80.

73. Okino K, Weibert RT. 1986. Warfarin-griseofulvin interaction. Drug Intell Clin Pharm 20:291-293.

74. Baciewicz AM, Morgan PJ. 1990. Ranitidine-warfarine interaction. Ann Inter Med 112:76-77.

75. MacDonald MG, Robinson DS. 1968. Clinical observations of possible barbiturate interference with anticoagulation. JAMA 204:97-100.

76. Denbow CE, Fraser HS. 1990. Clinically significant hemorrhage due to warfarin-carbamazepine interaction. South Med J 83:981.

77. Cucinell SA, Odessky L, Weiss M, Dayton PG. 1966. The effect of chloral hydrate on bishydroxycoumarin metabolism. JAMA 197:366-368.

78. Jahnchen E, Meinertz T, Gilfrich HJ, Kersting F, Groth U. 1978. Enhanced elimination of warfarin during treatment with cholestyramine. Br J Clin Pharmacol 5:437-440.

79. Rothstein E. 1972. Warfarin effect enhanced by disulfiram (Antabuse). JAMA 221:1052-1053.

80. McLeod AD, Burgess C. 1988. Drug interaction between warfarin and enoxacin. NZ Med J 101:216.

81. Udall JA. 1975. Clinical implications of warfarin interactions with five sedatives. Am J Cardiol 35:67-71.

82. Bachmann K, Shwartz JI, Forney R Jr, Frogameni A, Jauregui LE. 1984. The effect of erythromycin on the disposition kinetics of warfarin. Pharmacology 28:171-176.

83. Leor J, Levartowsky D, Sharon C. 1987. Interaction between nalidixic acid and warfarin. Ann Intern Med 107:601.

84. Fagan SC, Kertland HR, Tietjen GE. 1995. Safety of combination aspirin and anticoagulation in acute ischemic stroke. Ann Pharmacother 28:441-443.

85. Panegyres PK, Rischbieth RH. 1991. Fatal phenytoin warfarin interaction. Postgrad Med J 67:98.

86. Kates RE, Yee YG, Kirsten EB. 1987. Interaction between warfarin and propafenone in healthy volunteer subjects. Clin Pharmacol Ther 42:305-311.

87. O'Reilly RA. 1980. Spironolactone and warfarin interaction. Clin Pharmacol Ther 27:198-201.

88. Sioris LJ, Weibert RT, Pentel PR. 1980. Potentiation of warfarin anticoagulation by sulfisoxazole. Arch Intern Med 140:546-547.

89. Danos EA. 1992. Apparent potentiation of warfarin activity by tetracycline. Clin Pharm 11:806-808.

90. Hansten PD. 1980. Oral anticoagulants and drugs which alter thyroid function. Drug Intell Clin Pharm 14:331-337.

91. Guthrie SK, Stoysich AM, Bader G, Hilleman DE. 1995. Hypothesized interaction between valproic acid and warfarin. J Clin Psychopharmacol 15:138-139.

92. Schrogie JJ. 1975. Coagulopathy and fat-soluble vitamins. JAMA 232:19.

93. Rosenthal G.1971.Interaction of ascorbic acid and warfarin. JAMA 215:1671.

94. Kim JM, White RH. 1996. Effect of vitamin E on the anticoagulant response to warfarin. Am J Cardiol 77:545-546.

95. Krasinski K, Kusmiesz H, Nelson JD. 1982. Pharmacologic interactions among chloramphenicol, phenytoin and phenobarbital. Pediatr Infect Dis 1:232-235.

96. Back D J, Bates M, Bowden A, Breckenridge A M, Hall M J Jones H, MacIver M, Orme M, Perucca E, Richens A, Rowe P H, Smith E. 1980. The interaction of phenobarbital and other anticonvulsants with oral contraceptive steroid therapy. Contraception 22:495-503.

97. Brooks PM, Buchanan WW, Grove M, Downie WW. 1976. Effects of enzyme induction on metabolism of prednisolone. Ann Rheum Dis 35:339-343.

98. Stambaugh JE, Hemphill DM, Wainer IW, Schwartz I. 1977. A potentially toxic drug interaction between pethidine (meperidine) and phenobarbitone. Lancet 1:398-399.

99. Amabeoku GJ, Chikuni O, Akino C, Mutetwa S. 1993. Pharmacokinetic interaction of single doses of quinine and carbamazepine, phenobarbitone and phenytoin in healthy volunteers. East Afr Med J 70:90-93.

100. Bernus I, Dickinson RG, Hooper WD, Eadie MJ. 1994. Inhibition of phenobarbitone N-glucosidation by valproate. Br J Clin Pharmacol 38:411-416.

101. Brunova E, Slabochova Z, Platilova H, Pavlik F, Grafnetterova J, Dvoracek K 1977. Interaction of tolbutamide and chloramphenicol in diabetic patients. Int J Clin Pharmacol 15:7-12.

102. Vincent FM, Mills L, Sullivan JK. 1978. Chloramphenicol-induced phenytoin intoxication. Ann Neurol 3:469.

103. Achumba JI, Ette EI, Thomas WO, Essien EE. 1988. Chloroquine-induced acute dystonic reactions in the presence of metronidazole. Drug Intell Clin Pharm 22:308-310.

104. Makanjuola RO, Dixon PA, Ohorah E. 1988. Effects of antimalarial agents on plasma levels of chlorpromazine and its metabolites in schizophrenic patients. Trop Geographic Med 40:31-33.

105. Kuhlmann J. 1985. Effects of verapamil, diltiazem, and nifedipine on plasma levels and renal excretion of digitoxin. Clin Pharmacol Ther 38:667-673.

106. Brater DC, Morelli HF. 1977. Digitoxin toxicity in patients with normokalemic potassium depletion. Clin Pharmacol Ther 22:21-23.

107. Ascione FJ. 1977. Digitalis glycosides with potassium-depleting diuretics. Drug Ther (Hosp) 7:5.

108. Bartolone RS, Rao TLK.1983. Dysrhythmias following muscle relaxant administration in patients receiving digitalis. Anesthesiology 58:567-569.

109. Ascione FJ. 1977. Digitalis glycosides with potassium-depleting diuretics. Drug Ther (Hosp) 8:12.

110. Notelovitz M, Tjapkes J, Ware M. 1981. Interaction between estrogen and Dilantin in a menopausal woman. N Engl J Med 304:788-789.

111. Back DJ, Breckenridge AM, MacIver M, Orme M l'E, Purba H, Rowe PH. 1981. Interaction of ethinylestradiol with ascorbic acid in man. Br Med J 283:503.

112. Lehto P, Kivisto KT. 1994. Different effects of products containing metal ions on the absorption of lomefloxacin. Clin Pharmacol Ther 56:477-481.

113. Lehto P, Kivisto KT, Neuvonen PJ. 1994. The effect of ferrous sulphate on the absorption of norfloxacin, ciprofloxacin and ofloxacin. Br J Clin Pharmacol 37:82-85.

114. Barriere SL, Catlin, Donald H, Orlando PL, Noe A, Frost RW. 1990. Alteration in the pharmacokinetic disposition of ciprofloxacin by simultaneous administration of azlocillin. Antimicrob Agents Chemother 34:823-826.

115. Davis RL, Quenzer RW, Kelly HW, Powell JR. 1992. Effect of the addition of ciprofloxacin on theophylline pharmacokinetics in subjects inhibited by cimetidine. Ann Pharmacother 26:11-13.

116. Campbell NRC, Kara M, Hanschoff BB, Haddara WM, McKay DW. 1992. Norfloxacin interaction with antacids and minerals. Br J Clin Pharmacol 33:115-116.

117. Pupeschi G, Agenet C, Levron J-C, Barges-Bertocchio M-H. 1994. Do enzyme inducers modify haloperidol decanoate rate or release? Prog Neuro-Psychopharmacol Biol Psychiat 18:1323-1332.

118. Takeda M, Nishinuma K, Yamashita S, Matsubayashi T, Tanino S, Nishimura T. 1986. Serum haloperidol of schizophrenics receiving treatment for tuberculosis. Clin Neuropharmacol 9:386-397.

119. Schaffer CB, Batra K, Garvey MJ, Mungas DM, Schaffer LC. 1984. The effect of haloperidol on serum levels of lithium in adult maniac patients. Biol Psychiatry 19:1495-1499.

120. Nadel I, Wallach M. 1979. Drug interaction between haloperidol and methyldopa. Br J Psychiatry 135:484.

121. Young A, Kehoe R. 1989. Two cases of agranulocytosis on addition of a butyrophenone (haloperidol) to a long-standing course of phenothiazine treatment. Br J Psychiatry 154:710-712.

122. McSwain MJ, Forman LM. 1995. Severe parkinsonian symptom development on combination treatment with tacrine and haloperidol. J Clin Psychopharmacol 15:284.

123. Herings RM, de Boer A, Stricker BH, Leufkens HG, Porsius A. 1995. Hypoglycemia associated with use of inhibitors of angiotensin converting enzyme. Lancet 345:1195-1198.

124. Marrazzi MA, Jacober S, Luby ED. 1994. A naltrexone-induced increase in insulin requirements. J Clin Psychopharmacol 14:363-365.

125. Richardson T, Foster J, Mawer GE. 1986. Enhancement by sodium salicylate of the blood glucose lowering effect of chlorpropamide – drug interaction or summation of similar effects? Br J Clin Pharmacol 22:43-48.

126. Blanco R, Martinez Taboada VM, GonzalezGay MA, Armona J, FernandezSueiro JL, GonzalezVela MC, Rodriguez Valverde V. 1996. Acute febrile toxic reaction in patients with refractory rheymatoid arthritis who are receiving combined therapy with methotrexate and azathioprine. Arthritis Rheum 39:1016-1020.

127. Crom WR, Pratt CB, Green AA, Champion JE, Crom DB, Steewart CF, Evans WE. 1984. The effect of prior cisplatin therapy on the pharmacokinetics of high-dose methotrexate. J Clin Oncol 2:655-661.

128. Wadhwa NK, Schroeder TJ, O'Flaherty E, Pesce AJ, Myre SA, First MR. 1987. The effect of oral metoclopramide on the absorption of cyclosporin. Transplantation 43:211-213.

129. Kremer JM, Hamilton RA. 1995. The effects of NSAIDs on methotrexate (MTX) pharmacokinetics: impairment of renal clearance of MTX at weekly maintenance doses but not at 7.5 mg. J Rheumatol 22:2072-2077.

130. Dean R, Nachman J, Lorenzana N. 1992. Possible methotrexate-mezlocillin interaction. Am J Pediatr Hematol Oncol 14:88-89.

131. Morgan SL, Baggott JE, Alarcon GS. 1993. Methotrexate and sulphasalazine combination therapy: is it worth to risk? Arthritis Rheum 36:281-282.

132. Turck M. 1984. Successful psoriasis treatment then sudden 'cytotoxicity'. Hosp Pract 19:175-176.

133. Govert JA, Patton S, Fine RL. 1992. Pancytopenia from using trimethoprim and methotrexate. Ann Intern Med 117:877-878.

134. Thatte UM, Shah SJ, Dalvi SS, Suraokar S, Temulkar P, Anklesaria P, Kshirsagar NA. 1988. Acute drug interaction between indomethacin and nifedipine in hypertensive patients. J Assoc Physicians India 36:695-698.

135. Ahmad S. 1984. Nifedipine-phenytoin interaction. J Am Coll Cardiol 3:1582.

136. Tada Y, Tsuda Y, Otsuka T, Nagasawa K, Kimura H, Kusaba T, Sakata T. 1992. Case report: Nifedipine-rifampicin interaction attenuates thr effect on blood presure in a patient with essential hypertension. Am J Med Sci 303:25-27.

137. Sternbach H. 1991. Fluoxetine-associated potentiation of calcium-channel blockers J Clin Psychopharmacol 11:390-391.

138. Cohen AF, Kroon R, Schoemaker R, Hoogkamer H, van Vliet A. 1991. Influence of gastric acidity on the bioavailability of digoxin. Ann Int Med 115:540-545.

139. Hajela R, Cunningham GM, Kapur BM, Peachey JE, Devenyi P. 1990. Catatonic reaction to omeprazole and disulfiram in a patient with alcohol dependence. Can Med Assoc J (CMAJ) 143:1207-1208.

140. Anderson T, Lagerstrom PO, Unge P. 1990. A study of the interaction between omeprazole and phenytoin in epileptic patients. Ther Drug Monit 12:329-333.

141. Taylor AT, Pritchard DC, Goldstein AO, Fletcher JL Jr. 1994. Continuation of warfarin-nafcillin interaction during dicloxacillin therapy. J Fam Pract 39:182-185.

142. Rodondi LC, Flahertyy JF, Schonfeld P, Barriere SL, Gambertoglio JG. 1989. Influence of coadministration on the pharmacokinetics of mezlocillin and cefotaxime in healthy volunteers and in patients with renal failure. Clin Pharmacol Ther 45:527-534.

143. Finch RA. 1981. Hypernatremia during lithium and ticarcillin therapy. South Med J 74:376-377.

144. Mackie K, Pavlin EG. 1990. Recurrent paralysis following piperacillin administration. Anesthesiology 72:561-563.

145. Smith DA, Chandler MHH, Shedlofsky SI, Wedlund PJ, Blouin RA. 1991. Age-dependent stereoselective increase in the oral clearance of hexobarbitone isomers caused by rifampicin. Br J Clin Pharmacol 32:735-739.

146. Schenkel EJ. 1995. Severe exacerbation of asthma secondary to rifampin in a steroid-dependent asthmatic. J Allergy Clin Immunol 95:314.

147. Kim YH, Cha IJ, Shim JC, Shin JG, Yoon YR, Kim YK, Kim JI, Park GH, Jang IJ, Woo JI, Shin SG. 1996. Effect of rifampin on the plasma concentration and the clinical effect of haloperidol concomitantly administered to schizophrenic patients. J Clin Psychopharmacol 16:247-252.

148. Askgaard DS, Wilcke T, Dossing M. 1995. Hepatotoxicity caused by the combined action of isoniazid and rifampicin. Thorax 50:213-214.

149. Bhatia RS, Uppal R, Malhi R, Behera D, Jundal SK. 1991. Drug interaction between rifampicin and cotrimoxazole in patients with tuberculosis. Hum Exp Toxicol 10:419-421.

150. Csogor SI, Papp I. 1970. Competition between sulphonamides and thiopental for the binding sites of plasma proteins. Arzneimittelforschung 20:1925-1927.

151. Jones DK, Hakim M; Wallwork J, Higenbottam TW, White DJ. 1986. Serious interaction between cyclosporin A and sulphadimidine. Br Med J 292:728-729.

152. Johnson JF, Dobmeier ME. 1990. Symptomatic hypoglycemia secondary to a glipizide-trimetroprim/sulphamethoxazole drug interaction. DICP 24:250-251.

153. Hansen JM, Kampmann JP, Siersbaek-Nielsen K, Lumholtz IB, Arroe M, Abildgaard U, Skovsted L. 1979. The effect of different sulfonamides on phenytoin metabolism in man. Acta Med Scand 624:106-110.

154. Lindebaum J, Rund DG, Butler VP Jr, Tse-Eng D, Saha JR. 1981. Inactivation of digoxin by the gut flora: reversal by antibiotic therapy. N Engl J Med 305:789-794.

155. Malt U. 1978. Lithium carbonate and tetracycline interaction. Br Med J 2:502.

156. Neuvonen PJ, Penttila O, Lehtovaara R, Aho K. 1975. Effect of antiepileptic drugs on the elimination of various tetracycline derivatives. Eur J Clin Pharmacol 9:147-154.

157. McCormack JP, Reid SE, Lawson LM. 1990. Theophylline toxicity induced by tetracycline. Clin Pharm 9:546-549.

158. Andersson KE, Bratt L, Dencker H, Kamme C, Lanner E. 1976. Inhibition of tetracycline absorption by zinc. Eur J Clin Pharmacol 10:59.

159. Bailey RR. 1984. Leukopenia due to a trimethoprim-azathioprine interaction. N Z Med J 97:739.

160. Ringden O, Myrenfors P, Klintmalm G, Tyden G, Ost L. 1984. Nephrotoxicity by co-trimoxazole and cyclosporin in transplanted patients. Lancet 1:1016-1017.

161. Lee BL, Medina I, Benowitz NL, Jacob P 3rd, Wofsy CB, Mills J. 1989. Dapsone, trimethoprim, and sulphamethoxazole plasma levels during treatment of Pneumocystis pneumonia in patients with the acquired immunodeficiency syndrome (AIDS). Ann Intern Med 110:606-611.

162. Petersen P, Kastrup J, Bartram R, Molholm Hansen J. 1985. Digoxin-trimethoprim interaction. Acta Med Scand 217:423-427.

163. Beattie B, Biller J, Mehlhaus B, Murray M. 1988. Verapamil-induced carbamazepine neurotoxicity; a report of two cases. Eur J Neurol 28:104-105.

164. Jaffe R, Livshits T, Bursztyn M. 1994. Adverse interaction between clonidine and verapamil. Ann Pharmacother 28:881-883.

165. Hedman A, Angelin B, Arvidsson A, Beck O, Dahlqvist R, Nilsson B, Olsson M, Schenck-Gustafsson K. 1991. Digoxin-verapamil interaction: reduction of biliary but not renal digoxin clearance in humans. Clin Pharmacol Ther 49:256-262.

166. Fleisher D, Burgunda SW, Ameeta P. 2004. Drug absorption with food. In: Handbook of Drug-Nutrient Interactions, pp 129-154.

167. Kane Gc, Lipsky JJ. 2000. Drug-grapefruit juice interactions. Mayo Clinic Proceedings 75:933-942.

168. Bailey DG. 2004. Grapefruit juice-drug interactions issues. In: Handbook of Drug-Nutrient Interactions, pp 175-194.

169. Rollins CJ. 2004. Drug-nutrient interactions in patients receiving enteral nutrition. In: Handbook of Drug-Nutrient Interactions, pp 515-552.

170. Btaiche IF, Kraft MD. 2004. Nutrients that may optimize drug effects. In: Handbook of Drug-Nutrient Interactions, pp 195-216.

171. Kater RM, Roggin G, Tobon F, Zieve P, Iber FL. 1969. Increased rate of clearance of drugs from the circulation of alcoholics. Am J Med Sci 258:35-39.

172. Pham NT, Weibert RT. 1995. Warfarin-alcohol interaction. ASHP Midyear Clinical Meeting 30:P-101(E).

173. Misra PS, Lefevre A, Ishii H, Rubin E, Lieber CS. 1971. Increase of ethanol, meprobamate and phenobarbital metabolism after chronic ethanol administration in man and in rats. Am J Med 51: 346-351.

174. Uemura K, Komura S. 1995. Death caused by triazolam and ethanol intoxication. Am J Forensic Med Pathol 16:66-68.

175. Paul MD, Prfrey PS, Smart M, Gault H. The effect of ethanol on serum cyclosporin A levels in renal transplant recipients. Am J Kidney Dis 10:133-135.

176. Pentikainen PJ. 1994. Acute alcohol intake increases the bioavailability of felodipine. Clin Pharmacol Ther 55:141-149.

177. Kopanoff DE, Snider DE Jr, Caras GJ. 1978. Isoniazid-related hepatitis. Am Rev Respir Dis 117:991-1001.

178. Pai SH, Werthamer S, Zak FG. 1973. Severe liver damage caused by treatment of psoriasis with methotrexate. NY State J Med 73:2585-2587.

179. Deykin D, Janson P, McMahon L. 1982. Ethanol potentiation of aspirin-induced prolongation of the bleeding time. N Engl J Med 306:852-854.

180. Linnoila M, Seppala T, Mattila MJ. 1974. Acute effect of of antipyretic analgesics, alone or in combination with alcohol, on human psychomotor skills related to driving. Br J Clin Pharmacol 1:477-484.

181. Margolese HC, Malchy L, Negrete JC, Tempier R, Gill K. 2004. Drug and alcohol use among patients with schizophrenia and related psychoses: levels and consequences. Schizoprenia Research 67:157-166.

182. Tanaka E. 2003. Toxicological interactions involving psychiatric drugs and alcohol: an update. J Clin Pharm Ther 28:81-95.

183. Wayne JA. 2004. Alcohol and drug interactions. In: Handbook of Drug Interactions, pp 395-462.

184. Schein JR. 1995. Cigarette smoking and clinically significant drug interactions. Anns Pharmacother 29:1139-1148.

185. Jusko JW. 1979. Influence of cigarette smoking on drug metabolism in man. Drug Metab Rev 9:221-236.

186. Scavone JM, Joseph M, Greenblatt DJ, LeDuc BW, Blyden GT, Luna BG, Harmatz, Herold S. 1990. Differential effect of cigarette smoking on antipyrine oxidation and acetaminophen conjugation. Pharmacology 40:77-84.

187. Cassidenti DL, Vijod AG, Vijod MA, Stanczyk FZ, Lobo RA. 1990. Short-term effects of smoking on the pharmacokinetic profiles of micronized estradiol in postmenopausal women. Am J Obstet Gynecol 163:1953-1960.

188. Perel JM, Hurwic MJ, Kanzler MB. 1975. Pharmacodynamics of imipramine in depressed patients. Psychopharmacol Bull 11:16-18.

189. Fox K, Jonathan A, Williams H, Selwyn A. 1980. Interaction between cigarettes and propranolol in treatment of angina pectoris. Br Med J 281:191-193.

190. Garg SK, Ravi Kiran TN. 1982. Effect of smoking on phenylbutazone disposition. Int J Clin Pharmacol Ther Toxicol 20:289-290.

191. Grech-Belanger O, Gilbert M, Turgeon J, LeBlanc PP. 1985. Effect of cigarette smoking on mexiletine kinetics. Clin Pharmacol Ther 37:638-643.

192. Benowitz NL, Peng M, Jacob P. 2003. Effects of cigarette smoking and carbon monoxide on chlorzoxazone and caffeine metabolism. Clin Pharmacol 74:468-474.

193. Shoshana Z, Benowitz NL. 1999. Drug interactions with tobacco smoking: an update. Clin Pharmacokin 36:425-438.

194. Ritter JM. 1999. Adverse Drug Reactions. In: Ritter JM, Lewis LD, Mant TGK,editors. A Textbook of Clinical Pharmacology.NewYork:Oxford University Press Inc.,pp 84-96.

195. Gordon GG, Skett p. 1994. Toxicological aspects of xenobiotic metabolism. In: Introduction to drug metabolism. London: Backie Academic & Professional, an imprint of Chapmann & Hall, pp 166-176.

196. Royer RJ. 2000. Side effects of drugs and pharmacovigilance. In: Sirtori CR, Kuhlmann J, Tillement J-P, Vrhovac B, Reidenberd M, editors. Clinical Pharmacology. London: McGraw-Hill International (UK) Ltd., pp 65-74.

197. Bates DW, Leape L. 2000. Adverse drug reactions. In: Carruthers SG, Hoffman BB, Melmon KL, Nierenberg DW, editors. Melmon and Morrelli's Clinical Pharmacology. Basic Principles in Therapeutics, 4th ed., New York: McGraw-Hill Medical Publishing Division, pp 1223-1256.

198. Jeffrey EH, Mannering GJ. 1983. Interaction of constitutive and phenobarbital-induced cytochrome P-450 isozymes during the sequential oxidation of benzphetamine. Mol Pharmacol 23:748-757.

199. Lindeke B, Paulsen-Sorman U. 1988. Nitrogenous compounds as ligands to hemoproteins – the concept of metabolic-intermediary complexes. In: Cho AK, Lindeke B, editors. Biotransformation of Organic Nitrogen Compounds.Karger,Basel,pp 63-102.

200. Lindeke B. 1982. The non-and postenzymatic chemistry of N-oxygenated molecules. Drug Metab Rev 13:71-121.

201. Mansuy D, Beaune P, Cresteil T, Bacot C, Chottard JC, Gans P.1978. Formation of complexes between microsomal cytochrome P-450-Fe(II) and nitrosoarenes obtained by oxidation of arylhydroxylamines or reduction of nitroarenes *in situ*. Eur J Biochem 86:573-579.

202. Ford GP, Herman PS. 1992. Relative stabilities of nitrenium ions derived from polycyclic aromatic amines. Relationships to mutagenity. Chem-Biol Interact 81:1-18.

203. Streeter AJ, Hoener BA. 1988. Evidence for the involvement of a nitrenium ion in the covalent binding of nitrofurazone to DNA. Pharm Res 5:434-436.

204. Hartman GD, Schlegel HB. 1981. The relationship of the carcinogenic/mutagenic potential of arylamines to their singlet-triplet nitrenium ion energies. Chem-Biol Interact 36:319-330.

205. Kerdar RS, Dehner D, Wild D. 1993. Reactivity and genotoxicity of arylnitrenium ions in bacterial and mammalian cells. Toxicol Lett 67:73-85.

206. Clement B, Kunze T. 1990. Hepatic microsomal N-hydroxylation of adenine to 6-hydrxyaminopurine. Biochem Pharmacol 39:925-933.

207. Ford GP, Griffin GR. 1992. Relative stabilities of nitrenium ions derived from heterocyclic amine food carcinogens. Relationship to mutagenity. Chem-Biol Interact 81:19-33.

208. Dipple A, Michejda CJ, Weisburger EK. 1985. Metabolism of chemical carcinogens. Pharmacol Ther 27:265-296.

209. Quon CY. 1988. Nitrogen oxidation in carcinogenesis. In: Cho AK, Lindeke B, editors. Biotransformation of Organic Nitrogen Compounds. Karger, Basel, pp 132-160.

210. Borch RF, Millard JA. 1987. The mechanism of activation of 4-hydroxy-cyclophosphamide. J Med Chem 30:427-431.

211. Hong PS, Chan KK. 1987. Identification and quantitation of alcophosphamide, a metabolite of cyclophosphamide, in the rat using CI / MS, Biomed Environ Mass Spectrom 14:167-172.

212. Lindeke B, Cho AK. 1982. N-Dealkylation and deamination. In: Jacoby WB, Bend JR, Caldwell J, editors. Metabolic Basis of Detoxication. New York: Academic Press, pp 105-126.

213. Heur YH, Streeter AJ, Nims RW, Keefer LK. 1989. The Fenton degradation as a nonenzymatic model for microsomal denitrosation of N-nitrosodimethylamine. Chem Res Toxicol 2:247-253.

214. Decker CJ, Rashed MS, Baillie TA, Maltby D, Correia MA. 1989. Oxidative metabolism of spironolactone: evidence for the involvement of electrophilic thiosteroid species in drug-mediated destruction of rat hepatic cytochrome P450. Biochemistry 28:5128-5136.

215. Bolton JL, Le Blanc JCY, Siu KWM. 1993. Reactions of quinone methides with proteins: analysis of myoglobin adduct formation by electrospray mass spectrometry. Biol Mass Spectrom 22:666-668.

216. Vamvakas S, Dekant W, Anders MW. 1989. Mutagenicity of benzyl S-haloalkyl and S-haloalkenyl sulfides in the Ames-test. Biochem Pharmacol 38:935-939.

217. Dekant W, Lash LH, Anders MW. 1988. Fate of glutathione conjugates and bioactivation of cysteine S-conjugates by cysteine conjugate β-lyase. In: Sies H,

Ketterer B, editors. Glutathione Conjugation. Mechanisms and Biological Significance. London: Academic Press, pp 415-447.

218. Blagbrough IS, Buckberry LD, Bycroft BW, Shaw PN. 1992. Structure-activity relationship studies of bovine C-S lyase enzymes. Pharm Pharmacol Lett 1:93-96.

219. Sabourin PJ, Hodgson E. 1984. Characterization of the purified microsomal FAD-containing monooxygenase from mouse and pig liver. Chem-Biol Interact 51:125-139.

220. Doerge DR, Corbett MD. 1984. Hydroperoxy-flavin-mediated oxidations of organosulfur compounds. Model studies for the flavin monooxygenase. Molec Pharmacol 26:348-352.

221. Guengerich FP. 2005. Principles of covalent binding of reactive metabolites and examples of activation of bis-electrophiles by conjugation. Arch Biochem Biophys 433:369-378.

222. Ferrari A, Coccia CPR, Bertolini A, Sternieri E. 2004. Methadone – metabolism, pharmacokinetics and interactions. Pharmacol Res 50:551-559.

223. Gonzales FJ. 2005. Role of cythocromes P450 in chemical toxicity and oxidative stress: studies with CYP2E1. Mutat Res - Envir Muta 569:101-110.

224. Furst SM, Uetrecht JP. 1993. Carbamazepine metabolism to a reactive intermediate by the myeloperoxidase system of activated neutrophils.Biochem Pharmacol 45:1267-1275.

225. Reid JL, Rubin PC, Whiting B. 1989. In: Lecture Notes on Clinical Pharmacology, 3rd ed. London: Blackwell Scientific Publications, p 179.

226. Seligmann H, Podoshin L, Ben-David J, Fradis M; Goldsher M. 1996. Drug-induced tinnitus and other hearing disorders. Drug Saf 14:198-212.

227. Koegel L. 1985. Ototoxicity: a contemporary review of aminoglycosides, loop diuretics, acetylsalicylic acid, quinine, erythromycin, and cisplatin. Am J Otol 6:190-199.

228. Lien EJ, Lipsett LR, Lien LL. 1983. Structure side effect sorting of drugs: VI. Ototoxicities. J Clin Hosp Pharm 8:15-33.

229. Rybak LP. 1986. Drug ototoxicity. Annu Rev Pharmacol Toxicol 26:79-99.

230. Rybak LP. 1993. Ototoxicity of loop diuretics. Otolaryngol Clin North Am 26:829-844.

231. Abel R, Leopold IH. 1987. Drug-induced ocular diseases. In: Speight TM, editor. Avery's Drug Treatment, 3rd ed. Hong Kong: Adis Press, pp 409-417.

232. Schichi H, Nebert DW. 1982. Genetic differences in drug metabolism associated with ocular toxicity. Environ Health Perspect 44:107-117.

233. Whelton A, Hamilton CW. 1991. Non-steroidal anti-inflammatory drugs: effects on kidney function. J Clin Pharmacol 31:588-598.

234. Adler SG, Cohen AH, Border WA. 1986. Hypersensitivity phenomena and the kidney: role of drugs and environmental agents. Am J Kidney Dis 5:75-96.

235. Ali MZ, Goetz MB. 1997. A meta-analysis of the relative efficacy and toxicity of a single daily dosing versus multiple daily dosing of aminoglycosides. Clin Infect Dis 24:796-809.

236. De Marie S. 1996. Liposomal and lipid-based formulations of amphotericin B. Leukemia 10:S93-96.

237. Cole E, Keown P, Landsberg D. 1998. Safety and tolerability of cyclosporin and cyclosporin microemulsion during 18 months of follow-up in stable renal transplant recipients: a report of the Canadian Neoral Renal study Group. Transplantation 65:505-510.

238. Verstraete M, Boogaerts MA. 1987. Drug-induced haematological disorders. In: Speight T, editor. Avery's Drug Treatment, 3rded. Hong Kong: Adis Press, pp 1007-1022.

239. Farrell GC. 1997. Drug-induced hepatic injury. J Gastroenterol Hepatol 12:S242-250.

240. Sheikh NM, Philen RM, Love LA. 1997. Chaparral-associated hepato-toxicity. Ann Intern Med 157:913-919.

241. Cooper JAD Jr, White DA, Matthay RA. 1986. Drug-induced pulmonary disease. Am Rev Respir Dis 133:321-340.

242. Loebstein R, Lalkin A, Koren G. 1997. Pharmacokinetic changes during pregnancy and their clinical relevance. Clin Pharmacokinet 33:328-343.

243. Spielberg SP. 1982. Pharmacokinetics and the fetus. N Engl J Med 307:115-116.

244. Martz F, Failinger C, Blake DA. 1977. Phenytoin teratogenesis: Correlation between embryopathic effect and covalent binding of putative arene oxide metabolite in gestational tissue. J Pharmacol Exp Ther 203:231-239.

245. Loebstein R. Vohra S, Koren G. 2000. Drug theraphy in pregnant and breast-feeding women. In: Carruthers SG, Hoffman BB, Melmon KL, Nierenberg DW, editors. Clinical Pharmacology. Basic Principles in Therapeutics, 4thed. New York: Mc Graw-Hill (Medical Publishing Devision), pp 1117-1142.

246. Morselli PL. 1976. Clinical pharmacokinetics in neonates. Clin Pharmacokinet 1:81-98.

247. Loebstein R. Vohra S, Koren G. 2000. Drug theraphy in pediatric patients. In: Carruthers SG, Hoffman BB, Melmon KL, Nierenberg DW, editors. Clinical Pharmacology. Basic Principles in Therapeutics, 4thed. New York: Mc Graw-Hill (Medical Publishing Devision), pp 1145.

248. Loebstein R. Vohra S, Koren G. 2000. Drug theraphy in pediatric patients. In: Carruthers SG, Hoffman BB, Melmon KL, Nierenberg DW, editors. Clinical Pharmacology. Basic Principles in Therapeutics, 4thed. New York: Mc Graw-Hill (Medical Publishing Devision), pp 1147-1149.

249. Bhanthumnavin K, Schuster MM. 1077. Aging and gastrointestinal function. In: Finch CE, Hayflick L, editors. Handbook of the Biology of Aging. New York: Van Nostrand Reinhold, pp 709-723.

250. Shock NW, Watkin DM, Yiengst MJ, Norris AH, Gaffney GW. Gregerman RIO, Falzone JA. 1963. Age differences in the water content of the body as related to basal oxygen consumption in males. J Gerontol 18:1-8.

251. Homer TD, Stansky DR. 1985. The effects of increasing age on thiopental disposition and anesthetic requirement. Anesthesiology 62:714-724.

252. Woodhouse KW, Wynne HA. 1988. Age-related changes in liver size and hepatic flow: The influence of drug metabolism in the elderly. Clin Pharmacokinet 15:287-294.

253. Vestal RE, Gurwitz JH. 2000. Geriatric pharmacology. In: Carruthers SG, Hoffman BB, Melmon KL, Nierenberg DW, editors. Melmon and Morrelli's Clinical Pharmacology. Basic Principles in Therapeutics, 4th ed., New York: McGraw-Hill Medical Publishing Division, pp 1151-1177.

254. Turnheim K. 2003. When drug therapy gets old: pharmacokinetics and pharmacodynamics in the elderly. Exp Gerontol 38:843-853.

255. Turnheim K. 2004. Drug therapy in the elderly. Exp Gerontol 39:1731-1738.

256. Cusack BJ, Vestal RE. 1986. Clinical pharmacology: Special considerations in the elderly. In: Calkins E, Davis PJ, Ford AB, editors. Practice of Geriatric Medicine. Philadelphia: WB Saunders, pp 115-134.

257. Delafuente JC. 2003. Understanding and preventing drug interactions in elderly patients. Crit Rev Oncol Hemat 48:133-143.

258. Wong DD.1985. Phenytoin-dexamethasone: a possible drug-drug interaction. JAMA, 254:2062-2066.

259. Sugita O, Sawada Y, Sugiyama Y, Iga T, Hanano M. 1981. Kinetic analysis of tolbutamide-sulfonamide interaction in rabbits based on clearance concept: Prediction of species difference from in vitro plasma protein binding and metabolism. Drug Metab Dispos 12:131-138.

260. Honig PK, Wortham DC, Zamani K, Conner DP, Mullin JC, Cantilena LR. 1993. Terfenadine-ketoconazole interaction. JAMA 269:1513:1518.

261. Stille W, Harder S, Mieke S, Beer C, Shah PM, Staib AH. 1987 Decrease of caffeine elimination in man during co-administration of 4-quinolones. J Antimicrob Chemother 20:729-734.

262. Vereerstraeten P, Thiry P, Kinnaert P, Toussaint C. 1987. Influence of erythromycine of cyclosporin pharamacokinetics. Transplantation 44:155-156.

263. Siverman RB. 1988. Mechanism-base enzyme inactivation, Chem Enzimol, 1:3-30.

264. Okuda H, Nishiyama T, ogura K, Nagayama S, Ikea K, Yamaguchi S, NakamuraY, Kawaguchi Y, Watabe T. 1995. Mechanism of lethal toxicity exerted by simultaneous administration of the new antiviral, sorivudine, and the antitumor agent Tegafur. Xenobio Metabol Dispos 10:166-169.

265. Periti P, Mazzei T, Mini E, Novelli A. 1992. Pharmacokinetic drug interactions of macrolides. Clin Pharmacokinet 23:106-131.

266. Murray M, Reiy GF. 1990. Selectivity in the inhibition of mammalian cytochromes P450 by chemical agents. Pharmacol Rev 42:85-101.

267. Kunze LF, Trager WF. 1993. Isoform-selective mechanism-based inhibition of human cytochrome P450 1A2 by furafylline. Chem Res Toxicol 6:649-656.

268. Thummel KE, O'Shea , Paine MF, Shen DD, Kunze KL, Perkins JD, Wilkinson GR. 1996. Oral first-pass elimination of midazolam involves both gastrointestinal and hepatic CYP3A-mediated metabolism. Clin Pharmacol Ther 59:491-502.

269. Hebert FM, Roberts JP, Prueksaritanont T, Benet LZ. 1992. Bioavailability of cyclosporin with concomitant rifampicin administration is markedly less than predicted by hepatic enxyme induction. Clin Pharmacol Ther 52:453-457.

270. Gomez DY, Wacher VJ, Tomlanovich Sj, Hebert MF, Benet LZ. 1995. The effects of ketoconazole on the intestinal metabolism and bioavailability of cyclosporin. Clin Pharmacol Ther 58:15-19.

271. Davis AM, Riley RJ. 2004. Predictive ADMET studies, the challenges and the opportunities. Curr Opin Chem Biol 8:378-386.

272. Bugrim A, Nikolskaya T, Nikolsky Y. 2004. Early predictions of drug metabolism and toxicity: systems biology approach and modelling. Drug Discov To, 9:127-135.

273. Rostami-Hodjegan A, Tucker G. 2004. '*In silico*' simulations to assess the '*in vivo*'consequences of '*in vitro*' metabolic drug-drug interactions. Drug Discov To: Tech 1:441-448.

Chapter 9

STRATEGIES FOR DRUG DESIGN

9.1 INTRODUCTION

The previous chapters in this book dealt with the major aspects of the metabolism of drugs as well as some basic pharmacokinetic principles. This chapter, aimed primarily at the aspiring medicinal chemist, attempts to illustrate how considerations of pharmacokinetics and metabolism serve as invaluable input to the process of drug design and optimisation of drug *in vivo* activity. There is a vast literature on this subject and the treatment below is necessarily selective. However, the aim here is to highlight the main principles and popular strategies that are applied to overcome pharmacokinetic problems and to use metabolism to advantage in the discovery and development of drugs, as well as indicate useful recent literature sources for further study.

9.2 PHARMACOKINETICS AND METABOLISM IN DRUG RESEARCH

9.2.1 General overview

It is now widely accepted that while structure-activity relationships (SAR) have an important place in drug discovery and design, in particular to identify ligands with optimum affinities for their receptors, the most effective way to increase the therapeutic index of a new drug candidate intended for a specific application is to complement SAR-based approaches with additional data on its metabolites, its pharmacodynamic and pharmacokinetic properties and toxicological implications [1]. In other words, optimisation of *in vitro* activity through the employment of SAR-guided synthesis alone is no assurance of favourable *in vivo* activity, since the latter is subject to pharmacokinetics and metabolism that determine e.g. the drug bioavailability, duration of action, biotransformation into active/inactive/toxic metabolites, and so on [2].

An earlier survey [3] indicated that some 40% of a sample of ~300 new drug candidates investigated in humans were subsequently withdrawn due to serious shortcomings in their pharmacokinetics, as reflected in e.g. poor oral absorption, extensive first-pass metabolism, unfavourable distribution or clearance, or a combination of these. This emphasises the need for understanding the principal factors affecting pharmacokinetics *viz.* drug lipophilicity and solubility (see Chapter 1). These properties can be manipulated by chemical modification of the active compound or *via* formulation approaches so as to overcome the above problems, ideally without compromising the intrinsic pharmacological activity of the pharmacophore.

From a historical perspective, the rational use of metabolism input to the drug discovery process is a relatively recent innovation [4]. Frequently in the past, such information has mainly been used to explain the failure of a molecule to achieve its expected performance. During the last two decades however, the explosive growth of knowledge in the area of drug-metabolising enzymes coupled with technological advances in analytical instrumentation has allowed medicinal chemists to acquire valuable information on the metabolic fates of new drug candidates at an early stage of their development [2]. In addition, as shown in Chapters 1-3, based on a wealth of accumulated data, rules exist for predicting both the pharmacokinetic behaviour of a compound as well as its likely major routes of metabolism from a knowledge of its molecular structure and physicochemical properties [4]. During the last decade, there has been a growing emphasis on rapid metabolism assessment in the discovery phase [5] and numerous *in silico* tools have been developed to predict the metabolic properties of candidate drugs, e.g. their metabolic stability, likely sites of metabolism and ensuing metabolites, rates of metabolism, drug-drug interactions, clearance and toxicology. The status of such computational models has recently been reviewed [6,7]. Exploitation of the existing knowledge bases and responsible use of computerised resources can aid the medicinal chemist in optimising drug *in vivo* activity.

As is evident from earlier chapters, nature has evolved a formidable array of metabolic mechanisms to handle both endogenous and xenobiotic substances in humans. One feature of the metabolism of xenobiotics is the prevalence of oxidative processes, which may not only detoxify them, but also generate toxic, reactive intermediates such as epoxides and radicals. Mention has been made earlier of the possible negative consequences that can ensue from reaction of such intermediates with endogenous macromolecules. Therefore, as regards drug design, one principle that serves as a guideline is that oxidative pathways for the biotransformation of candidate drugs should generally be avoided. One way to achieve this is to rely on inactivation of designed drugs *via* hydrolytic mechanisms such as those effected by esterases that are widespread in the body.

In summary, consideration of pharmacokinetic and metabolic factors indicates that, in principle: (a) rational synthetic modifications can be made to a drug candidate to ensure its favourable absorption, distribution and clearance; (b) at the same time, appropriate functional groups, or other moieties such as carrier groups that undergo predictable metabolism can be attached to the pharmacophore to direct the specific routes of activation or deactivation as needed [8]. These guiding principles should lead to the development of drugs with high therapeutic indices.

An elegant and current application of (b) above is represented by metabolism-based drug-targeting, whereby advantage is taken of the prevalence of specific, known enzymes in an organ, body compartment or diseased tissue, to design a molecule that is metabolised only at that site, where it subsequently releases the active drug. Such an approach (see section 9.2.5 below) has in recent years led to the development of safer, more effective drugs with site-specificity and hence displaying fewer side effects.

In the sections that follow, an attempt is made to describe some of the main approaches to chemical modification that may result in improved drug pharmacokinetics, favourable metabolism or both. In addition, some developments in drug targeting are described, where they might depend on predictable metabolism. In each case, the design concept is briefly explained and illustrated with one or more pertinent examples. In keeping with the title of this book, we have attempted to include primarily recent case studies, though for didactic purposes reference is occasionally made to older examples in the literature.

Subsections 9.2.2-9.2.4 focus on chemical manipulation of drugs, highlighting the rationale behind the discovery of prodrugs, hard drugs and soft drugs respectively, with examples. Strategies based on more sophisticated 'chemical delivery systems' that deliberately include drug-targeting as their goal are described in section 9.2.5. For completeness, section 9.3 includes a brief interlude on aspects of formulation approaches that are mainly aimed at improving oral absorption of poorly soluble drugs. Some of these approaches involve physical modification of new chemical entities and may not rely on the creation of covalent bonds between the active drug and the matrix or carrier moiety. Nevertheless, the authors believe that medicinal chemists should be aware that alternatives to synthesis may sometimes be the route to meeting their objectives.

Finally, in section 9.4 we underscore the crucial roles of pharmacokinetics and metabolism in drug design in the hope that their consideration by aspiring medicinal chemists will result in the development of safer and more effective drugs in the future.

9.2.2 The prodrug approach

According to the original definition by Albert [9], a prodrug is a chemical with little or no pharmacological activity that undergoes biotransformation to the therapeutically active metabolite. Actually, the 'activation' of the prodrug i.e. its conversion to the pharmacologically active form, may proceed under enzyme control, by non-enzymatic reaction, or by each of these in sequence. In general, the intention of the prodrug approach is to improve the efficacy of an established drug [10].

Prodrugs are developed to address numerous shortcomings, but probably most frequently to improve oral bioavailability, either by enhancing oral absorption or by reducing pre-systemic metabolism. As indicated in Chapter 1, transport of a drug through membranes, and hence its absorption, depends critically on the balance between the drug's aqueous solubility and its lipophilicity. Optimisation of this balance is often achieved by attaching a 'carrier moiety' to a polar group such as an acidic, alcoholic, phenolic or amino-function of the active species to yield the prodrug, which should then undergo predictable metabolism to release the active form in the body. Chemical derivatisation used to improve lipophilicity often involves conversion of acidic, phenolic and alcoholic functions into appropriate esters that are metabolised to the corresponding active drugs by esterases, which are ubiquitous. Aldehydes and ketones may be converted into acetals, and amines into quaternary ammonium species, amino acid peptides and imines [11].

More recent developments involving prodrugs relate to their activation in two-step targeting therapies such as ADEPT (antibody-directed enzyme prodrug therapy) and GDEPT (gene-directed enzyme prodrug therapy). These approaches hold particular promise in the area of cancer treatment through selective liberation of anticancer drugs at the surface of tumour cells. In contrast to the prodrugs described above, which are developed primarily to overcome pharmacokinetic problems, those used in ADEPT and GDEPT therapies are associated with site-specific drug delivery. Prodrug activation again relies on specific enzymes but in this case these are 'pre-delivered' to the desired sites of action [12].

Several examples of relatively simple prodrugs are described first in this section. This is followed by a description of more complex systems that utilise prodrugs with the specific intention of site-specific delivery.

A recent example of a successful prodrug is ximelagatran (Figure 9.1), which upon absorption is converted into its active metabolite melagatran [13], a potent competitive inhibitor of human α-thrombin.

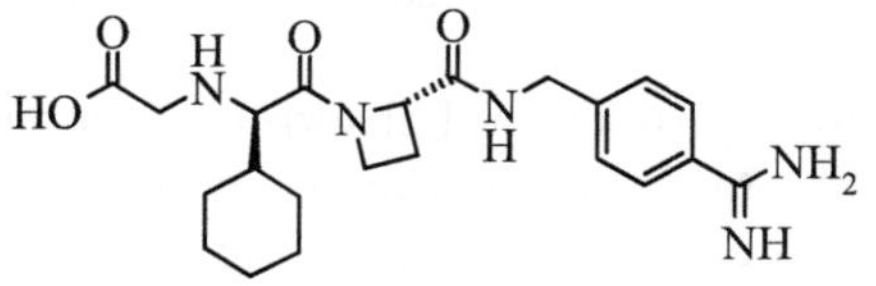

Melagatran

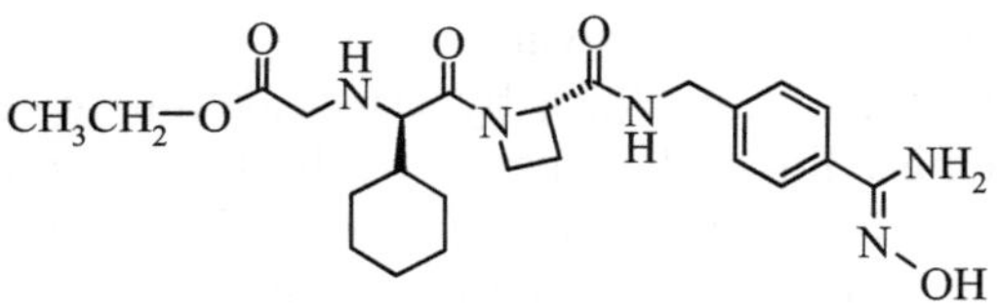

Ximelagatran

Fig.9.1 The active drug melagatran and its prodrug ximelagatran [13,14]

Melagatran was developed in the search for a new generation of oral anticoagulants with more predictable pharmacokinetic and pharmacodynamic properties than those of drugs in previous use, such as dicoumarol and warfarin. But despite having the necessary pharmacodynamic properties of a new antithrombotic agent, the oral bioavailability of melagatran was found to be only ~5%, which precluded its oral administration. This led to the development of its prodrug ximelagatran, produced by ethylation of the –COOH group and hydroxylation of the amidine group of the active compound. Poor bioavailability was attributed to the strong basic amidine functionality, originally selected in the design phase to fit the arginine side-pocket of thrombin [14]. Hence this was replaced by the less basic N-hydroxylated amidine. In addition, an ethyl ester protecting group was introduced. Biotransformation of the prodrug to melagatran, involving ester cleavage and reduction of the amidoxime function, was demonstrated *in vitro* using microsomes and mitochondria from liver and kidney of pig and human.

These chemical modifications resulted in significant reduction in the hydrophilicity of the active molecule, the prodrug having an apparent *in vitro* permeability coefficient around 80-fold higher than that of melagatran. Following oral administration of ximelagatran, it is thus rapidly absorbed and converted to melagatran which has a bioavailability of ~20% i.e. significantly greater than that following oral administration of the active drug. The successful performance of the prodrug ximelagatran led to its full-scale clinical evaluation in 2003. Further details of the pharmacodynamics, pharmacokinetics and metabolism of ximelagatran and melagatran have been published [13,14].

A similar approach to improving oral absorption has been applied for some time to many antibiotics. Ampicillin is poorly absorbed when administered orally. Large doses are thus required to achieve the necessary therapeutic level, leading to toxic effects in the GI tract. The prodrug strategy was employed to develop derivatives such as bacampicillin, pivampicillin and talampicillin (Figure 9.2), by esterification of the polar carboxylate group to yield these lipophilic, enzymatically labile prodrugs, all of which are metabolised to the active antibiotic ampicillin. Whereas the absoprtion of ampicillin is less than 50%, the above prodrugs are absorbed to the extent of 98-99% and continue to be widely used [2].

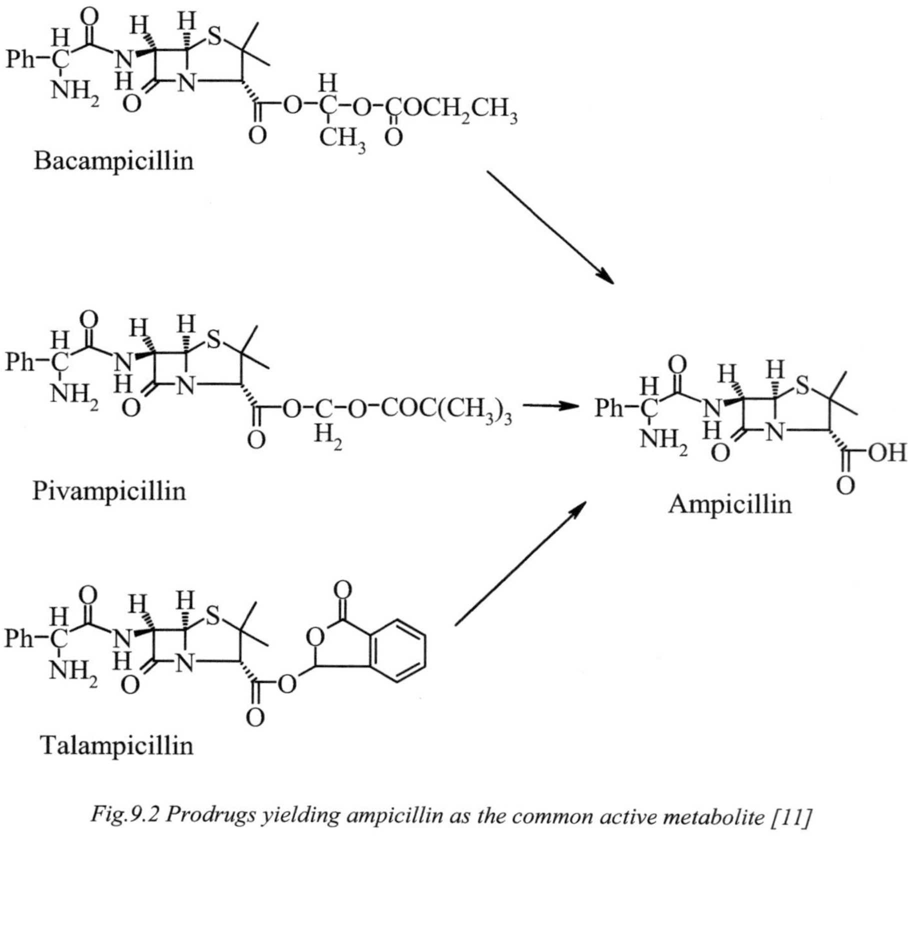

Fig.9.2 Prodrugs yielding ampicillin as the common active metabolite [11]

These compounds fall into the class of 'tripartate carrier-linked prodrugs' [10,11,15], characterised by the presence of three distinct moieties, namely the active drug, a linking structure and a carrier group. In the case of bacampicillin, these moieties are respectively the ampicillin 'core', -OCH(CH_3)O- and –COEt. In the first phase of metabolism of bacampicillin, the latter moieties are enzymatically hydrolysed to yield carbon dioxide and ethanol (Figure 9.3), and in the second phase, spontaneous loss of acetaldehyde from the resulting intermediate releases the active drug, ampicillin [11].

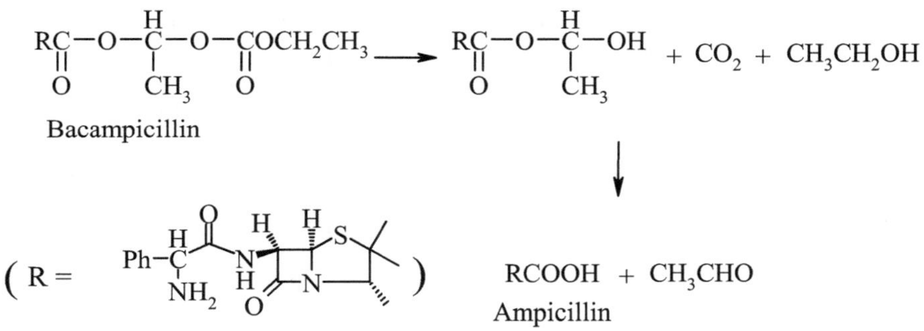

Fig.9.3 Metabolism of bacampicillin [11]

Analogous synthetic strategies have been employed in the development of prodrugs of biologically active phosphate esters, as detailed in a recent review [15]. Here, both improvement in bioavailability as well as some degree of site-specificity, reflected in elevation of the concentration of the active species within cells, have been achieved. At physiological pH in the range 7.0-7.4, phosphate esters, $O{=}P(OR)(O^-)_2$, are in the deprotonated state and therefore generally do not readily permeate cellular membranes. In order to increase bioavailability and cell permeability, various masking groups (MG) have been developed that convert the charged phosphate esters into neutral molecules, rendering the resulting prodrugs more permeable to cell membranes.

As shown schematically in Figure 9.4, following diffusion of the prodrugs through cellular membranes, the masking groups are removed by hydrolysis to yield the charged phosphate ester [15]. Since the active, charged species has thus been regenerated within the cell, it is effectively 'trapped' there, where it can carry out its medicinal function. Within the cell, conversion of the prodrug into the phosphate ester may occur either by chemical or enzymatic hydrolysis.

Though elegant as a design concept, this prodrug approach does nevertheless present significant synthetic challenges as regards the type of masking group to be employed to achieve appropriate stability for optimum chemical/enzymatic hydrolysis. In the case of anti-HIV nucleosides [16], successful derivatives have been based on the use of an unsymmetrical, cyclic bio-activatable protecting group [17] that undergoes a tandem reaction in its hydrolysis to yield the active drug.

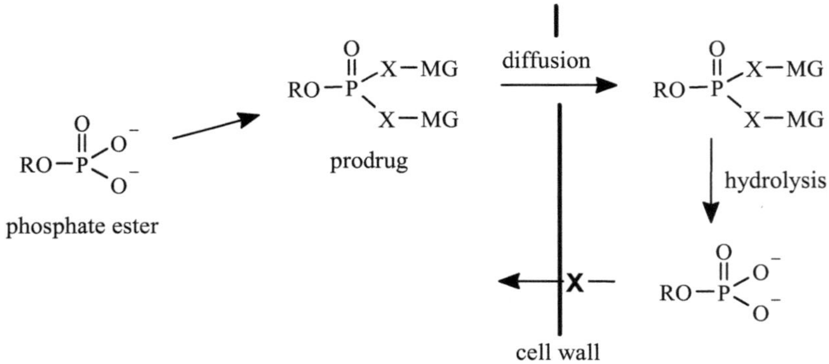

Fig.9.4 Prodrug concept for biologically active phosphate esters [15]

The same 'tripartate' principle employed in the development of ampicillin prodrugs described above has also been applied to phosphate esters [15]. Here, the protecting moiety attached to an oxygen atom of the active phosphate ester is of the form –X-Y. The residues X and Y are chosen such that enzymatic cleavage initially splits off the terminal group Y.

The resulting intermediate is unstable, the linking group X leaving spontaneously to yield finally the charged phosphate group. Acyloxyalkyl ester prodrugs of this type, for example, are hydrolysed by the enzyme carboxyesterase, yielding the active phosphate, an aldehyde, a carboxylate and two protons.

One concern relating to the use of prodrugs is the possible toxicity of the degradation products, in particular formaldehyde, which is associated with carcinogenicity. More recent work does, however, suggest that the human body can tolerate low levels of formaldehyde better then previously believed [18]. The medicinal chemist needs to be aware of the possibility of toxic metabolites or byproducts arising during the metabolism of the administered compound [19].

The reader is referred to the informative review [15] for further details on strategies used to synthesise analogous prodrugs including those of nucleotides and inositol phosphates. Interestingly, despite intensive efforts in the area of phosphate-containing prodrug candidates, clinical success has been very limited, and therefore this endeavour poses ongoing challenges to the medicinal chemist.

Another challenging area for the medicinal chemist is the delivery of peptides [9,20]. The two-step activation strategy referred to above for phosphate ester prodrugs has been employed in the case of peptide prodrugs, with the difference that, following the first step (enzymatically mediated hydrolysis of the ester), the second involves non-enzymatic intramolecular nucleophilic attack which cleaves the amide bond, releasing the peptide and a cyclic by-product (Figure 9.5a). An example of such a prodrug is shown in Figure 9.5b.

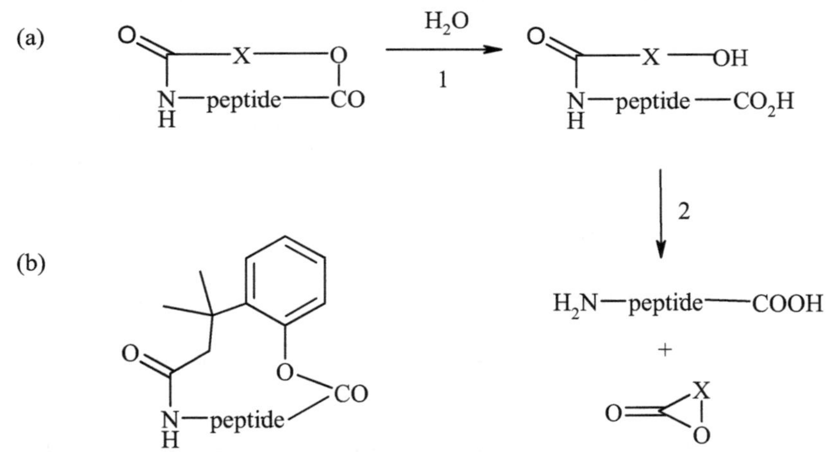

Fig.9.5 (a) Peptide prodrug undergoing two-step activation and (b) an example of a prodrug based on the design concept in (a) [9, and refs. therein]

In the metabolism of the model compound, step 1 unmasks a phenolic –OH group, which in step 2 is the nucleophilic centre that attacks the amide bond.

Other pharmacokinetic objectives that may be addressed by the prodrug approach include improvement in absorption *via* parenteral routes and extending the duration of drug action by slow metabolic release. An example of a prodrug fulfilling the second objective is bambuterol, derived from the active β_2-adrenergic agonist terbutaline [21]. Chemical modification in this case involved conversion of the two phenolic groups on the terbutaline molecule into their diethylcarbamato ($(CH_3)_2$-N-CO-) esters. Activation of the prodrug involves hydrolysis in blood serum (mediated by a cholinesterase) as well as oxidation involving monooxygenase present in various organs and tissues. The extended duration of action in this case, however, is novel, relying on prolonged inhibition of the cholinesterase by covalent attachment of the diethylcarbamato group of bambuterol. Consequently, bambuterol produces a more sustained bronchodilating effect than the parent terbutaline and can thus be administered less frequently.

Organ- or tissue-selective delivery is also possible *via* the prodrug approach, as exemplified recently by capecitabine, an orally active prodrug of the antineoplastic 5-fluorouracil [22]. The prodrug (Figure 9.6) is activated sequentially by carboxylesterase present in the liver, by cytidine deaminase in liver and tumour cells, and finally by thymidine phosphorylase present in tumours. The crucial feature of this metabolism that accounts for the successful 'targeting' performance of the prodrug is that the final step, release of 5-fluorouracil, occurs selectively within tumour cells.

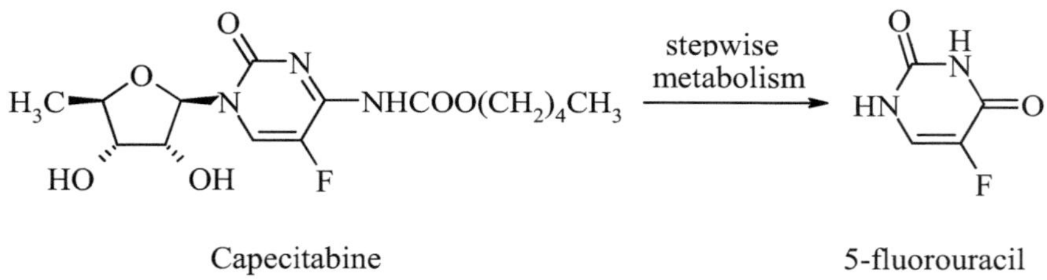

Fig.9.6 The prodrug capecitabine and its active metabolite 5-fluorouracil [22]

Another topical area in which prodrugs feature prominently relates to the development of anti-HIV nucleosides and their analogues. FDA-approved drugs in this class include zidovudine (AZT), didanosine, zalcitabine, stavudine and lamivudine. These compounds are effective inhibitors of HIV reverse transcriptase (HIV-RT) and are active as antiretrovirals in the form of their triphosphate metabolites, whose sequential formation is catalysed by a variety of cellular kinases. The cellular pharmacology, structure-activity relationships and pharmacokinetics of a number of anti-HIV nucleosides and their prodrugs have been the subject of a recent review [23]. The CNS and the lymphatic system act as reservoirs for HIV, but the nucleosides as such have limited penetration into these areas. Hence, one motivation for prodrug design is to overcome this delivery problem. One example of the documented prodrug design strategies is cited here.

Phospholipid prodrugs (Figure 9.7) have been designed for 2', 3'-dideoxynucleoside analogues such as AZT (denoted X in Figure 9.7), to improve their therapeutic profiles [23]. With such prodrugs, intracellular release of the nucleoside analogue or its monophosphate occurs, in the latter case bypassing the monophosphorylation step in the further metabolism to the active triphosphate.

Fig.9.7 Two types of phospholipid prodrugs for anti-HIV agents such as AZT [23]

We mention here that another very important type of prodrug for 2',3'-dideoxynucleoside analogues is based on the redox chemistry of dihydropyridine (DHP)-pyridinium salt interconversion, a strategy which has been employed elsewhere to design drugs that can cross the blood-brain barrier (BBB) [24,25]. However, since this type of system more appropriately falls under the classification of a chemical delivery system, a discussion of this strategy is reserved for section 9.2.5 where greater focus is given to drug targeting approaches.

Simple derivatisation *via* ester or amide formation does not guarantee a successful prodrug. A case in point is the poorly soluble antiviral acyclovir, which though possessing a hydroxyl group that could be chemically modified to improve absorption, has not yielded successful ester prodrugs [2]. Instead, the prodrug desoxyacyclovir (Figure 9.8) does result in superior oral delivery of acyclovir. Here, advantage is taken of oxidative metabolism by the enzyme xanthine oxidase, present in the gut and the liver, for biotransformation of the prodrug to the active acyclovir.

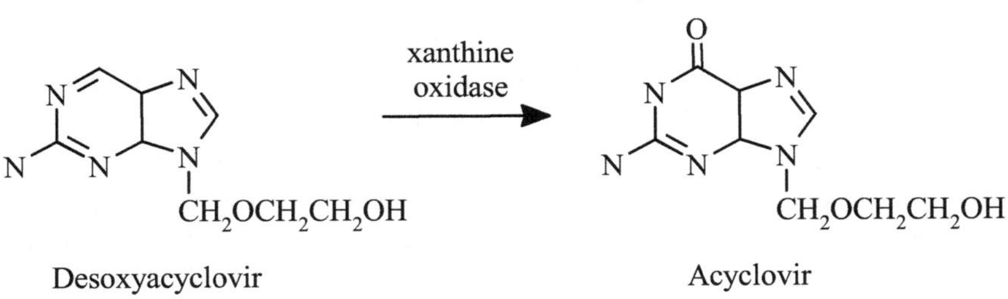

Fig.9.8 The first step in the metabolic activation of the prodrug desoxyacyclovir [2]

As with the nucleoside analogues described above, *in vivo* phosphorylation of acyclovir is necessary for the eventual activity to be manifested, which in this case is inhibition of the viral DNA polymerase.

A conceptually different category of prodrug that warrants discussion here is that used in novel therapies such as ADEPT (antibody-directed enzyme prodrug therapy) and GDEPT (gene-directed enzyme prodrug therapy), techniques that have developed through advances in molecular biology. This type of prodrug is non-toxic and instead of being activated by an endogenous enzyme, is metabolised in ADEPT to the active agent by an engineered enzyme-antibody conjugate that has been delivered site-specifically to a tumour cell in advance [26,27]. In the practice of ADEPT, for example, the monoclonal antibody-enzyme conjugate is administered intravenously, whereupon it localises in tumour cells. Several hours later, the anticancer prodrug is administered and is thus selectively activated at the tumour cells by the delivered enzyme. Normal cells, being devoid of the enzyme, are unaffected by this treatment. If only one prodrug is administered, the delivered enzyme must obviously catalyse a scission reaction for that specific compound. In general though, a desirable requirement of the enzyme should be versatility of catalytic action that may enable it to activate several anticancer agents. In GDEPT, a foreign gene encoding for the desired enzyme is delivered to the target tumour by means of a viral or liposomal vector. Following gene expression, the enzyme converts the non-toxic prodrug into the cytotoxic agent at the desired site.

Examples of enzymes that have recently been considered as suitable candidates for the activation of anticancer prodrugs in ADEPT and GDEPT approaches are β-galactosidase and β-glucuronidase [28]. The former is expressed in mammalian cells transduced with the *E. coli* LacZ gene. Figure 9.9 shows a designed prodrug 1 that has been shown to undergo conversion to the antitumour alkylating phosphoramide mustard 2 when incubated with *E. coli* β-galactosidase [29]. The first step in the proposed mechanism of activation involves enzymatic cleavage of the 4-β-D-galactopyranosyl unit to yield an intermediate phenol. The latter undergoes spontaneous 1, 6-elimination to release the cytotoxic mustard 2 as well as a quinone methide which spontaneously converts to 4-hydroxybenzylalcohol 3.

On the basis of these findings, prodrug 1 was considered to have good potential in conjunction with GDEPT to increase antitumour selectivity in cancer therapy. An analogous prodrug 4 (Figure 9.9) was designed with a β-D-glucuronic acid linked to a self-immolative spacer (in this case an N-(*o*-hydroxyphenyl)-N-carbamate) and a phenolic mustard 5 [30] as a candidate for use in conjunction with ADEPT. The presence of a spacer separating the drug and the enzyme substrate in systems of this type has been reported to produce more effective prodrugs [31].

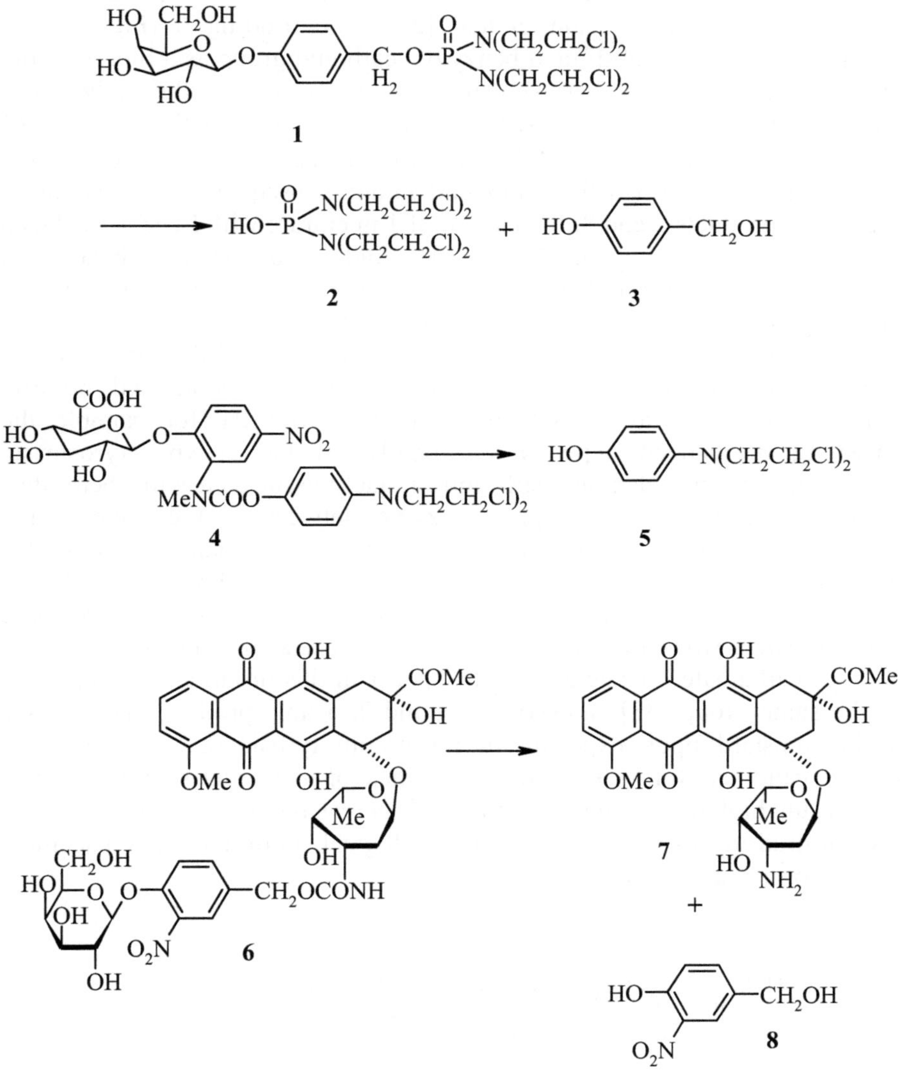

Fig.9.9 Activation of prodrugs designed for use in ADEPT and GDEPT therapies [28,29]

The cytotoxicities of the prodrug and the drug were compared, the former showing a reduced toxicity, which is a requirement for its application in ADEPT. In contrast to prodrug 1, compound 4 was designed to be cleaved by β-D-glucuronidase, and was indeed shown to release the drug 5 *in vitro* in the presence of *E.Coli* β-D-glucuronidase.

Analaogous prodrug design strategy has been applied to the anthracycline antibiotic daunorubicin [28]. The prodrug 6 (Figure 9.9), incorporating a *para*-substituted benzyloxycarbonyl group as a bioreversible amine protective group, was shown to be degraded into the active drug daunorubicin 7 and 3-nitro-4-hydroxybenzyl alcohol 8 in the presence of *E.coli* β-galactosidase. The first activation step is cleavage of the galactopyranosyl unit (as for compound 1) and subsequent decarboxylation to give the products 7 and 8. Incubation of 1 in culture with LacZ-transduced human cancer cells yielded 100-300 fold cytotoxicity enhancement compared to controls, confirming that 6 is a good substrate for the enzyme and has potential for use in ADEPT or GDEPT therapy.

Variations on the above methods include MDEPT (melanocyte-directed enzyme prodrug therapy) and VDEPT (virus-directed prodrug therapy). In contrast to ADEPT and GDEPT, in MDEPT, for example, the activating mechanism depends specifically on the enzyme tyrosinase, already present in melanoma cells and uniquely associated with them, thus circumventing the need for prior enzyme delivery to the tumour site. Prodrugs have been developed to treat malignant melanoma using this targeting technique. Examples are shown in Figure 9.10. Prodrug 1 [32] was designed to contain a catechol moiety to take advantage of tyrosinase oxidation, which would lead to release of the drug (a phenol mustard in this case).

A carbamate linkage acts as the spacer in this prodrug. A later study by the same group [33] showed compound 2, a urea prodrug that releases aniline mustard upon exposure to mushroom tyrosinase, to be a more suitable candidate for MDEPT than 1 because of its greater stability in sera. This result highlighted the possibility of increasing the stability and half-lives of prodrugs of this type under physiological conditions by replacement of urea for carbamate linkages.

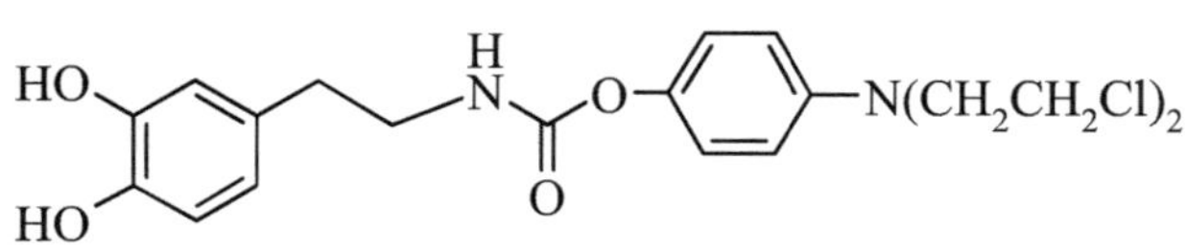

1

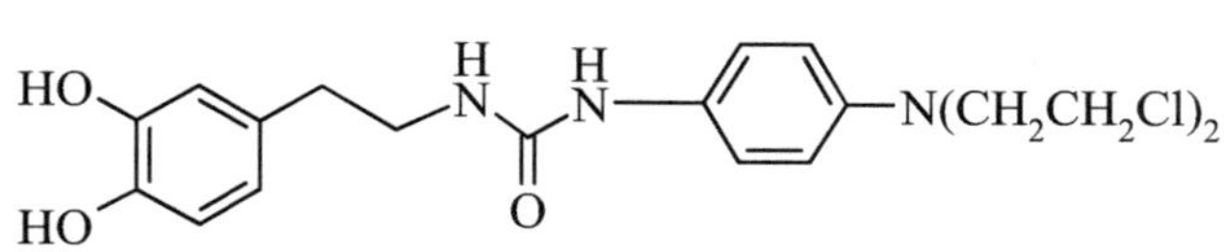

2

Fig.9.10 Carbamate and urea prodrugs, candidates in MDEPT [32,33]

In the context of prodrugs useful in anticancer treatment, we should mention also those that are hypoxia-selective. Relatively high proportions of hypoxic cells in tumours, compared with normal tissue, enable their targeting by prodrugs that can undergo bioreductive activation, releasing the cytotoxic agent intracellularly. Several classes of bioreductive drugs are known and have been reviewed recently [12]. An important example is tirapazamine (3-amino-1,2,4-benzotriazine 1,4-di-N-oxide), which is selectively toxic to hypoxic cells present in solid tumours. At low oxygen levels, intracellular metabolism converts the prodrug to a radical that is extremely toxic, causing DNA damage. The toxicity towards aerobic cells is lower because the toxic radical reverts to tirapazamine in the presence of oxygen. Evidently tirapazamine may be activated by multiple nuclear reductases [34].

Novel drug design strategies for targeted anticancer therapy were recently highlighted in the literature, with particular emphasis on effective chemical linkages between the targeting molecule and the cytotoxic agent [35]. Illustrative examples of conjugates with monoclonal antibody (mAb) include prodrugs containing calichimeacin (anti-leukemic) and paclitaxel derivatives. In the first case, conjugation is achieved by introducing a pH-sensitive bifunctional linker, which facilitates intracellular release of calicheamicin. In the second mAb conjugate, following its endocytosis, metabolic activation by non-specific cellular esterases releases free paclitaxel. The macromolecular nature of the antibody scaffold may present some problems with distribution and clearance and there is a trend towards the design of new anticancer agents with low molecular weight ligands.

Other macromolecules, including soluble polymers, have been employed in prodrug design to address drug solubility as well as stability issues, and problems such as enzyme degradation, hydrolysis and oxidative reduction that can occur in sera. Design strategies aimed at improving soluble macromolecular delivery systems have been reviewed [36]. While incorporation of macromolecules as constituents of prodrugs has in recent years contributed to increasing drug biovailability and reducing toxicity, there is a need for more effective constructs with additional capacities such as drug targeting and timed delivery.

Covalent coupling of drug molecules to polymers such as polyethylene glycol (PEG) or its derivatives could render them more soluble and more stable to systemic degradation. PEG is an approved material for medicinal application owing to its biocompatibility and non-toxic, non-antigenic and non-immunogenic nature. Figure 9.11 shows examples of prodrugs obtained by covalent coupling of activated PEG to the antiherpes agents valacyclovir and acyclovir [37]. In synthesising the prodrugs, appropriate derivatisation of PEG was necessary since its terminal –OH groups are not suitable for direct attachment of the drugs. The acyclovir-polymer conjugate was thus obtained by attaching the drug to carboxyl-PEG using an ester bond. Similarly, for the valacyclovir prodrug, the active agent was coupled to chloro-PEG using a covalent C-N bond. *In vitro* studies on the release of the drugs from these conjugates were performed in different media. It was concluded that PEG-valacyclovir is the more suitable prodrug for therapeutic use owing to its higher stability in various media and its larger percentage of drug release over time. In particular, this prodrug was considered suitable for oral administration, whereas the PEG-acyclovir prodrug, with the higher rate of hydrolysis, was considered suitable for other forms of administration and in cases where rapid drug release is required.

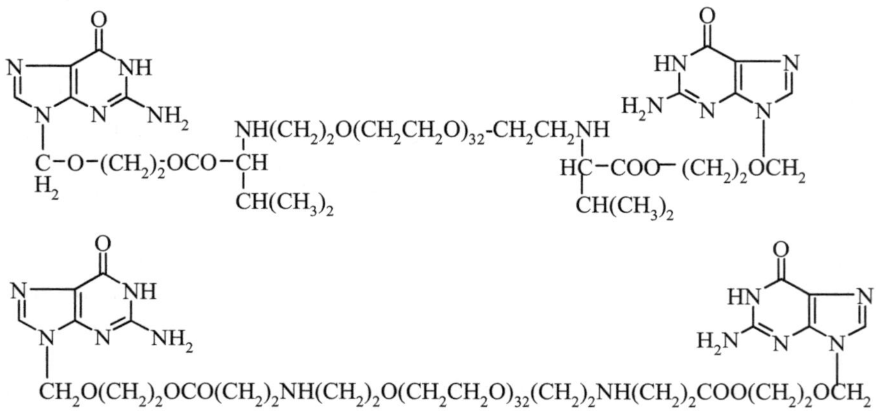

Fig.9.11 PEG-based prodrugs of valacyclovir (top) and acyclovir (bottom) [37]

Advances in the development of a number of acid-sensitive macromolecular anticancer drug delivery systems (DDSs), spanning the range from simple to site-specific antibody-targeted polymer-drug conjugates, have been reviewed [38]. A requirement for activation of these systems is that the spacer unit (separating the active drug from the polymeric backbone in the DDS) should be susceptible to lysosomal enzyme or chemical hydrolysis at physiological pH, or under conditions of environmental pH change. The principle behind this approach has proven to be valid, but much research is needed to optimise the performance of such DDSs. Because of their structural complexity, *in vitro* cytotoxicity studies are generally inadequate for assessing their activities and more feedback from *in vivo* studies in conjunction with computer-modelling are evidently necessary for their further development [38].

A considerable variety of prodrugs has been presented above with the intention of illustrating various design strategies for achieving different delivery objectives, in particular the improvement of pharmacokinetic properties and overcoming metabolic hurdles. Further examples can be found in the reviews quoted above. The study of an account of the development of fosphenytoin [39], a prodrug of the sparingly-soluble anticonvulsant phenytoin, is strongly recommended. This offers a detailed account of how practical problems associated with drug development may be addressed and is of interest not only from the chemical synthetic viewpoint, but also indicates the importance of understanding aspects of formulation, pharmacokinetics and metabolism in prodrug design.

The reader is especially referred to ref. 10, which is a recent commentary on the philosophy of prodrug research. One important conclusion that the author of the review draws is that the various objectives of prodrug development are interlinked, in the sense that the development of a prodrug may lead to beneficial properties over and above those originally intended by the structural modifications implemented. So, for example, improvement in drug solubility (a pharmaceutical objective) effected by prodrug formation may yield improved oral absorption (a pharmacokinetic objective). Similarly, unintended site-specificity might also result through improved chemical stability of the prodrug. A second observation that is made in ref. 10 regarding the prodrug approach is that it should be a promising chemical strategy in cases where there is a significant gap between the structural nature of the pharmacophore and other desirable pharmaceutical, pharmacokinetic or pharmacodynamic properties, since a traditional optimisation strategy would usually fail under these circumstances. Thus, many prominent medicinal chemists advise implementation of a prodrug strategy at an early stage of lead optimisation. On the other hand, some of the weaknesses of simple prodrugs stem from the single chemical conversion involved in their metabolism to yield the active species [1]. The potential advantage of multiple conversions is one factor that has contributed to the evolution of chemical delivery systems (see section 9.2.5).

9.2.3 The hard drug approach

Hard drugs are pharmacologically active compounds that undergo little or no metabolism i.e. the term 'hard' is synonymous with 'metabolically stabilised' [40,41]. The concept implies that after exerting their medicinal effect, hard drugs are excreted by the body unchanged. An important advantage is that since metabolism is absent or very limited, the risk from toxic metabolites (as might occur with prodrugs, for example) is minimised. However, the reader will appreciate that to design a hard drug, a formidable strategy would be required to ensure that the candidate drug could escape biotransformation by the plethora of enzymes that mediate Phase I and Phase II biotransformations of substrates of extremely wide chemical and structural diversity (Chapters 2-3). Thus, in practice, achieving a high degree of metabolic stabilisation is often fortuitous but structural modification can improve 'hardness', as indicated in examples that follow.

Enalaprilat (Figure 9.12) is a potent, orally active ACE inhibitor, regarded as an important example of a hard drug on account of its very limited metabolism and exclusive excretion *via* the kidney [2,42]. In contrast to the ACE inhibitor captopril, which contains a thiol function (believed to be the origin of its adverse side-effects through *in vivo* disulphide formation with endogenous proteins), enalaprilat is a carboxyalkyldipeptide whose structure was designed to effect significantly stronger inhibition of ACE than that displayed by captopril [43]. This goal was indeed achieved with concomitantly reduced side effects compared to captopril.

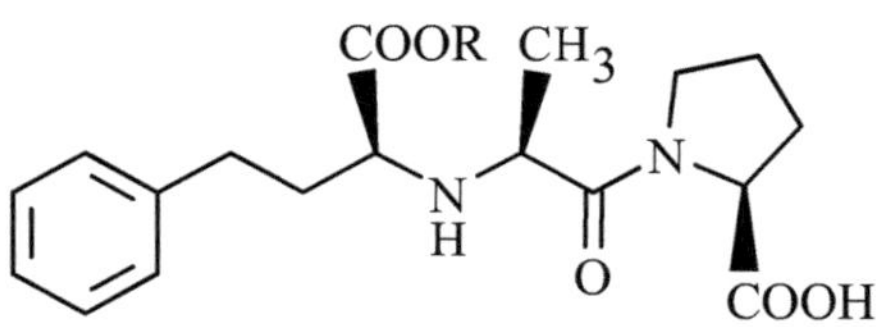

Enalaprilat: R = H

Enalapril: R = Et

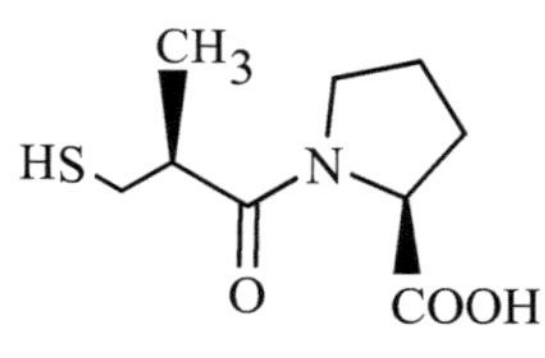

Captopril

Fig.9.12 Structures of ACE inhibitors [42]

As it happened, the poor lipophilicity of enalaprilat (octanol/water partition coefficient ~0.003 [44]) due to the presence of two carboxylic acid groups in the molecule, resulted in poor oral absorption (<10%) [2]. The prodrug strategy described in section 9.2.2 was therefore subsequently employed whereby one of the carboxylic acid groups was converted into its ethyl ester, to yield the widely prescribed drug enalapril (Figure 9.12), with significantly improved absorption (~60%) [2]. This prodrug is metabolised by hepatic esterolysis to enalaprilat as the major metabolite [45].

Bisphosphonates (Figure 9.13) were developed as inhibitors of bone resorption and display a remarkable metabolic stability, warranting their description as hard drugs [2,46]. The discovery of these compounds was based on earlier observations that inorganic pyrophosphates could bind to calcium phosphate, inhibiting the formation of calcium phosphate crystals and crystal dissolution *in vitro*, but lacking *in vivo* activity on bone resorption. Pyrophosphates (containing the P-O-P bond) were found to undergo rapid *in vivo* hydrolysis before reaching the site of bone destruction, whereas bisphosphonate analogues (characterised by P-C-P bonds) resisted biotransformation and successfully inhibited bone resorption. The bisphosphonates have high aqueous solubilities, lacking the typical substrate properties associated with metabolisable drugs. Consequently, they display simple pharmacokinetics and are exclusively excreted by the kidneys [46]. The high aqueous solubilities, however, result in extremely low oral bioavailability in humans (e.g. ~0.7% for alendronate).

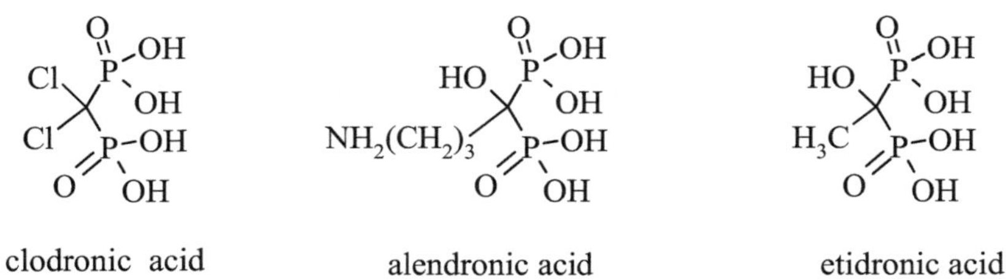

Fig.9.13 Representative parent acids of bisphosphonates with the ability to inhibit bone resorption

Drugs with metabolic stabilities approaching those of the biphosphonates are exceptional. More frequently, metabolic stability must be built in to lead compounds by appropriate systematic chemical synthesis. Furthermore, chemical modification should ideally not compromise pharmacological activity or potency. In some instances, other advantageous activities may fortuitously be gained during the iterative process of synthesis and biological evaluation.

An example illustrating the input of metabolism data to drug optimisation concerns the development of analogues of the potent cholesterol absorption inhibitor 1, (-) SCH 48461 (Figure 9.14) [47]. Some detail is presented here to illustrate the interplay between SAR-input and metabolism data input in the process of drug discovery/optimisation, in this case with the intention of metabolic stabilisation.

From metabolism studies, four primary sites of biotransformation in the molecule 1 had been identified (*a,b* – hydroxylation sites, *c,d* – demethylation sites) that led to no fewer than eleven metabolites. Since previous SAR-studies had indicated that the C-3 phenylpropyl group was an essential pharmacophore but could tolerate some modification, benzylic hydroxylation (site *b*) could be blocked *via* e.g. substitution of the benzylic C atom by an oxygen atom. Unexpectedly, this led to a significant reduction in potency of the product (±) 2 relative to (±) 1 (the racemate of (-) SCH 48461). It was then reasoned that (±) 2 might have been rendered metabolically labile by enhanced hydroxylation at site *a* relative to (±) 1, due to the presence of the electron-rich phenoxy group.

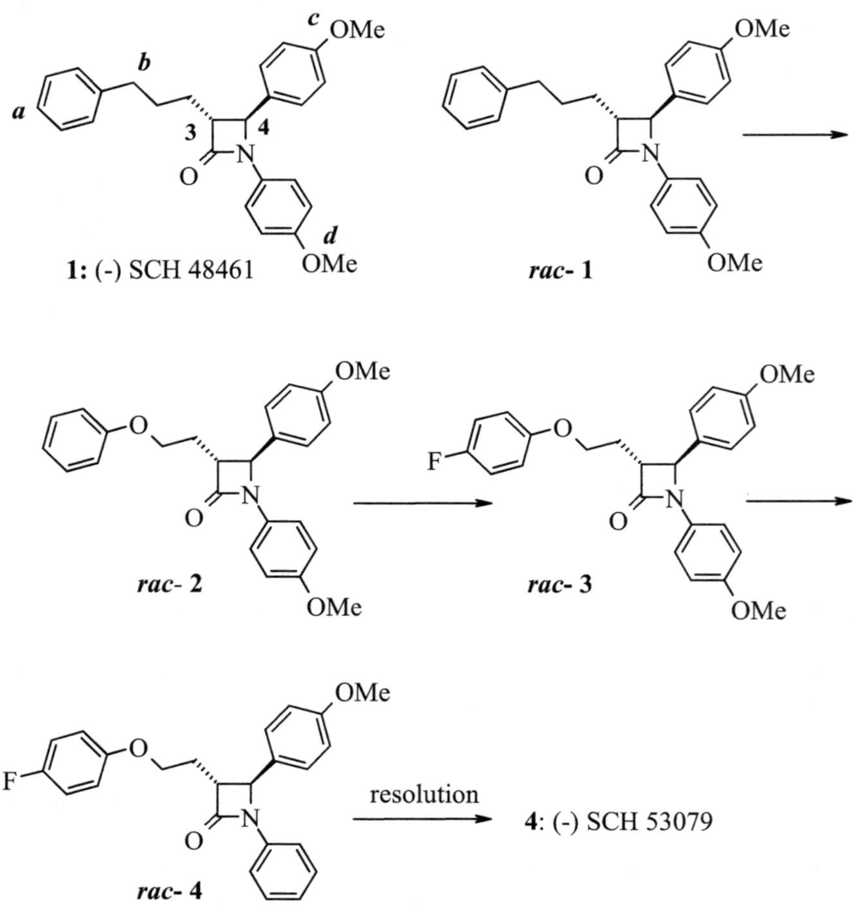

Fig.9.14 Stages in the metabolic stabilisation of a cholesterol absorption inhibitor [47]

Hence, (±) 3 was synthesised, site *a* being blocked by a fluorine atom. This did in fact restore activity comparable to that of (±) 1.

SAR considerations had also indicated that the methoxy group of the N-aryl moiety of 1 was not required for activity, a prediction confirmed by *in vivo* evaluation. Hence the *p*-methoxyphenyl moiety was replaced by a phenyl group and finally, resolution of the product (±) 4 yielded the eutomer (-) 4, designated (-) SCH 53079. The latter compound was found to be equipotent to (-) SCH 48461 and yielded only three metabolites as their glucuronide conjugates in animal studies.

Two examples in the more recent literature that illustrate the iterative process of synthesis and biological evaluation to improve the metabolic stability of lead compounds warrant further study. One of these relates to the metabolic stabilisation of inhibitors of TNF-α, whose over-expression is implicated in a number of diseases [48]. The other concerns the metabolic stabilisation of a benzylidene ketal M_2 muscarinic receptor antagonist [49]. The challenge in the latter case was to overcome (while maintaining M_2 selectivity and affinity) *in vivo* cytochrome P450-catalysed oxidative cleavage of the methylenedioxy group in the lead compound 1 (Figure 9.15), leading to a catechol intermediate. The latter, if further oxidised to an ortho-quinone, could induce toxic effects by DNA alkylation. In summary, the steps involved in the overall metabolic stabilisation and retention of M_2 activity of 1 included: (a) replacement of the metabolically labile methylenedioxy group with a *p*-methoxyphenyl group to yield 2 (with poor M_2 activity compared to 1, however), (b) replacement of the sulphonamide with a naphthamide moiety to yield 3 (with restored M_2 activity, but with the new moiety susceptible to undesirable metabolic oxidation to an arene oxide), (c) introduction of a fluorine atom at the 4-position of the naphthamide moiety (to optimise metabolic stability).

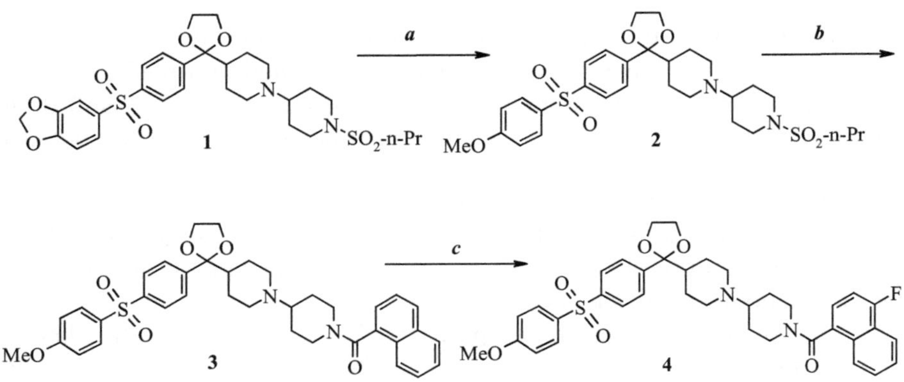

Fig.9.15 Stages in the metabolic stabilisation of an M_2 muscarinic antagonist [49]

Compound 4 demonstrated excellent M_2 affinity and selectivity, and human microsomal stability [49]. As a 'bonus', 4 displayed better bioavailability in rodents and primates than 1.

As a conclusion to this section, it is appropriate to comment on the question of metabolic stability in general and to reflect its current status in drug discovery and development. Ideally, metabolic stability is an issue addressed at the preclinical stage. Metabolism studies are thus used to identify candidate drugs with favourable pharmacokinetic and safety profiles for human administration. However, since new chemical entities cannot be administered directly to humans in the early stages of drug discovery, determination of their metabolic stabilities must rely on data from *in vivo* animal studies as well as *in vitro* cellular or subcellular (e.g. liver microsome) systems, and computational models. Part of this process involves the construction of libraries of compounds (typically based on combinatorial chemistry) and filtering them for metabolic stability in order to eliminate 'junk' leads [1]. The detailed protocol for preclinical metabolism studies and their utility in establishing the metabolic stabilities of candidate drugs, and hence predictions for human metabolism, have been described in a recent leading article [7]. In this context, computational models include those aimed at predicting specific enzyme-substrate binding affinity, the positions of metabolic attack in candidate drug molecules, and their rates of metabolism. While the success rates for prediction of *in vitro* properties from computational models appear to be improving, it remains true that accurate prediction from *in vitro* and animal studies to humans is still a significant challenge. Thus, qualitative and quantitative predictions of metabolic stability of new chemical entities in humans are still rather limited.

9.2.4 The soft drug approach

This concept was introduced in the late 1970's [50]. A soft drug is a pharmacologically active compound that is deactivated in a predictable and controllable way after it has fulfilled its therapeutic role [14]. A soft drug, therefore, not only possesses the required pharmacological activity, but also has built-in structural features that ensure its deactivation and detoxification in a desired way after it has carried out its biological action. Various subclasses of soft drugs have been defined and described in detail [24], the most successful of these being based on the inactive metabolite approach and the soft analogue approach [14].

In the inactive metabolite approach [24], one begins with a known inactive metabolite of the lead compound. This is used as the basis, in the so-called 'activation stage', for design of new molecules that are isosteric and/or isoelectronic analogues of the lead that gave rise to the inactive metabolite. Chemical modification of the inactive metabolite is thus intended to yield new molecules that would be metabolised to the inactive metabolite in a single step, thus ensuring the important requirement of 'predictable metabolism'. To ensure also 'controllable metabolism' (that could influence e.g. the rate and/or primary site of metabolism) appropriate chemical modification is performed during the activation stage.

An example from the class of soft β-blockers is illustrative [51]. Among the various metabolites of the well-known β-blockers atenolol and metoprolol (Figure 9.16) there is a common, significant metabolite (a phenylacetic acid derivative) that is known to be pharmacologically inactive. This molecule was therefore used to design new analogues of the β-blockers with predictable and controllable metabolism. Conversion of the carboxylic acid moiety of the inactive metabolite into an ester using a variety of substituents (R' in –COOR') thus yielded a family of soft β-blockers with variable transport properties and degradation rates, depending on the nature of R'. Thus, for example, with R' = $-CH_2SCH_3$, the derived compound, due to its rapid hydrolysis *in vivo*, displayed ultra-short antiarrhythmic activity on intravenous administration. On the other hand, the ethyl adamantanyl derivative has proven to be effective as a topical antiglaucoma agent, producing significant, prolonged reduction of intraocular pressure. At the same time, it undergoes rapid hydrolysis in human blood, a most important advantage, since this eliminates undesired systemic activity.

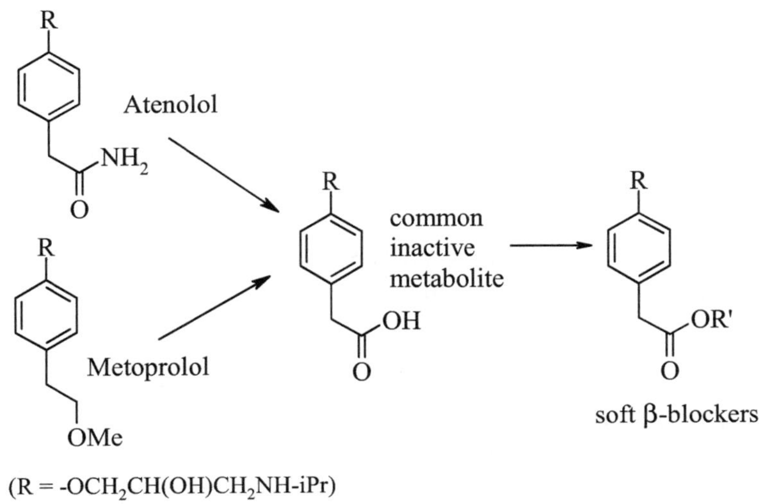

Fig.9.16 Design of soft β*-blockers based on the inactive metabolite approach [51]*

Essentially the same approach was adopted more recently to design ultra-short acting, soft bufuralol analogues, which are also β-blockers [52]. In this case the aromatic moiety of the lead compound, bufuralol, contains an ethyl substituent, which is metabolised to an acetic acid group in the inactive metabolite.

Esterification of the inactive metabolite with various reagents led to seven soft analogues of bufuralol, all of which were demonstrated to undergo extremely rapid metabolism by blood and tissue esterases to the common inactive acidic metabolite. Four members of the series displayed β-blocking potencies ranging between 25 and 50% that of the lead bufuralol.

An example of a highly successful corticosteroid designed on the above principle is the anti-inflammatory and anti-allergic loteprednol etabonate derived from prednisolone [1,8,53]. This compound is topically effective in the treatment of ocular inflammation, thereafter undergoing rapid biotransformation to inactive metabolites. Figure 9.17 shows the metabolism of loteprednol etabonate to its primary inactive metabolite [1,8]. In 1999, a turnover exceeding 250 million USD was reported for the drug [54].

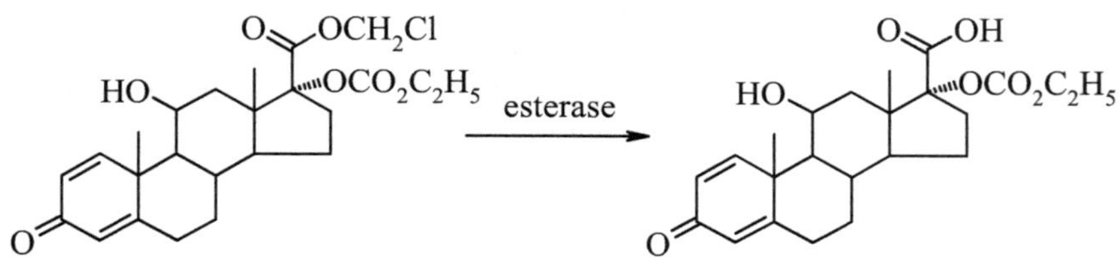

Loteprednol etabonate

Δ^1-cortienic acid 17α-ethylcarbonate

Fig.9.17 Metabolism of loteprednol etabonate to its primary inactive metabolite [1,8]

The clinical status and the success or otherwise of recently developed soft glucocorticoid steroids, including fluocortin-21 butyl ester, tipredane, butixocort propionate, itrocinonide, GW 215864 and loteprednol etabonate, have been reviewed [55].

In the search for safer dihydrofolate reductase (DHFR) inhibitors, the soft drug approach has recently been applied to synthesise a series of compounds in which the methylamino-bridge of non-classical inhibitors (e.g. trimethoprim) was replaced with an ester function [56]. These compounds were prepared as potential soft drugs intended for inhalation to treat *Pneumocystis carinii* pneumonia. This strategy anticipated rapid deactivation of these lipophilic esters by ubiquitously distributed esterases following their therapeutic action in the lungs. An interesting feature of this programme was the use of an automated docking and scoring procedure as well as molecular dynamics simulations to select the target compounds for synthesis. Meaningful data could be obtained from the docking routines since high-resolution X-ray structures of the human reductase with and without complexed inhibitors are available.

At this point, it is worth noting that the particular strategy employed in the examples above is a special case of the more general 'retrometabolic drug design' (RMDD) philosophy [8,24], manifested also in the design of more elaborate chemical delivery systems (CDSs) which are aimed at targeted drug delivery (see section 9.2.5). The general aim of the RMDD approach is thus to incorporate metabolism as well as targeting into the drug design process in a systematic way so as to derive safe, locally active compounds.

The second category of 'soft drug design' referred to earlier, namely the soft analogue approach, differs from the inactive metabolite approach in that the new soft compounds are close structural analogues of the selected lead compound into which a metabolically 'weak spot' has been deliberately incorporated by chemical synthesis. Again, ideally a one-step deactivation and non-toxic products are desirable, the first requirement often being achievable if the sensitive part of the molecule is susceptible to hydrolytic metabolism.

One area in which this approach has been employed is in the design of certain classes of long chain ammonium antibacterial agents [24,57-59]. In the case of cetyl pyridinium analogues, the 'hardness' of the parent cetyl pyridinium compound (containing the fully saturated $N^+–(CH_2)_{15}-CH_3$ chain) reflects that it requires several oxidative metabolic steps for its deactivation. As indicated earlier, oxidative metabolism generally leads to toxic intermediates. This hardness has been reduced significantly in the cetyl pyridinium analogue of Figure 9.18 by incorporation of the ester function, which lends itself to the facile and predictable metabolism shown [57,59]. The parent compound and the analogue display comparable activities as antimicrobials [24]. Because conventional long-chain quaternary ammonium compounds are used in massive quantities in a wide variety of pharmaceutical, domestic and industrial applications, there is serious concern regarding their effects on the environment. This is a strong motivation for producing novel soft quaternary ammonium compounds that can undergo facile degradation [59].

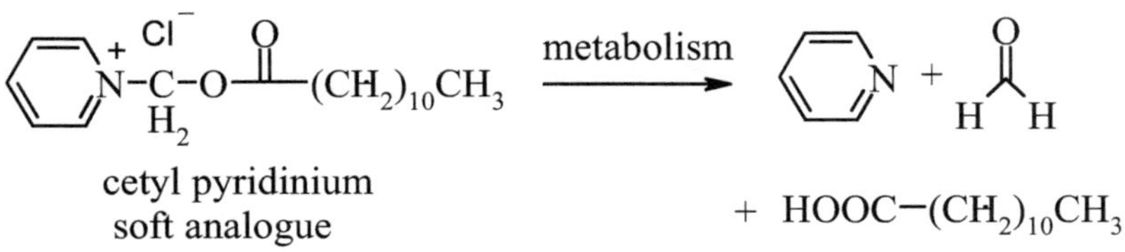

Fig.9.18 Metabolism of a soft analogue of an antibacterial [24,57,59]

Two categories of soft drug design were highlighted above to illustrate some of the principles involved. Other categories include controlled-release endogenous agents, activated soft compounds and active metabolite-based drugs [24]. Controlled-release endogenous agents are derived from e.g. natural hormones and neurotransmitters. Appropriate chemical modification can result in retardation of their typically rapid metabolism and thus yield soft drugs with prolonged action and/or site-specificity. To generate activated soft compounds, a pharmacophore is introduced into the structure of a non-toxic, inactive compound; the activated form loses the pharmacophore *in vivo*, yielding the original non-toxic species. Finally, active metabolite-based drugs are oxidative metabolites of a parent drug that still retain activity and are consequently more readily inactivated *in vivo*. In the last case, for drugs undergoing sequential oxidative metabolism to yield eventually an inactive metabolite, some previous metabolite (e.g. ideally that preceding deactivation) could represent a useful drug. Choosing a metabolite that appears earlier in the metabolic sequence would be counterproductive, raising complications of control due to the simultaneous presence of a number of its active metabolites. Many further instructive examples from all of the subclasses of soft drugs can be found in the references quoted in this section, as well as in a very recent review that also features the role of *in silico* tools in the design process [60].

9.2.5 Strategies based on Chemical Delivery Systems

As discussed briefly in the above section, soft drug design represents one extreme strategy in the overall scheme of retrometabolic drug design (RMDD) [1,8,24,61]. The complementing strategy in RMDD is based on the concept of the ‘chemical delivery system’ (CDS), which evolved from prodrugs (section 9.2.2) in the early 1980s. The essential difference between prodrugs and CDSs is that the latter rely on multi-step activation and contain targetor moieties [62]. Thus, in the nomenclature used to describe these concepts, a prodrug (as defined in section 9.2.2) consists essentially of the active drug (D) attached to 'modifier functions' (F_1 - F_n), which control the molecular properties of the prodrug (e.g. by altering its lipophilicity, acting as protecting groups) [8,62]. But whereas prodrugs are often designed to overcome problems such as poor absorption and rapid first-pass metabolism (see section 9.2.2), they are not necessarily designed to 'target' specific tissues, organs or other sites in the body. On the other hand, in the design of a CDS, an inactive derivative of a drug D, the intention instead is to incorporate the goal of targeting, so that ideally the drug is released from its CDS only at the intended site, but is present as an inactive species elsewhere in the body, from where it is safely eliminated.

In the general case, therefore, the CDS is designed as follows: an active molecule (the drug 'D') is synthetically transformed into an inactive molecule by attachment of not only modifiers (F_1 - F_n), but also a 'targetor function' (T) [8]. The role of the T function is to effect a specifically higher or sustained concentration of the drug at the site of interest and 'lock-in', while the F functions may control other molecular properties of the CDS (as in prodrugs). The RMDD approach should ensure that on administration of the CDS, the steps leading to drug activation are predictable, sequential metabolic reactions, generating inactive intermediates, disengaging the F functions first and eventually the T function (once the latter has performed its targeting role). As regards the targeting aspect, this may rely on the prevalence of specific enzymes at the target site (e.g. in ocular delivery), or on transport properties that are site-specific (e.g. for delivery to the brain).

One classification of chemical delivery systems includes the site-specific enzyme-activated CDS, the enzymatic physical-chemical based CDS, and the receptor-based CDS, and these have been illustrated with appropriate examples [24]. Some more recent examples of CDSs are described here, beginning with the simpler varieties and progressing towards more complicated ones. Many of these systems have been successful as ocular hypotensive agents and in the treatment of brain disorders (e.g. Alzheimer's disease). The design of CDSs for ophthalmic drugs takes advantage of the fact that the various compartments of the eye are regions having particularly high concentrations of metabolising enzymes of wide variety [8]. Bioactivation and targeting then rely on successive reduction-hydrolysis metabolic steps. In the case of brain-targeting, the 'lock-in' principle is based on the fact that a lipophilic precursor (e.g. T-DF) will readily cross the blood-brain barrier (BBB), but following e.g. local oxidative enzymatic conversion to a hydrophilic species (T^+-DF), the latter becomes trapped and the drug D is finally enzymatically released within the brain. The challenge of drug delivery to the brain stems from the very special nature of the BBB, as described later. Adequate delivery of a host of drugs that act on targets in the brain is an essential requirement. Such drugs include antidepressants, anaesthetics, anticonvulsants, antibiotics, anticancer and antiviral agents. The CDS approach represents an important recent advance in meeting this challenge [63]. These concepts are illustrated with several examples below.

The design, *in vitro* stability and ocular hypotensive activity of *t*-butalone CDSs have recently been described [64]. Here, the challenge was site-specific delivery to iris-ciliary body tissues of the active drug *t*-butaline 1 (Figure 9.19), a selective β_2-adrenoreceptor agonist, normally used to treat bronchorestrictive disorders. Two issues had to be addressed to prepare CDSs suitable for ocular delivery, namely the delivery aspect and the local bioactivation aspect.

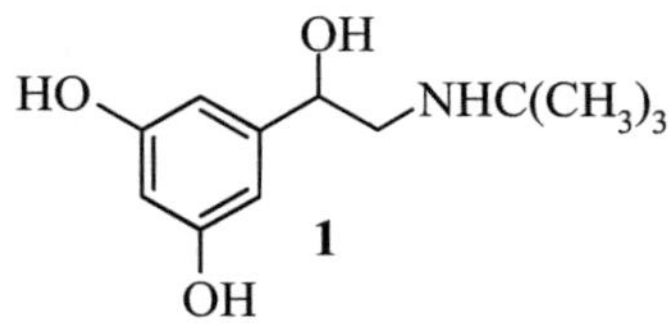

t-Butaline

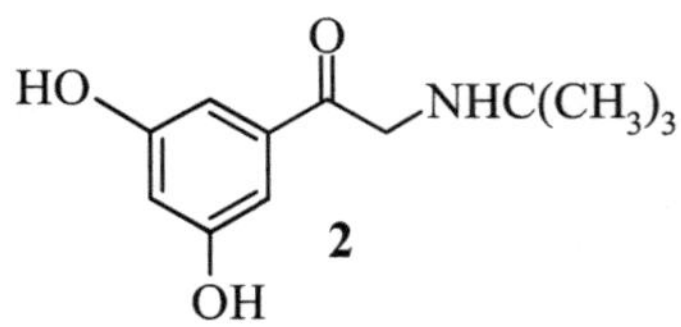

t-Butalone

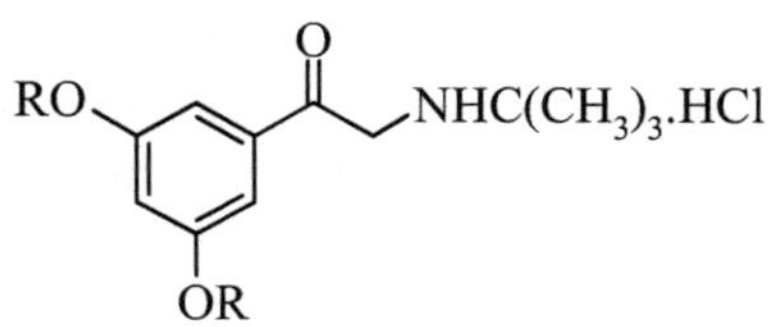

t-Butaline CDS

3. R = $-COCH_2CH(CH_3)_2$
4. R = $-COC(CH_3)_3$

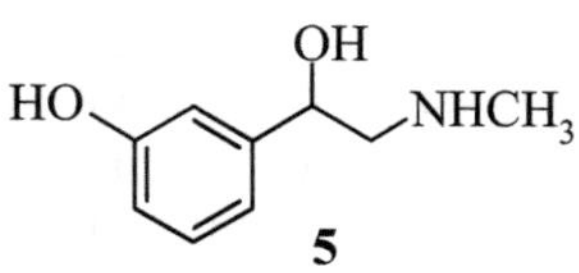

Phenylephrine

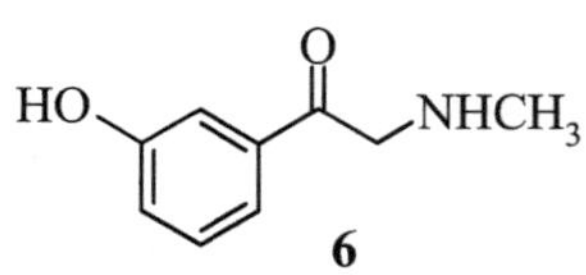

Phenylephrone

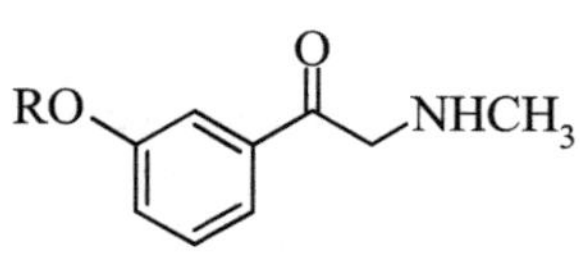

Phenylephrine CDS

7. R = $-COCH_2CH(CH_3)_2$
8. R = $-COCH_2Ph$
9. R = $-COC(CH_3)_3$

Fig.9.19 Examples of chemical delivery systems for ophthalmic application [64-67]

Since *t*-butaline is relatively hydrophilic, its permeability across biphasic corneal membrane is impaired, with the result that topical administration of its aqueous solution at 2% dose level does not alter the intraocular pressure (IOP) in test animals (normotensive rabbits) significantly. The lipophilicity of *t*-butaline was thus increased by esterification of the aromatic hydroxyl groups. The incorporated diisovaleryl- and dipivalyl-substituents thus correspond to the 'modifier' functions, which in this case facilitate corneal permeability. The 'targeting' aspect was addressed by converting the remaining hydroxyl group to a keto-function [64].

The rationale for the above approach was based on the analogy with esters of adrenolone, which were shown in previous work [65] to undergo a site-specific reduction-hydrolysis metabolic sequence by reductases and esterases present only in the iris-ciliary body, to the corresponding adrenaline derivatives. This is a key feature in the design strategy for ocular delivery. In the case of *t*-butaline, these chemical modifications thus produced the CDSs 3 and 4 (Figure 9.19), described as bioreversible diacyl derivatives of *t*-butalone 2. It is important to note that, in keeping with the CDS concept, t-butalone 2 is an inactive precursor of the active drug 1 and that the strategy being employed here is the site-specific CDS approach, based on predictable multi-step metabolic activation of bioreversible, inactive compounds at the site of action. These CDSs thus fall into the category of site-specific enzyme-activated CDSs [24].

Favourable results for the IOP-lowering and *in vivo* disposition of the dipivalyl terbutalone 4 in rabbits were reported [66] as were details of the synthetic procedures for the CDSs 3 and 4 and their comparative biological evaluation [64]. The outcomes of primary interest here were that both CDSs 3 and 4 exhibited a significant ocular hypotensive activity and that duration of action was found to be dependent on the 'modifier' function i.e. it can be controlled by judicious choice of steric bulk in the esterification step.

A wide variety of ester functions may be employed to alter the drug lipophilicity in CDSs of the type described above. For ocular delivery, however, strongly lipophilic functions have been employed for optimal corneal penetration. For example, in earlier studies of the design of soft β-blockers for ophthalmic use, the ester groups included the cyclohexylglycol, adamantylmethyl, adamantylethyl, *endo*- and *exo*-norbornyl, and isopinocamphyl functions [8].

The strategy described above for transforming *t*-butaline into a CDS has also recently been employed for phenylephrine 5 (an α_1-selective adrenergic agonist used in the eye for its mydriatic effect), by synthesising esters of the inactive ketone precursor phenylephrone 6 [67] (Figure 9.19).

In this case, esterification of the single phenolic group on the molecule was performed with the isovaleryl, phenylacetyl and pivalyl functions, and the mydriatic effect and ocular distribution/metabolism of the three resulting compounds 7-9 were investigated. Whereas phenylephrone showed no mydriatic activity whatsoever, the three esters displayed a significantly more potent mydriatic effect than phenylephrine, the phenylacetyl ester 8 being the most potent, with a short duration of action. It should be noted that for ocular delivery of the parent compound (phenylephrine hydrochloride), very high concentrations are usually employed because of the poor penetration of this hydrophilic drug into the epithelium of the cornea. This results in drainage of the drug into the nasolacrimal duct, with subsequent systemic distribution and primary systemic side effects. This is a common problem with administration of β-adrenergic antagonists directly into the eye and can result in heart-rate reduction (e.g. as found with the first widely used anti-glaucoma drug Timolol) or adverse respiratory events. A major advantage of the derived CDSs, such as that described for phenylephrine, is that the active drug is generated metabolically only in the iris-ciliary body tissues of the eye and is undetected in the systemic circulation.

The site-specific chemical delivery systems described above rely on ocular bioactivation for their drug targeting. One aspect that has been neglected thus far in describing the above systems is the question of stereospecificity. This important issue is often not explicitly mentioned in reported studies. It should be noted, however, that earlier studies based on ocular bioactivation of a CDS [68-72] showed that the released drug was the active *(S)*-isomer. Here the CDS was of the ketoxime-type i.e. an oxime or alkoxime, derived from the ketone corresponding to the β-adrenergic antagonist (Figure 9.20). In step 1, the CDS is metabolised by an oxime hydrolase present in the eye, reverting to the ketone from which it was chemically derived. The product is then metabolised in step 2 by a ketone reductase to give the active amino alcohol, In this case, the second step turned out to be stereospecific, yielding the active *(S)*-isomer [2,68].

One of the potential drug candidates that resulted from this strategy was alprenoxime, a CDS for the well-known β-blocker alprenolol. A shortcoming of the oximes, however, was their chemical instability in aqueous media, which shortened their shelf-lives. Subsequent conversion of the oximes (R = H in Figure 9.20) to the methoximes (R = CH_3) led to a significant increase in chemical stability and improved the overall performance of the CDS.

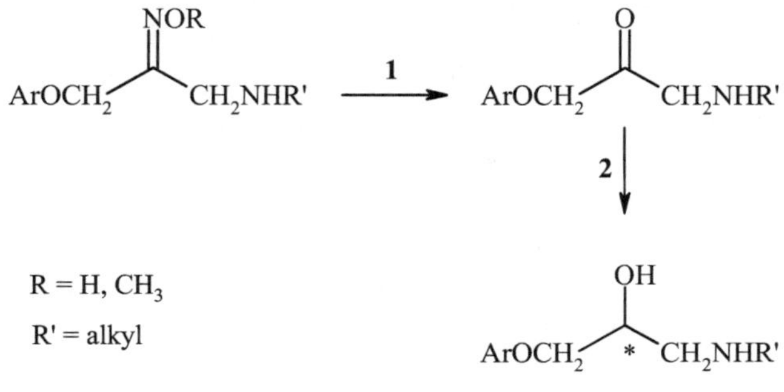

Fig.9.20 Sequential metabolism of a ketoxime with a stereospecific outcome [2,68]

The examples above illustrated design principles for the eye as the target organ for drug delivery. Here we discuss drug targeting to the central nervous system (CNS). Earlier, it was noted that the BBB represents a major obstacle for drug delivery to the brain. This is due to its unique lipoidal bilayer structure, which prevents the passage of hydrophilic drugs to the CNS. The use of a lipophilic prodrug may improve the level of drug uptake by the brain, but its efflux is likewise enhanced, resulting in low tissue retention. Poor selectivity (such prodrugs may enter other tissues as well) and the possibility of reactive catabolism of lipophilic prodrugs are additional factors that may not lead to an increase in the therapeutic index of a drug intended for delivery to the brain [1,63].

An essential concept in developing CDSs for brain-targeting is that of 'lock-in', i.e. ensuring that once the relatively lipophilic CDS has crossed the BBB, it is retained there. The strategy used to achieve this is to design the targetor moiety T to be susceptible to enzymatic transformation that converts it into a hydrophilic (commonly positively charged) moiety T^+-D (D = drug) that is then unable to exit the BBB. (The reader will no doubt recognize the conceptually analogous trapping of phosphate esters depicted in Figure 9.4, following diffusion of their lipophilic prodrugs into cells and subsequent hydrolysis to negatively charged species). Localisation of the modified CDS in the brain allows further, predictable metabolism that eventually releases the active agent. It should be emphasised that following administration and distribution of the CDS, the same enzymatic conversion to the strongly hydrophilic species T^+-D that occurs in the brain takes place elsewhere in the body, thus accelerating its peripheral elimination [24], in this way therefore contributing to brain-targeting.

The most successful brain-targeting approaches have incorporated, within the CDS, redox targetors, analogous to the NAD(P)H $\leftrightarrow$ NAD(P)$^+$ coenzyme system [24,11]. Systems based on the 1,4-dihydropyridine ring, whose chemistry has been extensively studied [73], have been found to be particularly versatile. Figure 9.21 shows schematically (1) the passive diffusion across the BBB of such a CDS containing both the drug D and a targetor moiety based on the lipophilic 1,4-dihydropyridine system, and (2) oxidation of the targetor moiety to the highly hydrophilic quaternary ammonium ion, mediated by oxidases in the brain, and resulting in 'lock-in'. Subsequent sustained, brain-specific release of the drug D is generally effected by hydrolysis mediated by appropriate esterases. The strategy above has been applied to drugs from a wide range of classes [24,63] that includes e.g. steroid hormones, anticancer agents, antiviral and antiretroviral agents.

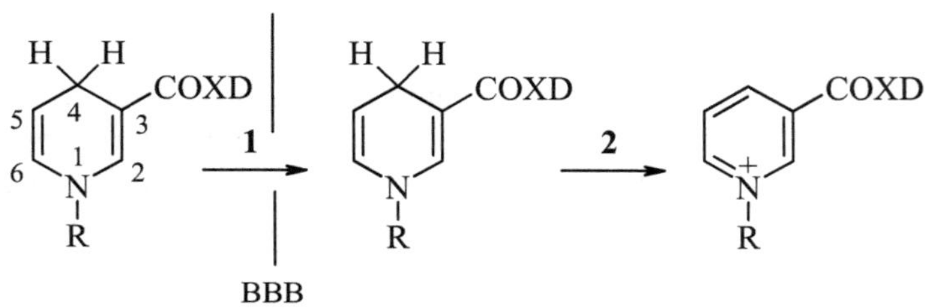

Fig.9.21 Targetor moiety for a drug D based on the 1,4-dihydropyridine system (X = N, O) [24,11]

Advantages of the employment of the 1,4-dihydropyridine ring as the targetor in such CDSs include the fact that it possesses a suitable degree of lipophilicity for penetrating both the BBB and other membranes, that its enzymatic oxidation to the T^{+}-D form proceeds at a reasonable rate, and that it may be suitably functionalised to link to a given drug D.

The use of phospholipid prodrugs for delivery of antiviral drugs such as AZT to the CNS was described in section 9.2.2 above. While such prodrugs may lead to improved transit into the CNS, extraction of these lipophilic compounds into other tissues may occur. This lack of selectivity may lead to serious side effects due to the potency of the antivirals. In a recent review describing the general problems associated with delivery of antiviral nucleosides to the CNS [25], the merits of chemical delivery systems based on redox trapping in the brain have also been discussed. A strategy based on the targeting methodology shown in Figure 9.21 above has been employed for delivery of drugs such as AZT to the CNS. Figure 9.22 illustrates such a CDS for AZT. As the drug molecule contains a single primary alcohol group, this is a convenient site for placement of the targetor moiety. This CDS relies on the versatile 1,4-dihydropyridine system for effective targeting, and various derivatives (with R = e.g. Me, Et, Pr, *i*-Pr, Bz) have been investigated [74-75]. Such compounds easily penetrate the BBB and their bioactivation involves (a) conversion to the hydrophilic quaternary ammonium species by oxidoreductases, effecting 'lock-in', and (b) subsequent hydrolysis by esterases, releasing the AZT in a sustained manner.

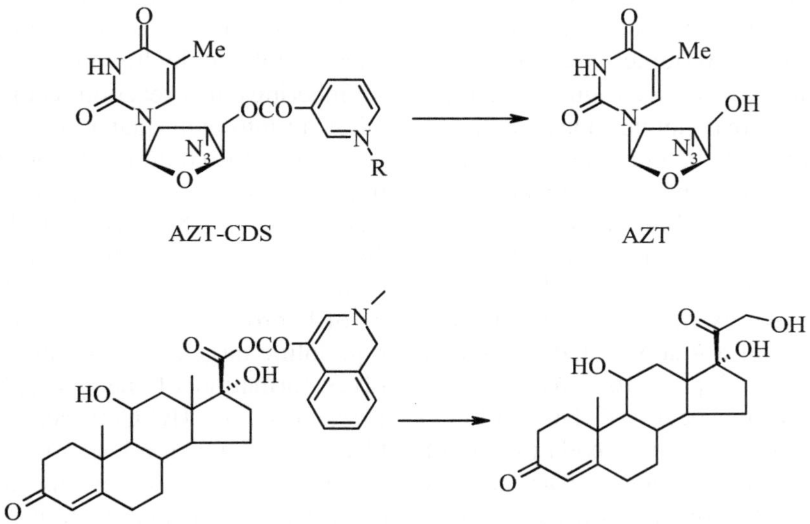

Fig.9.22 Examples of chemical delivery systems incorporating redox targetors [50,73-75]

The polarity of the AZT-T^+ species formed in the brain is orders of magnitude greater than that of the AZT-CDS, accounting for rapid peripheral elimination and effective 'lock-in' after its formation in the CNS [25]. Administration of the CDS consequently produces significantly higher AZT levels in the brain than does dosing of unmodified AZT. A systematic study of the effect of dihydronicotinate N-substitution on the brain-targeting efficacy [75] showed the N-propyl CDS to be the most efficient compound of the series examined.

Brain-targeting of pharmacologically active steroids can be effected using an analogous strategy. An important example of a very promising CDS designed for the delivery of estradiol (E_2) to the CNS has been described [62, 63]. This CDS was obtained by attaching the 1,4-dihydrotrigonelline targetor moiety to the 17-hydroxy function of the steroid. Intravenous administration of the E_2-CDS to rats led to confirmation of the lock-in mechanism, sustained release of E_2, and significantly elevated levels of the drug compared to those after simple E_2 treatment. Potential applications of the E_2-CDS include treatment of Alzheimer's disease and menopausal hot flushes. Phase I and Phase II clinical trials were reported as being in progress in 2001 [62].

Certain chemical shortcomings of the 1,4-dihydropyridine system were, however, noted recently in connection with the development of some steroidal CDSs [73]. While 1-alkyl-1,4-dihydropyridines easily undergo oxidation to their corresponding quaternary ammonium salts, 3-substituted-1,4-dihydropyridines are known to be susceptible to hydration at the 5, 6-double bond (see Figure 9.21 for ring-numbering), this reaction leading to products that can no longer undergo metabolism into their quaternary forms. Hydration is favoured under acidic conditions and this could have a negative impact on pharmaceutical formulation, leading to products with unacceptably short shelf-life. Thus, replacement of the hydrolytically labile 1,4-dihydropyridine ring with less reactive systems, such as those based on 1,4-dihydroquinoline and 1,2-dihydroisoquinoline, were investigated [73].

An example of a CDS for delivery of hydrocortisone (HC) based on the use of a 4-substituted-1,2-dihydroisoquinoline targetor is also shown in Figure 9.22 [73]. Its bioactivation involves formation of the hydrophilic, 'locked-in' species T^+-HC (following diffusion of the HC-CDS across the BBB and enzymatic oxidation) and final hydrolysis, which releases HC. The testosterone analogue was also synthesised and likewise evaluated for chemical stability studies and *in vivo* animal distribution studies. In the case of the HC-CDS, metabolism leading to prolonged release of HC in the brain of Sprague-Dawley rats was evident, whereas the blood levels of the CDS, its quaternary intermediate and the parent drug fell to undetectable values after a much shorter period.

Interestingly, the analogous CDS for testosterone behaved differently, no testosterone being detected in the brain. This was attributed to its excessively slow release rate by hydrolysis. The general conclusion of these studies, however, was that the 4-substituted-1,2-dihydroisoquinoline targetor moiety shows promise for brain-specific CDSs owing to favourable rates of metabolic activation as well as its chemical stability [73]. In particular, the desired result, a significantly reduced tendency to undergo hydration compared with 1,4-dihydropyridine, was confirmed, as was the greater stability of the alternative targetor ring-system towards aerial oxidation, these factors being favourable for formulation.

The same basic type of targetor moiety used in the above examples has been employed in considerably more elaborate CDSs designed for neuropeptide delivery to the brain, a topic that has recently been reviewed [63]. Many CNS disease states have potential for treatment with neuropeptides, but compounds in this class are notoriously difficult candidates for delivery to the BBB due to their rapid degradation by peptidases. More than a decade ago, a strategy was reported for delivering peptides into the CNS *via* a CDS that relied on sequential metabolism [76]. The challenges that such delivery presents are considerable. In designing an appropriate peptide CDS, the strategy must ensure, at minimum, a suitable level of liphophilicity for BBB penetration, prevention of premature inactivation of the peptide, as well provision for targeting to allow controlled metabolic release of the peptide in the brain.

The actual strategy employed has been described as an extension of the CDS approach to a 'molecular packaging strategy' [63], since it involves 'disguising' the peptide entity within a bulky molecule that is both lipophilic (for passive diffusion through the BBB) and that will evade recognition by peptidases. Moreover, the bioactivation of such a molecular package might involve as many as five or six metabolic steps, whose timing is crucial to successful peptide delivery. Part of the strategy therefore requires incorporation of a spacer unit between the peptide and the targetor to control the timing for targetor release.

Much research using this approach has been aimed at delivery of thyrotropin-releasing hormone (TRH), or Pyr-His-Pro-NH_2 and its analogues [63]. TRH is the primary neurotrophic hormone for secretion of thyroid-stimulating hormone and has beneficial effects in the treatment of e.g. memory impairment and amyotrophic lateral sclerosis. To put this into context, it should be mentioned that the simpler prodrug approach has also been employed to enhance delivery of TRH. One such study specifically focused on improving the lipophilicity of TRH and reducing its susceptibility to rapid enzymatic inactivation in the systemic circulation [77]. The prodrug strategy adopted involved N-acylation of the imidazole ring of the histidine residue with various chloroformates. N-alkoxycarbonyl prodrug derivatives were found to be resistant to enzymatic cleavage but underwent the desired facile bioreversal quantitatively to TRH *via* spontaneous hydrolysis or by plasma esterase-catalysed hydrolysis. These prodrugs were also significantly more lipophilic than the parent TRH.

An example of the 'molecular packaging strategy' CDS approach to peptide delivery is illustrated in Figure 9.23 and refers to a system designed to transport a pyroglutamyl peptide amide to the CNS, namely the TRH-analogue, Pyr-Leu-Pro-NH_2.

This CDS incorporated a Gln-Leu-Pro-Gly progenitor sequence of the above analogue [78]. Whereas previous methodology used by the same group was applicable only to neuropeptides containing free amino and carboxylic acid terminal functions [76], the example cited thus involved an extension to peptides with N-terminal pyroglutamyl (Pyr) and C-terminal carboxamide functions. In designing the CDS, the free carboxylic acid function of the glycine residue was rendered lipophilic by esterification with the bulky cholesterol molecule.

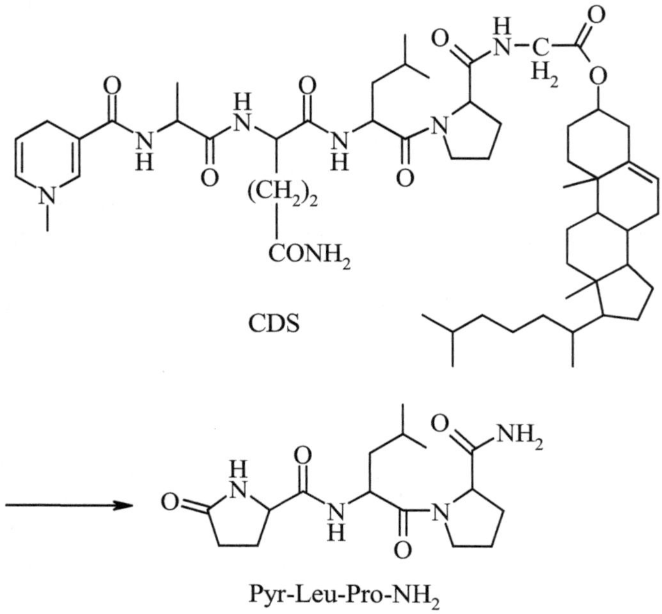

Fig.9.23 Example of a CDS designed to deliver a peptide to the CNS [78]

The targetor, a 1,4-dihydrotrigonellyl unit, was appended to the progenitor *via* a spacer unit, namely alanine. This elaborate CDS, comprising four units (a targetor, a peptide, a spacer unit and a lipophilic moiety) was shown to undergo a series of metabolic reactions, eventually releasing pharmacologically significant quantities of Pyr-Leu-Pro-NH_2 in the CNS of mice, following intravenous injection.

It should be noted that, as with the HC-CDS described above, the first step in the bioactivation of the peptide-CDS following passive transport across the BBB is oxidation (NAD $\leftrightarrow$ NADH) in the brain. The remaining bioactivation steps include cleavage of the cholesterol moiety by an esterase or lipase, and subsequent stepwise metabolic processes that are respectively mediated by (a) peptidyl glycine alpha-amidating monooxygenase, (b) dipeptidyl peptidase and (c) glutaminyl cyclase, the latter enzyme generating the peptide Pyr-Leu-Pro-NH_2. The intermediate structures in this sequence can be found in ref. 78. The application described for this particular compound thus highlighted the potential of incorporating within the CDS a suitable peptide progenitor with predictable bioactivation, as an extension to the existing methodology for synthesising peptide-CDSs.

The use of macromolecules in prodrug development to improve drug stability and bioavailability was described earlier. However, targeting of tissues using drugs attached to macromolecules such as neoglycoproteins and synthetic polymers has also been explored. This approach has been reviewed, with emphasis on anti-HIV therapy [79]. Many of the recently developed systems are based on a model comprising the drug molecule, solubiliser moieties (e.g. carboxylic, hydroxyl), spacer units that link drug molecules to the macromolecular carrier, and targeting moieties. The function of the spacers is to undergo chemical or enzymatic hydrolysis to release the drug. In contrast to simple prodrugs, drug-polymer conjugates can only be taken up by cells *via* pinocytosis.

Neoglycoprotein carriers have been successfully conjugated to anti-HIV drugs to produce prodrugs [80]. An example is the drug zidovudine, in the form of its 5'-monophosphate, AZTMP. Brain-targeting of AZT has also been effected by linking the drug in the form of its succinate to e.g. the anti-transferrin receptor antibody OX-26 [81]. *In vivo* studies indicated that the conjugate did indeed target brain capillaries and released the drug rapidly *in situ*. Conjugation of AZT by means of a succinate spacer with the water-soluble synthetic polymer α,β-poly(N-2-hydroxyethyl)-DL-aspartamide (PHEA) has also been achieved [82]. Release of AZT from the resulting macromolecular prodrug was tested under a range of pH conditions. Efficient bioactivation by plasmatic enzymes was evident from the fact that more than 60% of linked AZT was released from the conjugate in plasma.

The intention in the above section was to highlight the range of ideas currently being employed in the creation and development of novel chemical delivery systems, with an emphasis on those that employ targeting and that take advantage of predictable metabolism for their activation. Some of the challenges encountered in these areas of research have also been mentioned.

9.3 THE ROLE OF FORMULATION

All of the approaches to drug design described in the previous sections involved creation of new entities from existing, active agents in an attempt to overcome problems relating to pharmacokinetics, or to take advantage of controlled or predictable metabolism for bioactivation, or to achieve drug targeting, or a combination of these. In all cases, chemical synthesis, and hence the formation of new covalent bonds, was required to effect the necessary molecular modifications to arrive at prodrugs, hard drugs, soft drugs and chemical delivery systems.

In concluding this chapter it is appropriate to remind the reader that several negative aspects associated with drug delivery may be satisfactorily addressed by alternative, 'softer' approaches. In Chapter 1, a number of methods for improving oral absorption (e.g. by increasing aqueous solubility and drug stability, by reducing first-pass metabolism) were described. To mention a few approaches, these included e.g. selection of the appropriate solid form of the compound for formulation, the use of proliposomes to effect controlled drug release, enteric-coating to reduce first-pass metabolism, achieve tissue targeting and improve drug safety profiles, bioadhesive nanoparticles to reduce pre-systemic metabolism, and cyclodextrin inclusion to promote drug stability, increase bioavailability and reduce gastrointestinal irritation caused by NSAIDs. The medicinal chemist needs to keep such alternative formulation approaches in mind, since they might well serve in overcoming specific problems presented by new drug candidates. To stress the significance of formulation issues, we highlight just two of the above aspects, which happen to fall within our own research interests, namely drug polymorphism and cyclodextrin inclusion of drugs.

A very fundamental issue that is sometimes overlooked by synthetic chemists involved in drug discovery and design is the crucial importance of selecting the 'correct' solid form of the drug candidate intended for further development in an oral preparation (e.g. tablet, capsule). Often, this is the most stable (least soluble) solid form, but there may be good reasons for developing a metastable form of higher solubility. The recognition that every organic compound can potentially exist in multiple solid forms (polymorphs, solvates), each with a unique thermodynamic stability and solubility, and that physical and chemical factors during the manufacturing and processing stages can effect interconversion of solid forms, has alerted the pharmaceutical industry in recent years to the need to monitor the integrity of a drug substance from the point of its initial crystallization through to the finished product [83-85]. A notorious recent case involving the antiretroviral drug ritonavir illustrates the dire consequences of unexpected solid-state transformation in the pharmaceutical industry. In short, two years after successful marketing of this drug, several lots of capsules failed dissolution testing due to precipitation of the drug from semi-solid dosage forms. This was traced to conversion of the original crystalline form I of ritonavir to a thermodynamically more stable form II, with a solubility only ~25% that of form I. Rapid pervasion of form II followed and attempts to recover form I failed initially, necessitating temporary reformulation. A detailed, expensive and time-consuming investigation later revealed that the dramatic difference in solubility was due to significant differences in the hydrogen bonding arrangements in the two crystalline modifications of ritonavir. The source of form II was attributed to probable heterogeneous nucleation by a degradation product. A detailed account of this unwelcome occurrence of polymorphism has been published [86] and its perusal is highly recommended.

The chemical stabilities and bioavailabilities of many drugs have been improved by their encapsulation within cyclodextrins [87-88] and this is consequently one formulation technology that continues to be widely applied in the pharmaceutical industry. What may not be widely evident, however, is that the significant benefits that this technology brings may be applied not only to active drugs, but may even enhance the properties of e.g. prodrugs that have themselves been derived as carefully designed chemical drug delivery systems. In support of this statement, we cite the recent use of cyclodextrins in attempts to increase the aqueous solubility and stability of the designed soft corticosteroid loteprednol etabonate [89], whose structure, design strategy and bioactivation were described in section 9.2.3. In fact, a recent, authoritative review [90] reminds us that in the process of retrometabolic drug design of chemical delivery systems (CDSs), aimed at 'identifying new drug candidates with improved therapeutic indices based on predictable/controlled metabolism and/or site-targeted delivery', issues such as dosage form stability, solubility and dissolution properties are crucial for successful drug performance and pharmaceutical acceptability. Examples quoted include the use of the highly soluble 2-hydroxypropyl-β-cyclodextrin that has been successfully employed in stabilising and improving the water-solubility of a number of chemical delivery systems for parenteral administration, including, for example, the AZT-CDS and the estradiol-CDS described earlier.

9.4 CONCLUDING REMARKS

In this chapter, the authors attempted to show how considerations of pharmacokinetics and metabolism guide the process of developing drugs with improved delivery characteristics and the ability to target specific organs or tissues so as to maximise therapeutic efficacy. In reviewing some of the main approaches adopted to achieve these ends, we have deliberately omitted detailed chemical methodology, which we consider as secondary to the design concepts. Aspects of the synthetic methodology feature in many of the papers and reviews cited above. Compilation of this modest review owes everything to the eminent scientists whose original, imaginative concepts outlined above have been responsible for the creation of new generations of effective drugs. One message that is implicit in the above review is that due recognition of pharmacokinetic issues and a deeper insight into the nature of drug metabolism will contribute to more successful application of the principles outlined above to the design of new drug molecules.

The major thrust of this monograph has in fact been to elucidate the complex nature of drug metabolism and its ramifications in medicine. In aiming to reflect the richness and vitality of this subject, as well as justify the term 'current' in the title of this monograph, we have included many references to various types of studies performed during the last five years and attempted to explain their significance in modern medicine. We trust that the resulting document will cater for a wide readership, ranging from students of pharmacy, chemistry, biochemistry and pharmacology to established professionals in the health sciences.

References

1. Prokai L and Prokai-Tatrai K. 1999. Metabolism-based drug design and drug targeting. PSTT 2:457-462.
2. Lin JH, Lu AYH. 1997. Role of pharmacokinetics and metabolism in drug discovery and development. Pharmacol Rev 49:403-449.
3. Prentis RA, Lis Y, Walker SR. 1988. Pharmaceutical innovation by seven UK-owned pharmaceutical companies (1964-1985). Br J Clin Pharmacol 25:387-396.
4. Smith DA, Jones BC, Walker DK. 1996. Design of drugs involving the concepts and theories of drug metabolism and pharmacokinetics. Med Res Rev 16:243-266.
5. Bertrand M, Jackson P, Walther B. 2000. Rapid assessment of drug metabolism in the drug discovery process. Eur J Pharm Sci 11:S61-S72.
6. Ekins S. 2003. *In silico* approaches to predicting drug metabolism, toxicology and beyond. Biochem Soc Trans 31:611-614
7. Masimirembwa CM, Bredberg U, Andersson TB. 2003. Metabolic stability for drug discovery and development: Pharmacokinetic and biochemical challenges. Clin Pharmacokinet 42:515-528.
8. Bodor N. 1995. Retrometabolic drug design concepts in ophthalmic target-specific drug delivery. Adv Drug Deliv Rev 16:21-38.
9. Albert A. 1958. Chemical aspects of selective toxicity. Nature 182:421-422.
10. Testa B. 2004. Prodrug research: futile or fertile? Biochem Pharmacol 68:2097-2106.
11. Thomas G. 2000. Medicinal Chemistry. Chichester: John Wiley & Sons Ltd, Chapter 9, pp 327-374.
12. Naylor MA, Thomson P. 2001. Recent advances in bioreductive drug targeting. Mini-reviews Med Chem 1:17-29.

13. Gustafsson D, Elg M. 2003. The pharmacodynamics and pharmacokinetics of the oral direct thrombin inhibitor ximelagatran and its active metabolite melagatran: a mini-review. Thromb Res 109:S9-S15.

14. Clement B, Lopian K. 2003. Characterization of *in vitro* biotransformation of new, orally active, direct thrombin inhibitor ximelagatran, an amidoxime and ester prodrug. Drug Metab Dispos 31:645-651.

15. Schultz C. 2003. Prodrugs of biologically active phosphate esters. Bioorg Med Chem 11:885-898.

16. Meier C, Knispel T, De Clerq E, Balzarini J. 1999. CycloSal-pronucleotides of 2',3'-dideoxyadenosine and 2',3'-didehydroadenosine: synthesis and antiviral evaluation of a highly efficient nucleotide delivery system. J Med Chem 42:1604-1614.

17. Meier C. 1996. 4H-1.3.2-Benzodioxaphosphorin-2-nucleosyl-2-oxide – a new concept for lipophilic, potential prodrugs of biologically active nucleoside monophosphates. Angew Chem 108:77-79.

18. Sanghani PC, Stone CL, Ray BD, Pindel EV, Hurley TD, Bosron WF. 2000. Kinetic mechanism of human glutathione-dependent formaldehyde dehydrogenase. Biochem 39:10720-10786.

19. Testa B, Mayer J. 1998. Molecular toxicology and the medicinal chemist. Il Farmaco 53:287-291.

20. Pauletti GM, Gangwar S, Siahaan TJ, Aube J, Borchardt RT. 1997. Improvement of oral peptide bioavailability: peptidomimetics and prodrug strategies. Adv Drug Deliv Rev 27:235-256.

21. Persson G, Pahlm O, Gnosspelius Y. 1995. Oral bambuterol versus terbutaline in patients with asthma. Curr Therap Res 56:457-465.

22. Hwang JJ, Marshall JL. 2002. Capecitabine: fulfilling the promose of oral chemotherapy. Exp Opin Pharmacother 3:733-743.

23. Tan X, Chu CK, Boudinot D. 1999. Development and optimisation of anti-HIV nucleoside analogs and prodrugs: a review of their cellular pharmacology, structure-activity relationships and pharmacokinetics. Adv Drug Deliv Rev 39:117-151.

24. Bodor N, Buchwald P. 1997. Drug targeting *via* retrometabolic approaches. Pharmacol Ther 76:1-27.

25. Hasegawa T, Kawaguchi T. 2002. Delivery of anti-viral nucleoside analogues to the central nervous system. Curr Med Chem 1:55-63.

26. Uchegbu IF. 1999. (a) Parenteral Drug Delivery:1. Pharm J 263:309-318 (b) Parenteral Drug Delivery: 2. *ibid.* 263:355-358.

27. Li Z, Han J, Jiang Y, Browne P, Knox RJ, Hu L. 2003. Nitrobenzocyclophosphamides as potential prodrugs for bioreductive activation: synthesis, stability, enzymatic reduction, and antiproliferative activity in cell culture. Biorg Med Chem 11:4171-4178.

28. Ghosh AK, Khan S, Marini F, Nelson JA, Farquhar D. 2000. A daunorubicin β-galactoside prodrug for use in conjunction with gene-directed enzyme prodrug therapy. Tetrahedron Lett 41:4871-4874.

29. Ghosh AK, Khan S, Farquhar D. 1999. A β-galactoside phosphoramide mustard prodrug for use in conjunction with gene-directed enzyme prodrug therapy. Chem Commun 2527-2528.

30. Schmidt F, Florent J-C, Monnoret C, Straub R, Czech J, Gerken M, Bosslet K. 1997. Glucuronide prodrugs of hydroxy compounds for antibody directed enzyme prodrug therapy (ADEPT): a phenol nitrogen mustard carbamate. Bioorg Med Chem Lett 7:1071-1076.

31. Azoulay M, Florent J-C, Monneret C, Gesson J-P, Jacquesy J-C, Tillequin F, Koch M, Bosslet K, Czech J, Hoffmann D. 1995. Prodrugs of anthracycline antibiotics suited for tumor-specific activation. Anti-cancer Drug Des 10:441-450.

32. Jordan AM, Khan TH, Osborn HMI, Photiou A, Riley PA. 1999. Melanocyte-directed enzyme prodrug therapy (MDEPT): development of a targeted treatment for malignant melanoma. Bioorg Med Chem 7:1775-1780.

33. Jordan AM, Khan TH, Malkin H, Osborn HMI. 2002. Synthesis and analysis of urea and carbamate prodrugs as candidates for melanocyte-directed enzyme prodrug therapy (MDEPT). Bioorg Med Chem 10:2625-2633.

34. Delahoussaye YM, Evans JW, Brown JM. 2001. Metabolism of tirapazamine by multiple reductases in the nucleus. Biochem Pharmacol 62:1201-1209.

35. Maison W, Frangioni JV. 2004. Improved chemical strategies for the tatgeted therapy of cancer. Angew Chem Int Ed 42:4726-4728.

36. Christie RJ, Grainger DW. 2003. Design strategies to improve soluble macromolecular delivery constructs. Adv Drug Deliv Rev 55:421-437.

37. Zacchigna M, Di Luca G, Maurich V, Boccu E. 2002. Syntheses, chemical and enzymatic stability of new poly(ethylene glycol)-acyclovir prodrugs. Il Farmaco 57:207-214.

38. Ulbrich K, Šubr V. 2004. Polymeric anticancer drugs with pH-controlled activation. Adv Drug Deliv Rev 56:1023-1050.

39. Stella VJ. 1996. A case for prodrugs: fosphenytoin. Adv Drug Deliv Rev 19:311-330.

40. Ariëns EJ. 1972. Excursions in the field of SAR: a consideration of the past, the present and the future. In: Keverling Buiman JA, editor. Biological Activity and Chemical Structure. Amsterdam: Elsevier pp 1-35.

41. Ariëns EJ, Simonis AM. 1982. Optimalisation of pharmacokinetics: an essential aspect of drug development-by 'metabolic stabilisation'. In: Keverling Buiman JA, editor. Strategy in Drug Research. Amsterdam: Elsevier pp 165-178.

42. Lin JH, Chen I-W, Ulm EH, Duggan DE. 1988. Differential renal handling of angiotensin converting enzyme inhibitors in enalaprilat and lisinopril in rats. Drug Metab Dispos 16:392-396.

43. Thomas G. 2000. Medicinal Chemistry. Chichester: John Wiley & Sons Ltd, Chapter 6, pp 245-247.

44. Ondetti MA. 1988. Structural relationships of angiotensin-converting enzyme inhibitors to pharmacologic activity. Circulation 77:174-178.

45. Kelly JG, O'Malley K. 1990. Clinical pharmacokinetics of the newer ACE inhibitors. Clin Pharmacokinet 19:177-196.

46. Lin JH. 1996. Bisphosphonates: a review of their pharmacokinetic properties. Bone 18:75-85.

47. Dugar S, Yumibe N, Clader JW, Vizziano M, Huie K, Van Hek M, Compton DS, Davis Jr HR. 1996. Metabolism and structure activity data based drug design: discovery of (-) SCH 53079 an analog of the potent cholesterol absorption inhibitor (-)SCH 48461. Bioorg Med Chem 6:1271-1274.

48. Matsui T, Kondo T, Nishita Y, Itadani S, Tsuruta H, Fujita S, Omawari N, Sakai M, Nakazawa S, Ogata A, Mori H, Kamoshima W, Terai K, Ohno H, Obata T, Nakai H, Toda M. 2002. Highly potent inhibitors of tnf-α production. Part 2: metabolic stabilization of a newly found chemical lead and conformational analysis of an active diastereoisomer. Bioorg Med Chem 10:3787-3805.

49. Boyle CD, Chackalamannil S, Clader JW, Greenlee WJ, Josien HB, Kaminski JJ, Kozlowski JA, McCombie SW, Nazareno DV, Tagat JR, Wang Y, Zhou G, Billard W, Binch H III, Crosby G, Cohen-Williams M, Coffin VL, Cox KA, Grotz DE, Duffy RA, Ruperto V, Lachowicz JE. 2001. Metabolic stabilization of benzylidene ketal M_2-muscrinic receptor antagonists *via* halonaphthoic acid substitution. Biorg Med Chem Lett 11:2311-2314.

50. Bodor N, Kaminski JJ. 1980. Soft drugs. 2. Soft alkylating compounds as potential antitumor agents. J Med Chem 23:566-569.

51. Bodor N, El Koussi A, Kano M, Khalifa M. 1988. Soft drugs. 7. β-blockers for systemic and ophthalmic use. J Med Chem 31:1651-1656.

52. Hwang S-K, Juhasz A, Yoon S-H, Bodor N. 2000. Soft drugs.12. Design, synthesis, and evaluation of soft bufuralol analogues. J Med Chem 43:1525-1532.

53. Bodor N. 1993. In: Korting H, editor. Topical glucocorticoids.with increased benefit-risk ratio, current problems in dermatology. Basel: Karger AG pp 11-19.

54. Annual Report: Top 500 drugs, Pharma business, 2000, May p 58.

55. Belvisi MG, Hele DJ. 2003. Soft steroids: a new approach to the treatment of inflammatory airways diseases. Pulm Pharmacol Ther 16:321-325.

56. Graffner-Nordberg M, Kolmodin K, Åqvist K, Queener SF, Hallberg A. 2004. Design, synthesis and computational affinity prediction of ester soft drugs as inhibitors of dihydrofolate reductase from *Pneumocystis carinii*. Eur J Pharm Sci 22:43-53.

57. Bodor N, Kaminski JJ, Selk S. 1980. Soft drugs.1. Labile quaternary ammonium salts as soft antimicrobials. J Med Chem 23:469-474.

58. Pavlikova-Moricka M, Lacko I, Devinsky F, Masarova L, Mlynarcik D. 1994. Quaternary ammonium salts. 45. Quantitative relationships between structure and antimicrobial activity of new "soft" bisquaternary ammonium salts. Folia Microbiol 39:176-180.

59. Thorsteinsson T, Loftsson T, Masson M. 2003. Soft antibacterial agents. Curr Med Chem 10:1129-1136.

60. Bodor N, Buchwald P. 2004. Designing safer (soft) drugs by avoiding the formation of toxic and oxidative metabolites. Mol Biotechnol 26:123-132.

61. Bodor N. 1992. In: Sarel S, Mechoulam R, Agrana I, editors. Trends in Medicinal Chemistry. Proceedings of the XIth International Symposium on Medicinal Chemistry. Oxford: Blackwell Science pp 35-44.

62. Bodor N, Buchwald P. 2001. Drug targeting by retrometabolic design: soft drugs and chemical delivery systems. In: Schreier H, editor. Drug Targeting Technology: Physical·Chemical·Biological Methods. Drugs and the Pharmaceutical Sciences. New York: Marcel Dekker Inc pp 163-187.

63. Bodor N, Buchwald P. 2002. Barriers to remember: brain-targeting chemical delivery systems and Alzheimer's disease. DDT 7:766-774.

64. Reddy IK, Vaithiyalingam SR, Khan MA, Bodor NS. 2001. Design, *in vitro* stability, and ocular hypotensive activity of *t*-butalone chemical delivery systems. J Pharm Sci 90:1026-1033.

65. Bodor NS, Visor G. 1984. Improved delivery through biological membranes XVII. A site-specific chemical delivery system as a short acting mydriatic agent. Pharm Res 4:168-173.

66. Reddy IK, Vaithiyalingam SR, Khan MA, Bodor NS. 2001. Intraocular pressure-lowering activity and *in vivo* disposition of dipivalyl terbutalone in rabbits. Drug Dev Ind Pharm 27:137-141.

67. Goskonda VR, Ghandehari H, Reddy IK. 2001. Novel site-specific chemical delivery system as a potential mydriatic agent: formation of phenylephrine in the iris-ciliary body of phenylephrone chemical delivery systems. J Pharm Sci 90:12-22.

68. Bodor N, Prokai L. 1990. Site- and stereospecific ocular drug delivery by sequential enzyme bioactivation. Pharm Res 7:723-725.

69. Bodor N, El Koussi A, Kano M, Nakamura T. 1988. Improved delivery through biological membranes. 26. Design, synthesis, and pharmacological activity of a novel chemical delivery system for β-adrenergic blocking agents. J Med Chem 31:100-106.

70. Bodor N, Elkoussi A. 1991. Improved delivery through biological membranes. LVI. Pharmacological evaluation of alprenoxime. A new potential antiglaucoma agent. Pharm Res 8:1389-1395.

71. Polgar P, Bodor N. 1995. Minimal cardiac electrophysiological activity of alprenoxime, a site-activated ocular beta-blocker, in dogs. Life Sci 56:1207-1213.

72. Prokai L, Wu W-M, Somogyi G, Bodor N. 1995. Ocular delivery of the β-adrenergic antagonist alprenolol by sequential bioactivation of its methoxime analog. J Med Chem 38:2018-2020.

73. Bodor N, Farag HH, Barros MDC, Wu W-M, Buchwald P. 2002. *In vitro* and *in vivo* evaluations of dihydroquinoline- and dihydroisoquinoline-based targetor moieties for brain-specific chemical delivery systems. J Drug Target 10:63-71.

74. Brewster ME, Anderson W, Bodor N. 1990. Brain, blood, and cerebrospinal fluid distribution of a zidovudine chemical delivery system in rabbits, J Pharm Sci 80:843-846.

75. Brewster ME, Pop E, Braunstein J, Pop AC, Druzgala P, Dinculescu A, Anderson W, Elkoussi A, Bodor N. 1993. The effect of dihydronicotinate N-substitution on the brain-targeting efficacy of a zidovudine chemical delivery system. Pharm Res 10:1356-1362.

76. Bodor N, Prokai L, Wu W-M, Farag H, Jonnalagadda S, Kawamura M, Simpkins J. 1992. Strategy for delivering peptides into the central nervous system by sequential metabolism. Science 257:1698-1700.

77. Bundgaard H, Møss J. 1990. Prodrugs of peptides. 6. Bioreversible derivatives of thyrotropin-releasing hormone (TRH) with increased lipophilicity and resistance to cleavage by the TRH-specific serum enzyme. Pharm Res 7:885-892.

78. Prokai L, Ouyang X-D, Wu W-M, Bodor N. 1994. Chemical delivery system to transport a pyroglutamyl peptide amide to the central nervous system. J Am Chem Soc 116:2643-2644.

79. Giammona G, Cavallaro G, Pitarresi G. 1999. Studies of macromolecular prodrugs of zidovudine. Adv Drug Deliv Rev 39:153-167.

80. Molema G, Jansen RW, Visser J, Herdewijn P, Moolenaar F, Meijer DKF. 1991. Neoglycoproteins as carriers for antiviral drugs: synthesis and analysis of protein-drug conjugates. J Med Chem 34:1137-1141.

81. Tadayoni BM, Friden PM, Wallus LR, Musso GF. 1993. Synthesis, *in vitro* kinetics and *in vivo* studies on protein conjugates of AZT: evaluation as a transport system to increase brain delivery. Bioconjugate Chem 4:139-145.

82. Giammona G, Cavallaro G, Fontana G, Pitarresi G, Carlisi B. 1998. Coupling of the antiviral agent zidovudine to polyaspartamide and *in vitro* drug release studies. J Control Release 54:321-331.

83. Bernstein J. 2002. Polymorphism in Molecular Crystals-IUCr monographs on Crystallography, No.14. Oxford: Clarendon Press.

84. Caira MR. 1998. Crystalline polymorphism of organic compounds. In: Weber E (editor). Design of Organic Solids. Berlin: Springer; pp 163-208.

85. Byrn SR, Pfeiffer RR, Stowell JG. 1999. Solid-State Chemistry of Drugs. 2nd Edition. West Lafayette, Indiana: SSCI, Inc.

86. Chemburkar SR, Bauer J, Deming K, Spiwek H, Patel K, Morris J, Henry R, Spanton S, Dziki W, Porter W, Quick J, Bauer P, Donaubauer J, Narayanan BA, Soldani M, Riley D,

McFarland K. 2000. Dealing with the impact of Ritonavir Polymorphs on the late stages of bulk process development. Org Process Res Dev 4:413-417.

87. Frömming K-H, Szejtli J. 1988. In: Davies JED (editor). Topics in Inclusion Science Vol 5 – Cyclodextrin Technology. Dordrecht: Kluwer Academic Publishers.

88. Loftsson T, Brewster ME. 1996. Pharmaceutical applications of Cyclodextrins. 1. Drug solubilization and stabilization. J Pharm Sci 85:1017-1025.

89. Bodor N, Drustrup J, Wu W. 2000. Effect of cyclodextrins on the solubility and stability of a novel soft corticosteroid loteprednol etabonate. Pharmazie 55:206-209.

90. Brewster ME, Loftsson T. 2002. The use of chemically modified cyclodextrins in the development of formulations for chemical delivery systems. Pharmazie 57:94-101.

Index

MIX
Papier aus verantwortungsvollen Quellen
Paper from responsible sources
FSC® C105338

If you have any concerns about our products,
you can contact us on
ProductSafety@springernature.com

In case Publisher is established outside the EU,
the EU authorized representative is:
Springer Nature Customer Service Center GmbH
Europaplatz 3, 69115 Heidelberg, Germany

Printed by Libri Plureos GmbH
in Hamburg, Germany